CIVIL VIOLENCE
IN THE
URBAN
COMMUNITY

Edited with an introduction by

LOUIS H. MASOTTI
DON R. BOWEN

SAGE PUBLICATIONS, INC. / BEVERLY HILLS, CALIFORNIA

For information address:

SAGE PUBLICATIONS, INC.
275 South Beverly Drive
Beverly Hills, California 90212

First Printing

Printed in the United States of America

Standard Book Number 8039-1003-4

Library of Congress Catalog Card No. 68-57145

To The Memory Of

JOSEPH D. LOHMAN

Criminologist

CONTENTS

INTRODUCTION

CIVIL VIOLENCE
A Theoretical Overview

Don R. Bowen

Louis H. Masotti

We open this collection of theoretical and empirical investigations of urban rioting in the United States by attempting to place these phenomena in a somewhat more general framework. It is certainly the case that rioting, by itself, is not the only kind of civil violence. One thinks immediately, for example, of such things as guerrilla warfare, coups d'etat, insurrections, rebellions, revolutions, and related events. Indeed, in the literature on civil disorder, not rioting, but revolution, has long occupied the central focus of students of social conflict. However, it is also obvious that civil disorder or civil violence as a category is not exhaustive of all possible kinds of violence and conflict. Homicide, for example, is not normally considered a case of civil violence. Therefore we address ourselves to the questions of defining and delimiting the terms for various kinds of violence, conflict, and disorder employed in this universe of discourse, if for no other reason than that the reader may follow our analysis.

We begin with the observation that violence and conflict appear to be so widespread in human societies that they occur in all but a minute handful. It may be that there actually have been or will be societies, such as those contemplated from utterly different perspectives by Plato and Karl Marx, where no violence occurs. But we need not grapple with that issue in order to note that such societies are so rare that the possibility is generally described as utopian (meaning literally, from the Greek, "not of this place"). We also need not join the controversy on whether the propensity to violent behavior is innate in man *qua* man, or whether it is socially derived. Either point of view is compatible with our observation about the extreme incidence, frequency, and latitude of violent acts.

The problem that we do face is of a somewhat more technical nature: namely, how can we distinguish among the various types of violence and what are the conditions under which one type of violence may lead to—or become—another? The problem arises because from the viewpoint of the outside observer (perhaps the hypothetical strict behaviorist), one can distinguish acts of violence one from another most easily in terms of hard, observable characteristics such as the number of people involved, the number of people killed or injured, the amount of damage done, and the like. Now it may be that such an inquiry would be extremely valuable. It might be shown, for example, that what we commonly call civil disorder is almost always associated with large numbers of persons engaged, many casualties, much damage, and so on. We would therefore be inclined to say that what distinguishes civil violence from other violence is the magnitude or scope of the disturbance. Even more important would be some answers concerning whether large-scale disorders are preceded by a cumulative buildup of smaller disturbances. Thus it has been suggested that riots in the United States were preceded by a general breakdown of law and order. To our knowledge, nobody has yet systematically investigated this proposition, but an analysis of the frequency and distribution of violent acts would yield the information necessary to explore the asserted relation.

Although we think distinguishing civil violence from other violence in terms of magnitude is a valuable, and for certain purposes methodologically compelling, line of inquiry, nevertheless the difficulties we foresee are so formidable that we reject that categorization. The general problem is that events which seem to us very different phenomena may have a similar magnitude. The secondary problem is that a given event may change over time from one category to another, but its magnitude may not correspondingly alter. Let us illustrate what is implied in these statements with some examples.

Political assassination, to take a simple case, seems unlikely to appear very different from a murder growing out of a lovers' quarrel, on any scale of magnitude. Yet social scientists and laymen alike would agree that they are different things, different enough that it is worth distinguishing between them. At the other end of the continuum, there is a distinction between riots and revolutions, but the magnitude of both may be the same from a global point of view. Even more disturbing is that what appears initially to be a riot may become a revolution without altering significantly the magnitude of the disorder, as in St. Petersburg in 1917. Finally, the same kind of event involving the same kinds of levels of violence may or may not be a case of civil disorder. Consider banditry, an activity not normally considered politically relevant—unless the leader of the gang is Joseph Stalin robbing banks in the province of Georgia in order to buy arms.

This latter example invokes most clearly the distinction which we submit must be made between civil violence and other kinds of violence. That distinction rests on the *purposes or intentions* of the participants. If their purpose is to direct violence against the people or things which are symbols or agents of the political or civil order, then the violence is civil violence. We recognize that such a distinction opens a methodological Pandora's box. Was Stalin robbing banks perhaps to aggrandize himself, and merely hiding his private purpose when he said he used the money to support revolutionary activities? Conversely, when Black rioters in this country loot White-owned stores, are we justified in imputing to them the goal of striking at White social and political domination?

Ultimately this question is unanswerable, because the social scientist, no matter how subtle or understanding his techniques, must necessarily rely on some kind of overt behavior from which he infers the intentions or purposes of those he is observing. And in that process of inference, from what is observable to what is not, lies the possibility of error. Given the impossibility of absolutely comprehending the intentions of other persons (if for no other reason than that they occupy unique points in space-time), we fall back on a somewhat less rigorous standard of categorization and evidence. We take civil violence to be a subset of the more general category of violence. What marks off civil violence is the intent of those engaged in it to strike out at the animate or inanimate representatives of the civil order. Such intent may be directly expressed by the participants themselves, or inferred from their actions by the observer.

Either of these criteria of acceptability is subject to error. The expressed intent of the participants is not absolutely reliable, because respondents may, knowingly or unknowingly, incorrectly report their purposes. Obviously, such statements have to be evaluated in the same way all statements of intent are evaluated: namely, is it consistent with what we know of the individual, the situation, the time, etc.? Inferences by the observer based on the participants' actions run the danger of imputing intentions to the participants which they themselves never even thought of and might possibly reject if they did. Again, such inferences must be evaluated in the same manner all such statements are treated—do they seem credible given time, place, circumstance?

Given our distinction between civil violence and other kinds of violence, we face two additional questions not easily answered. These are: (1) Within the subset civil violence, what, if anything, distinguishes riots from other events such as revolutions, coups d'etat, etc., which also fall in this category? And (2) are riots, in fact, a case of civil violence as defined above? Taking these issues in order, we submit there are two crucial distinctions between riots and other forms of civil violence. The first is that riots do not present themselves as an attempt to seize state power or to throw it off. Revolutions, coups d'etat, guerrilla war, civil war, are all examples of civil violence which do have this component. They aim at seizing, overthrowing, or withdrawing from the instruments of political and social control. Riots, in our view, do not.

This is not to say that riots do not develop into revolutionary movements. Indeed, quite the contrary, they may, and often do. It is only to say that there is a difference in the initial intentions of rioters and revolutionaries. It is also not to deny that a series of riots like those which have taken place in the United States in recent years may give rise to a revolutionary rhetoric or ideology which describes, justifies, or explains that phenomenon we are here calling "rioting" in terms of revolution, insurrection, civil war, and the like. In fact, one of the contributors to this edition, T. M. Tomlinson, has traced the growth of what he terms a "riot ideology." Such a rhetoric arises because terms like "revolution" have taken on a hortatory meaning. To fulminate against the appropriation of technical terms for rhetorical purposes is a futile endeavor. We can only note that when urban riots in this country are described by some as "revolutions," they are using the term in a different sense than we here employ.

The second crucial distinction between riots and other kinds of civil violence is that riots appear to be more spontaneous, unplanned, and disorganized. To be sure, what appears to us as a relatively spontaneous and disorganized outburst may in fact have been secretly planned long in advance. In that event we, the observers, will make an error in classification. But there is virtually no evidence to support such a belief about Negro rioting in the United States, pre-1930 communal rioting in India, the New York draft riots of 1863, or, as we go to press, student rioting in most of the western world.

We are further confirmed in this view by the work of another of the contributors to this volume, Douglas P. Bwy. Bwy factor-analyzed a large number of instances of civil violence in Latin America for different time periods, and two basic factors emerged which he terms, respectively, "turmoil" and "internal war." The difference between them is that the former is characterized by a high degree of spontaneity and relative disorganization. And riots load very high on this turmoil factor, in distinction to such activities as revolutions, guerrilla war, revolutionary invasion, etc., which load on the more planned and organized dimension of internal war.

We turn now to the second question raised earlier, concerning whether riots are to be termed a case of civil violence. We raise this question because there are views current in high quarters which deny the civil character of the violence in question, at least as we have used the term here. Specifically, there are two radically different perspectives, both of which tend to deny or de-emphasize what in our view are the intentions of rioters to attack the symbols and agencies of political domination and social control. The first of these is the more easily dismissed. It is the view that riots are simply another form of lawless criminal activity, no different in kind but merely different in scale from murder, robbery, arson, and other criminal activities.

This view falters, as it were, on three shoals. First, it fails to take account of the expressed intentions of the rioters themselves, as revealed, for example, in such statements as "Burn, Baby, Burn" or "Get Whitey" or "Black Power." To the editors and contributors represented here, these statements reveal not just an extension of normal (in the statistical sense) criminal activity, but an attack on White political, social, and economic dominance. Second, the criminality hypothesis fails to explain why all kinds of criminals are not given to riot, only Black ones. It may be, of course, that an unarticulated major premise of this view is that Black Americans are innately more prone to criminality, and given that premise it follows that the participants in this kind of crime will be overwhelmingly Black. Needless to say, there is not a shred of evidence to support that proposition. The third and perhaps most decisive reason for rejecting this view is that it fails to tell us why the targets of violence are representatives of a political and social order. After all, if one is simply an arsonist, why pick on schools or precinct stations—which, generally being constructed of relatively noncombustible materials, can hardly yield as heart-warming a blaze as, say, a forest fire? Or, if one simply wants to shoot somebody, why pick White policemen, who are harder to hit with impunity and, in any event, are trained and equipped to shoot back?

The second view that tends to deny the civil or political character of rioting as a form of violence is not so easily dealt with, both because it is more subtle and because there is at least some evidence to support it. It is the view that rioting is a form of pathology characterized by the breakdown of superego controls on libidinal impulses. This failure of internal control is particularly likely to occur, according to this notion, where the individual is caught up in the "maddening crowd" and where the agencies of external social control, such as the police, appear to be intimidated or even absent. Needless to say, this view is a particular favorite of those whose training is in psychiatry.

As admitted above, there is some evidence to support this view. There are numerous reported instances of looted merchandise being returned by individuals claiming they did not know how they got it. Presumably the superego norms have reasserted themselves strongly enough by the time this happens that they work to repress even the memory of having looted the goods in the first place. There are equally numerous reports by rioters of a sense of almost orgiastic release in the early stages of a riot. Presumably such feelings of "freedom" or "joy" are the result of the release of previously inhibited libidinal energy in a kind of free-floating destructive aggressivity. In fact, such reports are so widespread that Dr. John Spiegel of the Lemburg Center for the Study of Violence has characterized the crowd-gathering and looting stages of a riot as a "Roman holiday" or "carnival."

Despite the persuasiveness of this view, which we tentatively label "rioting for kicks," there are persistent nagging difficulties which cannot be reconciled with it. In the first place, why are most of the rioters Black and poor? Must we assume that the psyches of poor Blacks are somehow so different that this particular segment of the population is more prone to failures of inhibition, more prone to be swept up in behavioral contagion, more prone to engage in riots as an irrational, impulsive release, rather than some other equally obvious channel? It is, we submit, as plausible, if not more so, that there are social and political realities which account for why the Black poor riot, which have little to do directly with their psyches and which have the additional advantage of accounting for why the White rich do not riot.

A second problem in the "rioting for kicks" hypothesis is that it fails to account for the direction and selectivity of the riots. It does not explain why White-owned property is particularly singled out for attack. It does not account for why White policemen become particular targets. Presumably, an irrational release of libidinal energy could be as gratifying when directed against any store or any bystander. Why is it necessary to loot, say, a White-owned finance company or to shoot at safety forces?

As an example of this point: In Detroit, units of the 82nd Airborne, an elite paratrooper division of the United States Army, more than half of whose ranks are Negro, were able to pacify an area of the city in a few hours, inflicting no casualties and suffering none themselves. This in contrast to the police and National Guard, who were unable to quell the riot in five days and who suffered or inflicted 43 officially acknowledged deaths (unofficial estimates run as high as 200). We suggest that if rioting is simply for kicks, the question has to be asked why rioters didn't shoot at the paratroopers but did at the police and the Guard. Part of the answer to that question is, of course, that the paratroopers were a superbly trained and disciplined force, officered apparently by men of both skill and compassion. But another part of the answer is that the paratroopers have quite a number of conspicuously Black faces among them, and they are also not identified with the local police or local government, but rather with the federal government. And there is a wealth of evidence indicating that Negro dissatisfaction and alienation is—so far—directed against local institutions. In light of this kind of selectivity of targets on the part of rioters, we feel that, while viewing riotous behavior as a form of pathology has merit in the case of certain individuals at certain stages of riots, as an overall explanation it is both unfortunate and incorrect.

At this point, let us sum up our arguments and take stock of where we stand. Violent behavior we have taken to be a general category of human existence. Civil violence is a special case of the more general category, characterized by attacks on symbols of the civil order *because* they are symbolic of that order. Riots, then, are a special case of the more general category of civil violence, differentiated from other kinds of civil violence either because they are not a deliberate attempt to seize or throw off political power or because they are more spontaneous and disorganized, or both. Additionally, we have the impression that riots may be distinguished from certain other kinds of civil violence by some measure of magnitude. For example, there are undoubtedly more people involved in rioting than, say, assassination. But for reasons explained earlier, we believe measures of scope or magnitude are inadequate to distinguish rioting from all other kinds of civil violence

FACTORS LEADING TO RIOTS

Having at this point trapped the beast in our conceptual net—provisionally, at least—let us move to an examination of its origins. We have so far emphasized the differences between riots and other types of civil violence, because we were concerned with defining the term. We are now compelled to point out the similarities among these various phenomena, implied by our suggestion that they belong to a general category. The two decisive similarities in our view are that they are attacks on the civil order and that they arise because the attackers are for some reason dissatisfied, disenchanted, or alienated from the civil order.

Neither of these statements is exactly novel, but there are implications which should be carefully noted. The most important one is that all kinds of civil violence have common beginnings: namely, the dissatisfaction, disenchantment, or alienation from the civil and political order of at least some people. Bluntly, urban rioting in the United States and guerrilla war in Columbia have the same roots, even though they are patently different events. If this is so, it follows that the hypotheses contained in a very long intellectual tradition—the study of revolution—are equally applicable to riots. And in fact, even the most casual reader of the essays in this collection cannot miss the feeling that the assumptions, hypotheses, findings, and conclusions about riots in this country and elsewhere have a very familiar ring.

The proposition that all kinds of civil violence grow out of dissatisfaction with the civil order (and the social and psychic conditions which bring about such dissatisfaction) unfortunately creates another monumental problem. Why is it that, if the conditions which breed civil violence are essentially the same, one country has a revolution, another has general strikes, and yet another riots? Why a revolution in Vietnam, but riots in the United States? At this point we can only suggest the outline of an answer to that question, and only in very general terms.

This outline has three points. The first is what political scientists call "system affect," or legitimacy. We believe that civil violence aimed at capturing or throwing off state power, such as revolution, civil war, coup d'etat, etc., occurs in countries where the regime has lost or is losing its legitimacy. Conversely, we believe that in systems where there is a high overall positive affect for the system, civil violence, if it breaks out, will be of the kind which does not initially aim at seizing power, such as riots or general strikes.

The second point we may call "system capacity." It has to do with the willingness and ability of the regime to respond to threats of civil violence with force or with reform, or with both. The importance of force can be grasped when it is realized that few if any regimes have ever been overthrown if they maintained control of the army. We might also point out that reform as a technique for quieting dissidence has its limits. It may work where the demands of the dissidents are specific and divisible and do not go beyond that. For example, a massive public "Marshall plan" for American Negroes would probably stop most rioting, although at this writing even that seems extremely unlikely to be forthcoming. But when the demands of those in rebellion reach the point of threatening the very power base of the regime, concessions or reforms can no longer be granted without the regime's conceding its own existence. Incidentally, the real importance of force here is how it is perceived in the eyes of the dissenters. They may over-evaluate or underevaluate, but their actions are controlled by these evaluations, not by objective reality.

The third part of this outline concentrates on what we call facilitating agencies. The general preconditions of civil disorder and violence may lead to revolutions in one political system, but to demonstrations in another, because of the presence or absence of revolutionary leaders, the presence or absence of norms condoning certain kinds of civil violence as opposed to others, or the fact that a given kind of civil violence has been employed before in that system and there is a kind of built-in cultural bias toward repeating it. Thus, we experience riots in the United States in part because we have had widespread rioting every summer since 1964 and because the country has had a long history of riots prior to that. This list, parenthetically, is meant to be illustrative, not exhaustive. But we believe that the presence or absence of facilitating agencies, or perhaps they should be called channels, is a consideration the analyst of civil violence can ignore only at great peril.

To repeat, we argue that all kinds of civil violence grow out of conditions of unrest or dissatisfaction which differ perhaps in degree but not in kind. The question as to whether the conditions will result in a riot, an insurrection, a revolution, or whatever, we think can be answered by examining the variance in (1) system legitimacy, (2) system capacity, and (3) facilitating agencies. We have not undertaken such a project, nor to our knowledge has anyone else.[1] We can therefore do no more than suggest. However, as the above propositions are formulated they at least have the advantage of being testable, and therefore subject to empirical confirmation—or disconfirmation. At this point we can do no more.

One further implication of the interaction of these three variables with the general conditions of dissatisfaction is that they may also throw some light on the dynamics of civil violence. We have previously pointed out that a riot, for example, may turn into a revolution, or, conversely, what is initially a revolution may peter out into a series of more or less spontaneous and disorganized demonstrations. This question has been raised with respect to the recent riots in the United States. Some Black militants, for example, have spoken of urban insurrections or guerrilla war in the cities growing out of the current riotous situations. In fact, one of

the contributors to this volume, Martin Oppenheimer, has addressed himself precisely to this question.

In our view such a development is unlikely in the near future,
because the overall legitimacy of the political system remains very
high (Negro political alienation, at least for the time being, is
directed at specific institutions of local government, particularly
the police, rather than at the legitimacy of the American political
system itself) and because the ability of the system to employ
massive force to suppress such a development is very high. Here
enters a tricky question, however. It may be that the Black militants misperceive the willingness of the society to use that force.
They may unwittingly depend on liberal pusillanimity (although
given their own assumption that the society is basically racist
through-and-through, it is hard to see how such a misperception
could long stand). Therefore an urban guerrilla movement might
develop despite the force capabilities of the political system,
because in the eyes of the potential guerrillas the system is unwilling to use that force to its full effectiveness.

So far we have avoided discussing that set of conditions characterized by dissatisfaction or unrest of some kind which, we have
hypothesized, lies at the roots of all kinds of manifestations of
civil violence. That discussion can no longer be postponed; and so,
by way of introduction, let us repeat an observation made earlier.
That is, if different phenomena grow out of the same kinds of
social or psychic conditions, then the analysis of those conditions
relative to one of these phenomena is also applicable to others. In
the present context, if the same kinds of conditions breed riots as
breed civil war, guerrilla movements, revolution, and so on, then
the study of rioting in the United States will profit greatly from an
examination of the literature on revolution.

FOUR MAJOR THEMES OF INQUIRY

We cannot here review the entire history of this line of inquiry.
We can do no more than suggest some major themes, foci, or
emphases, and some of the methodological and theoretical differences intrinsic to this universe of discourse. There are four major
themes running through the entire tradition and equally throughout the essays represented in this collection. The first of these is

centered on the notion of deprivation. We call it the "deprivational hypothesis." Men engage in civil violence because the current or anticipated distribution of values in their societies is unfair or unjust by some standard. Usually the unfairness is felt with respect to the distribution of wealth. But that is obviously only normal, not necessary. One might as easily be dissatisfied with the distribution of any other valued thing or symbol such as status, power, learning, rectitude, leisure, etc.

Two corollaries to this proposition must be noted. First, the injustice or unfairness of the current or anticipated distribution must be perceived or felt by the individuals themselves. If that proposition is not included, one has a difficult time accounting for the numerous political systems which have remained enormously stable with a politically quiescent populace over long periods of time and yet, by all available indicators, have a radically unequal distribution of values. The second corollary is related to the first. It is not necessarily the perception of an unequal distribution of values that moves men to civil violence, but rather the perception that the inequality in question is also unjust. And it follows that a perfectly equal distribution of values, if perceived as unfair or unjust, may be equally a motivating force for violence, although it must be admitted that the theorists of revolution have seldom, if ever, suggested this conclusion.

At this point the unamimity ends and the literature is divided on two central related questions. They are: (1) whether men perceive themselves as unjustly dealt with, or they feel deprived, as measured against some absolute standard, or as measured in relation to some other segment of the population or their own past experience; and (2) whether men are primarily moved to violence by despair or by hope. The answers to these two questions have methodological and explanatory implications. They affect how the student of civil violence will measure felt deprivation and, further, they affect how he will explain why certain individuals participate in such activities while others do not. Let us illustrate these points with some examples.

Consider the "absolute deprivation" hypothesis. It suggests that those individuals who are the most deprived are those who are the most likely to rise up. If we are considering the distribution of wealth, those with the least amount of it should be, potentially at least, those who are the most prone to engage in civil violence.

They are acting out of despair with their lot. In the case of the United States, it would be argued that urban Negroes engage in riots because they are overwhelmingly disadvantaged by the distribution of wealth in this country.

There is much in the simple—some say "vulgar"—version of Marxism which smacks of this notion. One of the things which makes the proletariat an "oppressed class" is that they do not receive a proportionate share of the national wealth in a capitalist economic order. More recently, Seymour Martin Lipset has argued that polities with a very low degree of wealth are likely to be more politically unstable,[2] and Bruce Russett has shown that there is a positive relation between unequal land distribution and civil disorder.[3] Lipset's analysis implies that poor people generally are more prone to civil disorder, while Russett's suggests that in agricultural societies the presence of large numbers of virtually landless individuals is highly destabilizing.[4]

Methodologically, those who direct their efforts along these paths have shown a marked preference for hard, objective data on income distribution, or on land, or on whatever other standard of wealth is valued in the societies in question. This preference risks the ecological fallacy, because one must assume that those who appear disadvantaged in fact *feel* themselves disadvantaged. But the risk may not be very great, for it cannot be gainsaid that one is much more likely to find persons who feel themselves deprived among the people who objectively *are* deprived than one is to find such persons among the ranks of the fat, the happy, or the smug. After all, one of the single best predictors of riots in northern American cities is the proportion of Negro population, even when we know that not all Negroes feel themselves unfairly dealt with.

Let us turn our attention now to the second emphasis: that which moves men to violence is deprivation relative to something else. The answers to the question of what that "something else" is are varied. Some suggest that one feels deprived when invidious comparisons are possible with the achievements of other salient individuals or reference groups. Here, for example, is Marx on this point: "A house may be large or small; as long as the surrounding houses are equally small it satisfies all social demands for a dwelling. But if a palace arises beside the little house, the little house shrinks into a hut."[5]

In the United States, many have argued that urban Negroes riot because they compare their own standard of living with that of the White middle class, and find theirs wanting. In particular it is suggested that the transmission of White middle-class life styles via television creates among Black viewers a very great deal of resentment at the inevitable comparison between what they see on the screen and what they see around them. To our knowledge this relation has not been systematically investigated and,[6] while there are obvious difficulties in this simple version, it does rest squarely on the notion of relative deprivation.

Other theorists have suggested that men feel deprived when they make comparisons between their present and anticipated level of achievement, or between their present or anticipated level of achievement and their past experience. James Davies has argued in this vein that revolutions are set off not in times of general economic depression, but rather when a sudden sharp economic downturn follows a period of general improvement.[7] In the United States, many have pointed out that the civil rights movement served to raise hopes among Negroes which were dashed once the legal barriers to segregation were broken and the much more difficult problems of the ghetto in the North were confronted.

The second major theme locates the genesis of civil violence in the feeling of dissatisfaction which results from the comparison between what one currently enjoys and what one aspires to—variously defined as what one expects, what one thinks one ought to have, or what one regards as ideal. We term this the "expectational theory." The distance between current status and aspiration has in fact been termed the "revolutionary gap," or, in the Feierabends' nicely turned phrase, "the want-get ratio."[8] This is also precisely the notion involved in what is called the revolution of rising expectations, referring to the formation of expectations which outrun the capacity of the political system to satisfy them.

From the viewpoint of social peace, according to this hypothesis, it is neither the wholly downtrodden—who have no aspirations—nor the very well off—who can satisfy theirs—who represent a threat. It is the man in the middle who is dangerous. It was De Tocqueville who observed; "It would appear that the French found their condition more insupportable in proportion to its improvement." On this view then, it is hope, not despair, which generates civil violence and disorder. The reason why Black Americans riot is because there has been just enough improvement in their conditions to generate hopes, expectations, or aspirations beyond the capacities of the system to meet them.

Since there is a nice exchange model present in this relation, we think it incumbent to point out that there are only two ways for a political system to get itself out of this problem. The system can either restrict demand or increase supply. That is, riots or violence generally will go on until the expectations which are being formed are met or until they are severely repressed. The strategy of a little bit of reform mixed with a little bit of repression is inherently inefficacious, because current satisfaction and aspirations cannot be brought into equilibrium among the disaffected in that way. A little bit of reform only raises expectations higher, and a little bit of repression only increases frustration.

The third major theme running through the literature has been an emphasis on the breakdown of consensual norms and the inability or unwillingness of the agencies of social control to act in such a way as to restore those norms. We term this the "systemic hypothesis." From the standpoint of the individual, the hypothesis focuses on two related, albeit distinct, characteristics: anomie, or the inability to share such norms, and alienation, the rejection of the norms in favor of some other set. In particular, when we deal with civil disorder and violence, this view leads the investigator to look for instances of political alienation.

From this standpoint of the regime or the ruling elites, this hypothesis directs attention to the cohesiveness of the governing group, to their effectiveness in exercising what we have earlier called reform capabilities or repressive capabilities, and to their legitimacy. As an example, Crane Brinton has pointed to the disintegration of ruling elites as one of the factors present in four revolutions which he examined.[9] The disintegration of the elites was responsible in his view for a general failure of agencies of social control, typified by the failure of the Russian troops to fire on a crowd of workers who, after being locked out of a metal working factory, were marching to petition the Czar in St. Petersburg in March, 1917.

From the viewpoint of the political system, this approach looks to such things as large-scale changes in social structure and process. Trends such as industrialization, urbanization, or modernization create new classes or groups of the population with different perceptions and life styles.[10] In such situations, traditional elites are unable to enforce traditional norms, and such phenomena as anomie and alienation grow apace.[11] The northern migration of the American Negro and his accompanying urbanization have often been cited as a prime example of the phenomenon. Rioting, it is argued, may be understood as an attack on the traditional accommodative structure of race relations in the United States, growing out of these large-scale social changes.

The fourth major theme we can identify views civil violence as a product of a struggle for power among various groups within the society. We call this the "group conflict" hypothesis. It primarily directs attention to cleavages within a society, such as ethnic, racial, religious, or regional splits. In particular, those who take this tack are likely to emphasize the presence of reinforcing cleavages, as opposed to cross-cutting ones. Where lines of social distance are mutually reinforcing, civil violence or intergroup conflict is likely to arise. Thus, communal rioting in India breaks out because religious, linguistic, residential, and ethnic cleavages all reinforce each other.[12] In the United States and South Africa, social class and racial cleavages reinforce each other.

From the viewpoint of our analysis here, treating intergroup conflict as a major theme in the literature on civil violence runs certain dangers—primarily the possibility that such conflict is not necessarily an instance of *civil* violence, in that it may represent simply an attack by one group on another, without involving the agencies or symbols of the civil order. In this sense, it would be much more analogous to warfare among political systems than to what we have defined as civil violence. However, we have the impression that group struggles for power normally include a struggle over the instrumentalities of political and social control, and if and when that is so, such events do fall within our definition.

We mention this point because it becomes crucial in what we take to be the premier example of the "group struggle" hypothesis: namely, secession and civil war. There is a point at which successful secessionist movements become in effect independent or autonomous political systems. At that point, in our view, the conflict between the breakaway system and the former one is no longer a case of civil war but of international war. Exactly where or when that point is reached we do not know. Experts in international law and politics generally argue that a new system becomes autonomous when it is given fairly widespread diplomatic recognition and/or when it has effective political control over a given territory. We must leave the resolution of this problem to those who know far more than we do about it. At this time we can only note that when group conflict and violence moves to that point, it no longer falls within our understanding of civil violence.

The reader will quickly discover that all four of these theoretical perspectives are represented in this volume. However, what may not appear when one is engaged in the details of the individual contributions is that these different theories specify differing units of analysis and different research methodologies. For example, the "deprivational" and "expectational" hypotheses rest on individual perceptions, norms, attitudes, etc., and fundamentally the unit of analysis of such theories is the single individual's state of mind.[13] Quite obviously, therefore, such theories will lean heavily on survey research techniques, since those techniques are currently among the most reliable for ascertaining the data necessary to validate or invalidate the hypotheses in question.

The "systemic" and "group conflict" hypotheses, on the other hand, examine the interactions of various groups or segments of the population within entire political systems. They direct our efforts toward larger units of analysis and tend to rely somewhat more on aggregate social data. For this reason, such theories currently lend themselves more easily to cross-national comparisons, if for no other cause than the fact that cross-national survey research is still in its infancy. The obvious point we want to make here is that different theoretical perspectives are the major controlling factor in decisions concerning the conduct of research. And that is as true of research in civil violence as it is in all other scientific inquiry.

The above we think are the major points of view describing the general conditions which result in various manifestations of civil violence. To close this part of our discussion, we note that these propositions are generally meant to explain the actions of the mass following of participants in such activities. They do not speak primarily to the question of the motivations of revolutionary leaders—the Lenins, the Mao Tse Tungs, the Che Guevaras. It has very often been noted that leadership cadres tend to be recruited from middle- or even upper-class backgrounds. And the hypotheses to explain this phenomenon range from "a lust for power" to "status discrepancy." We pass over this question lightly because riots, the major subject of this collection, are not characterized by the presence of individuals of this type. Although we may take judicial note of the fact that among Black Americans the most militant violence-prone attitudes (as distinct from actual behavior) are to be found precisely among that segment of the population which in other times and places has provided fertile recruiting grounds for revolutionary leaders: namely, middle-class university students or unemployed—perhaps unemployable—recent university graduates.

PRECIPITATING EVENTS

We come to the final part of this paper with one great unresolved issue. We have used the language of causal relations without addressing ourselves to the underlying question. We have spoken of the general conditions which breed civil violence without asking whether those conditions are necessary and sufficient to produce violence. Intimately related to this question is the issue of prediction. Every time we uncover a set of conditions such as described above, could we with any confidence predict that civil violence will occur? In our view the answer to that question is no. But in order to explain why we take this position, it is necessary to introduce an additional proposition. That is the notion of a precipitating event.

A precipitating event is the occurrence which is the immediate cause of civil violence: the match which lands in dry tinder. But such events are in our view essentially random and therefore unpredictable. In the United States, such events have ranged from the refusal of a White waiter to serve a Negro a glass of water to a police raid on an after-hours bar. To be sure, if one knew enough about police movements, one might have been able to predict these specific precipitating events. But that line of argument ultimately lands one in an infinite chain of causal regress, or requires the assumption that the universe in which we live is finite, knowable, and totally determinant, so that the investigator may function as a kind of Laplacean devil.

Since our own aspirations do not run that high, and since explanation must stop somewhere, we submit that the kinds of events which serve as the immediate causes of civil violence are, from the viewpoint of the analyst, random and unpredictable. They are, however, necessary. At this point, then, we have two sets of causes of civil violence: (1) the general conditions of dissatisfaction and unrest described above, and (2) precipitating events. Both are necessary; neither by itself is sufficient; together they are both necessary and sufficient.

One of these two sets, the conditions of unrest, is knowable in advance. They exist, for example, in virtually every large American city and many small ones. What is not knowable in advance is what kind of event, in which city, when, will touch off a riot. Let us reason by analogy. It is known that lightning starts fires in dry forests. If the forest is dry and there is heavy atmospheric electrical disturbance, it is a reasonable prediction that there will be fires, even though one does not know precisely when, where, or how the lightning will strike.

Similarly, we suggest that whether one's theoretical commitments emphasize deprivation in the absolute or relative sense, expectations beyond the capacity of the system to perform, breakdown of normative consensus and social control, or a struggle for power among conflicting groups—in any event, the conditions specified as important or relevant in these different hypotheses are all knowable in advance. They are the dry forest and the electrical disturbances. We feel reasonably certain that riots will continue in American urban areas so long as the conditions outlined by the contributors to this volume continue to hold. But we have no confidence in predicting precisely when and where the next outbreak of civil disorder will occur. That depends on what we have called above a precipitating event—one which is utterly contingent on specific circumstances.[14] We do not know where or when the lightning will strike, only that it probably will.

NOTES

1. This was written before the additional work of one of our contributors, Douglas P. Bwy, came to our attention. Bwy has addressed himself to the question posed here by examining the relations between such things as force capacity, system legitimacy, elite coherence, and other relevant variables and the outbreak of relatively anomic types of violence such as riots, strikes, and demonstrations, as opposed to the outbreak of more purposeful, planned events such as civil war, revolution, and insurrection. See his "Discerning Causal Patterns Among Conflict Models: A Comparative Study of Political Instability in Latin America," a paper delivered at the Midwest Conference of Political Science, Chicago, May 2-4, 1968.
2. Seymour Martin Lipset, *Political Man* (N.Y.: Doubleday, 1963), in particular chap. 2.
3. Bruce Russett, "Inequality and Instability: The Relation of Land Tenure to Politics," *World Politics,* XVI (April, 1964), 442-454.
4. For a general statement of the relation between revolution and increasing misery, see Pitirim Sorokin, *The Sociology of Revolution* (Philadelphia: J. B. Lippincott, 1925), p. 367.
5. Karl Marx, "Wage Labour and Capital," in *Selected Works,* Vol. I (N.Y.: International Publishers, 1933), pp. 268-269.
6. Charles Wolf has made a similar suggestion when he argues that "political vulnerability" is related to the level of economic aspirations, which, in turn, "...depends on the amount and kind of information that a region or country is exposed to." See his "Economic Change and Political Behavior: An Experimental Approach in an Asian Setting," in Wolf (ed.), *Foreign Aid: Theory and Practice in Southern Asia* (Princeton, N.J.: Princeton Univ. Press, 1960), p. 311.
7. James C. Davies, "Towards a Theory of Revolution," *Am. Sociol. Rev.,* XXVII (Feb., 1962), 6.
8. Ivo K. and Rosalind L. Feierabend, "Aggressive Behaviors Within Polities, 1948-62: A Cross-National Study," *J. Conflict Resolution,* X (Sept., 1966), 256-257.
9. See,Crane Brinton, *The Anatomy of Revolution* (N.Y.: Vintage Books, 1965).
10. See, for example, Daniel Lerner's formulation of "strains" in developing societies, resulting from unequal rates of modernization, urbanization, and literacy which lead to political instability. Daniel Lerner, *The Passing of Traditional Society* (N.Y.: Free Press, 1964), in particular the comparisons of Turkey and Egypt, chaps. 5 and 7.
11. As an example of this point of view, together with some suggested measures of anomie and alienation as preconditions of revolution, see Edward Tiryakian, "A Model of Societal Change and Its Lead Indicators," in Samuel Klausner (ed.), *The Study of Total Societies* (N.Y.: Praeger, 1967), pp. 69-98.
12. The problems raised by social cleavages and some of the means of coping with them form the central theme of Myron Weiner's study of India. See Myron Weiner, *The Politics of Scarcity* (Chicago: Univ. of Chicago Press, 1962). For an analysis directed specifically to group conflict, see Arnold Rose, "The Comparative Study of Intergroup Conflict," *Sociol. Q.,* I (Jan., 1960), 57-66.

13. Ultimately we would argue that all four of these theoretical perspectives rest on the individual's state of mind, since we do not quite see how one can speak of groups doing something or systems doing something except as a shorthand way of saying individuals are making certain kinds of decisions and acting on them. This observation has also been made by, among others, Lawrence Stone: "Psychological responses to changes in wealth and power. . .are politically more significant than material changes themselves." Lawrence Stone, "Theories of Revolution," *World Politics,* XVIII (Jan., 1966), 173.

14. Chalmers Johnson takes essentially the same view about predicting the outbreak of revolutions as we do about rioting: namely, that revolutions are highly contingent upon very specific kinds of events and, while one can know the general conditions (which Johnson terms as "disequilibrated social system"), it would be virtually impossible to know enough about these specific events in order to make accurate predictions. See Chalmers Johnson, *Revolutionary Change* (Boston: Little, Brown, 1966), in particular chap. 5.

Part I
THEORETICAL APPROACHES TO THE STUDY OF CIVIL VIOLENCE

Introduction

Students of civil disorder have variously sought to explain both the reasons why individuals are motivated to participate in such activities, and the sequences or stages of events which seem to occur when disorder does break out. These different kinds of inquiry, while in no sense incompatible with one another, do lead to different kinds of emphasis and do require the researcher to examine different kinds of empirical phenomena. The essays contained in this first section reflect this difference precisely.

The opening piece, by psychologist Leonard Berkowitz, views civil disorder as a special case of the more general category of aggression. Hence he suggests that in trying to explain why individuals would engage in civil violence we should examine the degree to which individuals are thwarted or frustrated in their attempts to reach the social and economic goals which they desire. And, because the experience of being frustrated leads to aggressive action, the explanation of individual motivation to engage in civil violence and disorder lies in frustration. However, Berkowitz warns, frustration must not be seen simply as a case of objectively measured deprivation. It is vitally necessary to know what the goals or hopes of the individual are before any assessment of his degree of frustration can be made. In addition, his analysis also suggests that the degree or intensity with which the individual engages in violence is furthered by the presence (or retarded by the absence) of symbols of violence, such as guns. Therefore, in the case of civil violence, Berkowitz argues that the presence of armed forces representing agencies of social control (such as the police) may, in fact, increase violent actions on the part of the population.

Ted Gurr develops the frustration-aggression hypothesis further in the second essay in this group. Gurr notes that the goals which men seek include those things they would like to have, those things they expect to have, and those things to which they think they are entitled. Men may be frustrated in any of these different kinds of goal-seeking behavior. Thus, according to Gurr, frustration or felt deprivation is always relative to the individual and is determined by his perception of the gap between what he has attained or what he perceives to be attainable on the one hand, and what he aspires to, expects, or thinks he deserves on the other. Gurr argues that only by understanding the relativity of individual goal-seeking can we explain the seeming paradox that it is very often the relatively better off, rather than those who are very poor, who engage in civil violence. The relatively better off may have extremely high hopes for the future and the gap between what they have and what they aspire to is actually greater than for the utterly destitute, who entertain no expectations for the future beyond what they already experience.

However, in Gurr's model between the feelings of deprivation and the outbreak of civil violence there are intervening variables which relate to the political system. Three of these serve to deter aggression; one facilitates it. The former are: (1) legitimacy—if individuals regard the government as legitimate they are more likely to accept deprivations; (2) coercive potential—if individuals perceive that the government can bring overwhelming force against them, they are less likely to engage in violence; (3) institutionalization—if the individual perceives stable alternate sources of satisfaction he is less likely to engage in violence. The variable which facilitates violence (facilitation) is the presence in the society of channels for readily expressing it, such as revolutionary movements (recall Berkowitz's point that the presence of symbols of violence facilitates the outbreak of aggression).

In considering this model it is important to remember that Gurr is addressing himself always to the problem of individual motivation. Even where he speaks of "systemic" variables such as, say, coercive potential the crucial point is that the deterrent to individual aggression here is the individual's *perception* of the system's capacity to bring force into play. Needless to say it is perfectly possible for the individual to credit the system with a coercive potential wholly unrelated to the actual capacity that can in fact be employed. In a very real sense, then, both Berkowitz and Gurr are working with a decision-making model applied to the individual decision to engage in aggression, violence, and the like. They direct our attention to the conditions in which individuals exist and, more importantly, to those individuals' perceptions of those conditions and their probable reactions to them.

The last two papers in this section turn us away from the question of individual motivation and perception and lead us to examine civil conflict from the viewpoint of contending social forces. Both of these essays are essentially equilibrium models. Both concentrate on race relations in the United States and view the confrontation between Black and White as a confrontation between contending and, at least partially, incompatible social groups. This confrontation sets the stage for the outbreak of conflict and violence between the two groups. Then through a series of stages or sequences, the relations between the two become stabilized or return to equilibrium—awaiting a renewal of the confrontation. It is inherent in the equilibrium model that this recycling will occur over and over again until each side has reached the point where, by some mode of accommodation, they both enjoy their maximum gain and minimum loss. And, even if that point is reached, it is quite obvious that exogenous change may once again upset the stable equilibrium.

Marilyn Gittell and Sherman Krupp, respectively a political scientist and a sociologist, rest their analysis of racial tension on four basic variables. These are the more or less habitual patterns of discrimination by Whites against Negroes, Negro expectations concerning change in those patterns, Negro action to effect such change, and the environmental context within which these various attitudes and activities occur. The output of the interaction of these variables is racial tension or conflict. Changes in any one of them will affect the level of tension either upward or downward. For example, if the patterns of discrimination remain the same and if the environmental conditions are stable then a rise in Negro expectations of change and an increase in efforts to effect that change will yield an increase in racial conflict. Conversely if the first two conditions remain constant and Negro expectations and actions decline, racial tension will decrease.

Beginning with a given situation, most likely one in which the relations between Whites and Negroes are those of dominance and subordination, the interactions of these four variables specify the dynamics of the model. It is important to recall, however, that four variables are not independent of each other. An increase in

Negro actions to effect change may increase White discrimination. Despite this potential complexity the authors believe that enough of the conditions can be specified in advance and enough other relevant conditions be discovered in the course of the analysis that the levels of racial conflict and the equilibrium points between the two contending groups can be predicted.

The final essay in this section is James Laue's analysis of change in segregation patterns in the American South. In Laue's view such change comes about as an end product of a process he terms the "7 C's." These are stages or sequences of conflict initiated by Negro demands for change in the dominant accommodative patterns of race relations and terminating in some change in those patterns. It must be understood that Laue does not view the working out of this cycle as determinant. It may be short-circuited by skipping some of the intermediate steps or it may be stopped altogether by White refusal to change buttressed by overwhelming force. The energy for the dynamics of this process is power. For example, direct action techniques provided Negroes in the South with a hitherto unrealized or unused source of power. A peak of the cycle is reached in the crisis stage where elites who have power in the White community perceive the necessity for some kind of change.

Laue argues that this cycle may recur over and over again. Beginning at one point in time with a given pattern of racial accommodation the conflict process proceeds until a new pattern is established. This new pattern then becomes the takeoff point for another cycle of change-through-conflict. Like Gittell-Krupp, Laue's analysis rests fundamentally on the proposition that conflict between two contending social groups and the establishment of an equilibrium point where both come to a momentary rest is an ever-present condition of all social interaction and a particularly valuable starting point for the analysis of interracial conflict and violence in urban America.

—L. H. M. and D. R. B.

THE STUDY OF URBAN VIOLENCE
Some Implications of Laboratory Studies of Frustration and Aggression

Leonard Berkowitz

■ The frustration-aggression hypothesis is the easiest and by far the most popular explanation of social violence — whether political turmoil, the hot summers of riot and disorder, or robberies and juvenile delinquency. We are all familiar with this formulation, and there is no need to spell out once again the great number of economic, social, and psychological frustrations that have been indicted as the source of aggression and domestic instability. Espoused in the social world primarily by political and economic liberals, this notion contends that the cause of civil tranquility is best served by eliminating barriers to the satisfaction of human needs and wants. Indeed, in the version that has attracted the greatest attention, the one spelled out by Dollard and his colleagues at Yale in 1939, it is argued that "aggression is always the result of frustration."[1]

AUTHOR'S NOTE: *This is a slightly revised version of a paper delivered at the annual meeting of the American Political Science Association, Chicago, September, 1967. The research reported in this paper has been sponsored by grants GS 1228 and GS 1737 from the National Science Foundation.*

The widespread acceptance of the frustration-aggression hypothesis, however, has not kept this formula safe from criticism. Since we are here concerned with the roots of violence, it is important to look closely at the relationship between frustration and aggression and consider the objections that have been raised. These criticisms have different, sometimes radically divergent, implications for social policy decisions. Before beginning this discussion, two points should be made clear. One, I believe in the essential validity of the frustration-aggression hypothesis, although I would modify it somewhat and severely restrict its scope. Two, with the Yale psychologists I prefer to define a "frustration" as the blocking of ongoing, goal-directed activity, rather than as the emotional reaction to this blocking.

One type of criticism is today most clearly associated with the ideas and writings of the eminent ethnologist, Konrad Lorenz. Throughout much of his long and productive professional career Lorenz has emphasized that the behavior of organisms — humans as well as lower animals, fish, and birds — is largely endogenously motivated; the mainsprings of action presumably arise from within. Behavior, he says, results from the spontaneous accumulation of some excitation or substance in neural centers. The external stimulus that seems to produce the action theoretically only "unlocks" inhibitory processes, thereby "releasing" the response. The behavior is essentially not a reaction to this external stimulus, but is supposedly actually impelled by the internal force, drive, or something, and is only let loose by the stimulus. If a sufficient amount of the internal excitation or substance accumulates before the organism can encounter a releasing stimulus, the response will go off by itself. In his latest book, *On Aggression*, Lorenz interprets aggressive behavior in just this manner. "It is the spontaneity of the [aggressive] instinct," he maintains, "that makes it so dangerous"[2] (p. 50). The behavior "can 'explode' without demonstrable external stimulation" merely because the internal accumulating *something* had not been discharged through earlier aggression. He strongly believes that "present-day civilized man suffers from insufficient discharge of his aggressive drive . . ." (p. 243). Lorenz's position, then, is that frustrations are, at best, an unimportant source of aggression.

We will not here go into a detailed discussion of the logical and empirical status of the Lorenzian account of behavior. I should note, however, that a number of biologists and comparative psychologists have severely criticized his analysis of animal behavior. Among other things, they object to his vague and imprecise concepts, and his excessive tendency to reason by crude analogies. Moreover, since Lorenz's ideas have attracted considerable popular attention, both in his own writings and in *The Territorial Imperative* by Robert Ardrey, we should look at the evidence he presents for his interpretation of human behavior. Thus, as one example, he says his views are supported by the failures of "an American method of education" to produce less aggressive children, even though the youngsters have been supposedly "spared all disappointments and indulged in every way" (*On Aggression,* p. 50). Since excessively indulged children probably expect to be gratified most of the time, so that the inevitable occasional frustrations they encounter are actually relatively strong thwartings for them, Lorenz's observation must leave the frustration-aggression hypothesis unscathed. His anthropological documentation is equally crude. A psychiatrist is quoted who supposedly "proved" that the Ute Indians have an unusually high neurosis rate because they are not permitted to discharge the strong aggressive drive bred in them during their warlike past (p. 244). Nothing is said about their current economic and social frustrations. Again, we are told of a psychoanalyst who "showed" that the survival of some Bornean tribes is in jeopardy because they can no longer engage in head-hunting (p. 261). In this regard, the anthropologist Edmund Leach has commented that Lorenz's anthropology is "way off," and reports that these Bornean tribes are actually having a rapid growth in population.

Another citation also illustrates one of Lorenz's major cures for aggressive behavior. He tells us (p. 55) that quarrels and fights often tear apart polar expeditions or other isolated groups of men. These people, Lorenz explains, had experienced an unfortunate damming up of aggression because their isolation had kept them from discharging their aggressive drive in attacks on "strangers or people outside their own circle of friends" (p. 55). In such circumstances, according to Lorenz, "the man of perception finds an outlet by creeping out of the barracks (tent, igloo) and smashing a not too expensive object with as resounding a crash as the occasion merits" (p. 56). According to this formulation, then, one of the best ways to prevent people from fighting is to provide them with "safe" or innocuous ways of venting their aggressive urge. Efforts to minimize their frustrations would presumably be wasted or at least relatively ineffective.

In must strongly disagree with Lorenz's proposed remedy for conflict. Informal observations as well as carefully controlled laboratory experiments indicate that attacks upon supposedly safe targets do not lessen, and can even increase, the likelihood of later aggression. We know, for example, that some persons have a strong inclination to be prejudiced against almost everyone who is different from them. For these prejudiced personalities, the expression of hostility against some groups of outsiders does not make them any friendlier toward other persons. Angry people may perhaps feel better when they can attack some scapegoat, but this does not necessarily mean their aggressive tendencies have been lessened. The pogroms incited by the Czar's secret police were no more successful in preventing the Russian Revolution than were the Russo-Japanese and Russo-Germanic wars. Attacks on minority groups and foreigners did not drain away the hostility toward the frustrating central government. Aggression can stimulate further aggression, at least until physical exhaustion, fear, or guilt inhibits further violence. Rather than providing a calming effect, the destruction, burning, and looting that take place during the initial stages of a riot seem to provoke still more violence. Further, several recent laboratory studies have demonstrated that giving children an opportunity to play aggressive games does not decrease the attacks they later will make upon some peer, and has a good chance of heightening the strength of these subsequent attacks.[3]

These misgivings, it should be clear, are not based on objections to the notion of innate determinants of aggression. Some criticisms of the frustration-aggression hypothesis have argued against the assumption of a "built-in" relationship between frustration and aggression, but there is today a much greater recognition of the role of constitutional determinants in human behavior. However, we probably should not think of these innate factors as constantly active instinctive drives. Contemporary biological research suggests these innate determinants could be likened to a "built-in wiring diagram" instead of a goading force. The "wiring" or neural connections makes it easy for certain actions to occur, but only in response to particular stimuli.[4] The innate factors are linkages between stimuli and responses — and an appropriate stimulus must be present if the behavior is to be elicited. Frustrations, in other words, may inherently increase the likelihood of aggressive reactions. Man might well have a constitutional predisposition to become aggressive after being thwarted. Clearly, however, other factors — such as fear of punishment or learning to respond in non-aggressive ways to frustrations — could prevent this potential from being realized.

It is somewhat easier to accept this interpretation of the frustration-aggression hypothesis, if we do not look at frustration as an emotionally neutral event. Indeed, an increasing body of animal and human research suggests that the consequences of a severe thwarting can be decidedly similar to those produced by punishment and pain. In the language of the experimental psychologists, the frustration is an aversive stimulus, and aversive stimuli are very reliable sources of aggressive behavior. But setting aside the specific emotional quality of the frustration, more and more animal and human experimentation has provided us with valuable insights into the frustration-aggression relationship.

This relationship, first of all, is very widespread among the various forms of life; pigeons have been found to become aggressive following a thwarting much as human children and adults do. In a recent experiment by Azrin, Hutchinson, and Hake,[5] for example, pigeons were taught to peck at a key by providing them with food every time they carried out such an action. Then after the key-pressing response was well established, the investigators suddenly stopped giving the bird food for his behavior. If there was no other animal present in the experimental chamber at the time, the pigeon exhibited only a flurry of action. When another pigeon was nearby, however, this burst of responding did not take place and the thwarted bird instead attacked the other pigeon. The frustration led to aggression, but only when a suitable target was present. This last qualification dealing with the nature of the available target is very important.

Before getting to this matter of the stimulus qualities of the target, another aspect of frustrations should be made explicit. Some opponents of the frustration-aggression hypothesis have assumed a person is frustrated whenever he has been deprived of the ordinary goals of social life for a long period of time. This assumption is not compatible with the definition of "frustration" I put forth at the beginning of this paper or with the results of recent experimentation. Contrary to traditional motivational thinking and the motivational concepts of Freud and Lorenz, many psychologists now insist that deprivations alone are inadequate to account for most motivated behavior. According to this newer theorizing, much greater weight must be given to anticipations of the goal than merely to the duration or magnitude of deprivation per se. The stimulation arising from these anticipations — from anticipatory goal responses — is now held to be a major determinant of the vigor and persistence of goal-seeking activity. As one psychologist (Mowrer) put it, we cannot fully account for goal-striving unless we give some attention to "hope." Whether a person's goal is food, a sexual object, or a color TV set, his goal-seeking is most intense when he is thinking of the goal and anticipating the satisfactions the food, sexual object, or TV set will bring. But similarly, his frustration is most severe when the anticipated satisfactions are not achieved.[6]

The politico-social counterpart of this theoretical formulation is obvious; the phrase "revolution of rising expectations" refers to just this conception of frustration. Poverty-stricken groups who had never dreamed of having automobiles, washing machines, or new homes are not frustrated merely because they had been deprived of these things; they are frustrated only after they had begun to hope. If they had dared to think they might get these objects and had anticipated their satisfactions, the inability to fulfill their anticipations is a frustration. Privations in themselves are much less likely to breed violence than is the dashing of hopes.

James Davies has employed this type of reasoning in his theory of revolutions.[7] The American, French, and Russian

Revolutions did not arise because these people were subjected to prolonged, severe hardships, Davies suggests. In each of these revolutions, and others as well, the established order was overthrown when a sudden, sharp socioeconomic *decline* abruptly thwarted the hopes and expectations that had begun to develop in the course of gradually improving conditions. Some data recently reported by Feierabend and Feierabend[8] can also be understood in these terms. They applied the frustration-aggression hypothesis to the study of political instability in a very impressive cross-national investigation. Among other things, they observed that rapid change in modernization within a society (as indicated by changes in such measures as the percentage of people having a primary education and the per capita consumption of calories) was associated with a relatively great increase in political instability (p. 265). It could be that the rapid socioeconomic improvements produce more hopes and expectations than can be fulfilled. Hope outstrips reality, even though conditions are rapidly improving for the society as a whole, and many of the people in the society are frustrated. Some such process, of course, may be occurring in the case of our present Negro revolution.

Let me now return to the problem of the stimulus qualities of the target of aggression. Recall that in the experiment with the frustrated pigeons the thwarted birds did not display their characteristic aggressive behavior unless another pigeon was nearby. The presence of an appropriate stimulus object was evidently necessary to evoke aggression from the aroused animals. Essentially similar findings have been obtained in experiments in which painful electric shocks were administered to rats.[9] Here too the aroused animals only attacked certain targets; the shocked rats did not attack a doll placed in the experimental chamber, whether the doll was moving or stationary. Nor did they attack a recently deceased rat lying motionless in the cage. If the dead animal was moved, however, attacks were made. Comparable results have been obtained when electrical stimulation was applied to the hypothalamus of cats.[10] Objects having certain sizes or shapes were attacked, while other kinds of objects were left alone.

This tendency for aroused animals to attack only particular targets can perhaps be explained by means of Lorenz's concept of the releasing stimulus. The particular live and/or moving target "releases" the animal's aggressive response. But note that the action is not the product of some gradually accumulating excitation or instinctive aggressive drive. The pigeon, rat, or cat, we might say, was first emotionally aroused (by the frustration, pain, or hypothalamic stimulation) and the appropriate stimulus object then released or evoked the action.

Similar processes operate at the human level. A good many (but not all) aggressive acts are impulsive in nature. Strong emotional arousal creates a predisposition to aggression, and the impulsive violent behavior occurs when an appropriate aggressive stimulus is encountered. Several experiments carried out in our Wisconsin laboratory have tried to demonstrate just this. Simply put, our basic hypothesis is that external stimuli associated with aggression will elicit relatively strong attacks from people who, for one reason or another, are ready to act aggressively. A prime example of such an aggressive stimulus, of course, is a weapon. One of our experiments has shown that angered college students who were given an opportunity to attack their tormentor exhibited much more intense aggression (in the form of electric shocks to their frustrator) when a rifle and pistol were nearby than when a neutral object was present or when there were no irrelevant objects near them.[11] The sight of the weapons evidently drew stronger attacks from the subjects than otherwise would have occurred in the absence of these aggressive objects. Several other experiments, including studies of children playing with aggressive toys, have yielded findings consistent with this analysis.[12] In these investigations, the aggressive objects (guns) acquired their aggressive stimulus properties through the use to which they were put. These stimulus properties can also come about by having the object associated with aggression. Thus, in several of our experiments, people whose name associated them with violent films shown to our subjects later were attacked more strongly by the subjects than were other target-persons who did not have this name-mediated connection with the observed aggression.[13]

These findings are obviously relevant to contemporary America. They of course argue for gun-control legislation, but also have implications for the riots that have torn through our cities this past summer. Some of our political leaders seem to be looking for single causes, whether this is a firebrand extremist such as Stokely Carmichael or a history of severe social and economic frustrations. Each of these factors might well have contributed to this summer's rioting; the American Negroes' frustrations undoubtedly were very important. Nevertheless, a complete understanding of the violence, and especially the contagious spread from one city to another, requires consideration of a multiplicity of causes, all operating together. Some of these causes are motivational; rebellious Negroes may have sought revenge, or they may have wanted to assert their masculinity. Much more simply, a good deal of activity during these riots involved the looting of desirable goods and appliances. Not all of the violence was this purposive, however. Some of it arose through the automatic operation of aggressive stimuli in a highly emotional atmosphere.

This impulsive mob violence was clearly not part of a calculated war against the whites. Where a deliberate anti-white campaign would have dictated attacks upon whites in all-white bastions, it was often Negro property that was destroyed. Moreover, aggressive stimuli had an important role. A lifetime of cruel frustrations built up a readiness for aggression, but this readiness had to be activated and inhibitions had to be lowered in order to produce the impulsive behavior. Different types of aggressive stimuli contributed to the aggressive actions. Some of these stimuli originated in the news reports, photographs, and films from other cities; research in a number of laboratories throughout this country and Canada indicates that observed aggression can stimulate aggressive behavior. This media-stimulated aggression may not always be immediately apparent. Some aggressive responses may operate only internally, in the form of clenched fists and violent ideas, but they can increase the probability and strength of later open aggression. The news stories probably also lower restraints against this open violence. A person who is in doubt as to whether destruction and looting are safe and/or proper behavior might have his doubts resolved; if other people do this sort of thing, maybe it isn't so bad. Maybe it is a good way to act and not so dangerous after all. And again the likelihood of aggression is heightened.

Then a precipitating event occurs in the Negro ghetto. The instigating stimulus could be an attack by whites against Negroes — a report of police brutality against some Negro — or it might be the sight of aggressive objects such as weapons, or even police. Police probably can function as stimuli automatically eliciting aggression from angry Negroes. They are the "head thumpers," the all-too-often hostile enforcers of laws arbitrarily imposed upon Negroes by an alien world. Mayor Cavanagh of Detroit has testified to this aggression-evoking effect. Answering criticisms of the delay in sending in police reinforcements at the first sign of rioting, he said experience in various cities around the country indicates the presence of police can inflame angry mobs and actually increase violence (*Meet the Press,* July 30, 1967). Of course the events in Milwaukee the week after Mayor Cavanagh spoke suggest that an army of police and National Guardsmen swiftly applied can restrain and then weaken mob violence fairly effectively. This rapid, all-blanketing police action obviously produces strong inhibitions, allowing time for the emotions inflamed by the precipitating event to cool down. Emptying the streets also removes aggression-eliciting stimuli; there is no one around to see other people looting and burning. But unless this extremely expensive complete inhibition can be achieved quickly, city officials might be advised to employ other law-enforcement techniques. Too weak a display of police force might be worse than none at all. One possibility is to have Negroes from outside the regular police department attempt to disperse the highly charged crowds. There are disadvantages, of course. The use of such an extra-police organization might be interpreted as a weakening of the community authority or a sign of the breakdown of the duly constituted forces of law and order. But there is also at least one very real advantage. The amateur law enforcers do not have a strong association with aggression and arbitrary frustration, and thus are less likely to draw out aggressive reactions from the emotionally charged people.

There are no easy solutions to the violence in our cities' streets. The causes are complex and poorly understood, and the possible remedies challenge our intelligence, cherished beliefs, and pocketbooks. I am convinced, however, that the roots of this violence are not to be found in any instinctive aggressive drive, and that there is no easy cure in the provision of so-called "safe" aggressive outlets. The answers can only be found in careful, systematic research free of the shopworn, oversimplified analogies of the past.

NOTES

1. John Dollard et al., *Frustration and Aggression* (New Haven: Yale Univ. Press, 1939), p. 3.

2. Konrad Lorenz, *On Aggression* (N. Y.: Harcourt, Brace & World, 1966).

3. For example, S. K. Mallick and B. R. McCandless, "A Study of Catharsis of Aggression," *J. Pers. Soc. Psychol.*, IV (1966), 591–596.

4. See L. Berkowitz, "The Concept of Aggressive Drive," in L. Berkowitz (ed.), *Advances in Experimental Social Psychology*, Vol. II (N. Y.: Academic Press, 1965).

5. N. H. Azrin, R. R. Hutchinson, and D. F. Hake, "Extinction-Induced Aggression," *J. Exp. Anal. Behavior*, IX (1966), 191–204.

6. See Berkowitz, *op. cit.*, for a further discussion, and also L. Berkowitz (ed.), *Roots of Aggression: A Re-examination of the Frustration-Aggression Hypothesis* (N. Y.: Atherton Press, 1968).

7. J. C. Davies, "Toward a Theory of Revolution," *Am. Sociol. Rev.*, XXVII (1962), 5–19.

8. I. K. Feierabend and R. L. Feierabend, "Aggressive Behaviors Within Polities, 1948–1962: A Cross-National Study," *J. Conf. Resol.*, X (1966), 249–271.

9. R. E. Ulrich and N. H. Azrin, "Reflexive Fighting in Response to Aversive Stimulation," *J. Exp. Anal. Behavior*, V (1962), 511–520.

10. P. K. Levison and J. P. Flynn, "The Objects Attacked by Cats During Stimulation of the Hypothalamus," *Animal Behavior*, XIII, (1965), 217–220.

11. L. Berkowitz and A. Le Page, "Weapons as Aggression-Eliciting Stimuli," *J. Pers. Soc. Psychol.*, VII (1967), 202–207.

12. E.g., Mallick and McCandless, *op. cit.*

13. See Berkowitz, note 4, *op. cit.* for a summary of some of this research.

URBAN DISORDER
Perspectives from the Comparative Study of Civil Strife

Ted Gurr

■ I assume that the sources and dynamics of urban disorder in the United States are fundamentally comparable to those of civil violence throughout the world. American Negro rioters and their white antagonists seem to share one basic psychological dynamic with striking French farmers, Guatemalan guerrillas, and rioting Indonesian students: most of them feel frustrated in the pursuit of their goals, they are angered as a consequence, and because of their immediate social circumstances they feel free enough, or desperate enough, to act on that anger.

It is not a sufficient explanation to say that Negroes riot because all angry men have a propensity to violence, but it is a useful assumption on which to base explanation. There is considerable theoretical speculation about the psychological sources of aggression, some of which has been applied to civil

AUTHOR'S NOTE: *This research was supported in part by the Center for Research in Social Systems (formerly SORO) of the American University, and by the Advanced Research Projects Agency of the Department of Defense. This support implies neither sponsor approval of this article and its conclusions nor the author's approval of the policies of the U. S. government toward civil strife. The assistance of Charles Ruttenberg in research design, data collection, and analysis is gratefully acknowledged. Joel Prager and Lois Wasserspring assisted in data collection. Diana Russell provided useful criticisms of the article in draft form.*

disorder. The first section of this article summarizes briefly some concepts and propositions about psychological factors that dispose men to violence and suggests their implications for research on urban disorder. We also know from cross-national studies of civil strife that there are regularities in its occurrence and intensity, that certain kinds and patterns of social conditions seem regularly to lead to strife and that others tend to minimize it. In the second section of the article I report on a cross-national study that was designed to determine the effects of some social conditions on the outcome of discontent, and suggest some of the study's implications for the most recent phase of the American dilemma.

SOME PSYCHOLOGICAL SOURCES OF TURMOIL

The sociological and popular cliché is that "frustration" or "discontent" or "despair" is the root cause of rebellion. Cliché or not, the basic relationship appears to be as fundamental to understanding civil strife as the law of gravity is to atmospheric physics: relative deprivation, the phrase I have used, is a necessary precondition for civil strife of any kind, and the more severe is relative deprivation, the more likely and severe is strife.[1] Relative deprivation is not whatever the outside observer thinks people ought to be dissatisfied with, however; it is a state of mind that can be defined as a discrepancy between people's expectations about the goods and conditions of life to which they are justifiably entitled, on the one hand, and on the other their value capabilities — what they perceive to be their chances for getting and keeping those goods and conditions. This is not a complicated way of making the simplistic and probably inaccurate statement that people are deprived and therefore angry if they have less than what they want. At least two characteristics of value perceptions are more consequential: what people think they *deserve,* not just what they want in an ideal sense, and what they think they have a chance of getting, not just what they have. They can have relatively little, but are likely to be relatively satisfied so long as they feel they are making satisfactory progress toward their goals. They feel deprived, and become angry, when they encounter or anticipate increased resistance. Since men live mentally in the immediate future as much as the present, *anticipated* frustration may be as important a cause of deprivation as actual frustration.

Underlying this relative deprivation approach to civil strife is a frustration-aggression mechanism, apparently a fundamental part of our psychobiological makeup. When we feel thwarted in an attempt to get something we want, we are likely to get angry, and when we get angry the most satisfying inherent response is to strike out at the source of frustration. Relative deprivation is, in effect, a perception of thwarting circumstances. *How* angry men become in response to the perception of deprivation seems determined partly by the nature of the expectations affected and partly by the kind and seriousness of interference with capabilities. The following propositions suggest some variables that affect the intensity of emotional response to the perception of deprivation.

The first proposition is that the greater the extent of discrepancy that men see between what they seek and what seems to be attainable, the greater their anger and consequent disposition to aggression. This variable is highly susceptible to change, for any increase in men's expectation levels, or any decrease in perceived capabilities, can increase this discrepancy. It is also one of the keys to the supposed paradox that dissatisfied people often (but by no means always) revolt just when things begin to get better: a little improvement accompanied by promises that much improvement is to come raises expectations, and if much improvement does *not* come when expected, perceptions of capabilities may drop, sometimes abruptly and sharply.

The proposition suggests a basic question for research on urban disorder in the United States: What has happened to levels of Negro expectations and to their perceptions of capabilities? The expectations of many Negroes have clearly increased, but how much, and why? Capabilities also seem to have increased for the majority, as suggested by the Brink and Harris survey finding that two-thirds of American Negroes felt that they were better off in 1966 than in 1963. But it may be that perceived capabilities are increasing less rapidly than expectation levels, so that the discrepancy, and consequently Negro anger, is still growing. Moreover, 5% of Negroes feel that they are worse off now than they were. Who are they, and are they the potential extremists?[2]

A second proposition relates to "opportunities": men who feel they have many ways to attain their goals are less likely to become angry when one is blocked than those who have few alternatives. The educated man has more skills, hence usually has more opportunities, than the uneducated. The city-dweller usually has more economic opportunities than the peasant, and thus the discontented peasant throughout the world often responds to the lack of rural opportunity by migrating to the city. For the unemployed Negro, the creation of job-training programs is presumably an expansion of opportunities. But this way there also may be dragons, since those who invest their energies in what appear to be opportunities but fail to make progress toward their goals tend to become more bitterly angry than those who do nothing. It is not entirely coincidental that Fidel Castro rebelled after his brief conventional political career was ended by the cancellation of elections, or that the young man now called Ho Chi Minh trained for service in a bureaucracy whose upper ranks were largely closed to Vietnamese, or that Job Corps drop-outs are found among riot leaders. Theory, along with popular liberal belief, suggests that expanding opportunities for goal attainment can be a crucial means of minimizing discontent. The qualification suggested here is that if what was promoted as opportunity proves just another dead end, the effects can be explosive. What the psychological effects of public and private efforts to expand opportunities for various groups of American Negroes have been I at least cannot say, but social researchers certainly have survey techniques that can provide some answers.

A third general proposition is that the greater the *intensity* of men's expectations, the greater their anger when they meet unexpected or increased resistance. By "intensity" is meant how badly they want whatever it is that they are seeking. The questions this poses for an analysis of Negro discontent are: Which groups of American Negroes have become more strongly motivated toward their economic and social goals in the past decade or so, and how much more strongly? One would predict that expectations are most intensely held among the young, among Northern Negroes, and among Negro leaders, the groups that have the greatest personal investment in alleviating or escaping from discrimination. These are in fact the groups that are most dissatisfied with the pace of improvement in Negro rights.[3]

Another perceptual variable serves to control the effects of deprivation. The proposition is that if men think that deprivation is legitimate, i.e., justified by circumstances or by the need to attain some greater end, the intensity and perhaps the level of expectations decline and consequently deprivation tends to be accepted with less anger. This proposition seems applicable, for example, to the great personal sacrifices men sometimes will make in the service of revolutionary movements. It probably also helps explain why most Negroes, like most white Americans, accept tax increases and the risks of military service without open rebellion.

These comments are not designed to provide an explanation of urban disorder in the United States but to demonstrate that to assess any particular act of political violence, or the potential for it, social scientists should begin by getting answers to some specific questions about the extent, intensity, and character of discontent. We need to know which groups feel most deprived, for these are the people who are most likely to become alienated to the point of violence. We need to assess the intensity of their anger, because this is a major determinant of the intensity of their action. Most important for policy purposes we need to know — not just to assume that we know — the *content* of their deprivation, for effective solutions can be devised only on the basis of accurate diagnosis.

The four propositions outlined above specify one set of psychological variables that can be examined. There is empirical evidence for these propositions and a good deal of suggestive evidence, from studies of civil strife, that they operate among collectivities. But there is far too little evidence about the psyches of rebels for us to say generally how important one or another of the variables is, or how they interact, or whether they are the most consequential psychological factors in civil strife. Some others bear mentioning: civil violence is sometimes assumed to be a calculated strategy for getting what the participants want — which it certainly is for some men, angry or not. Some scholars suggest that discontented men rebel only when faced with a specific threat.

Some, notably Fanon, interpret violence as a therapeutic assertion of self for the oppressed. Others of a Freudian or Lorenzian disposition suggest that aggression is a fundamental human drive, not necessarily a response to a specific kind of
circumstance but a predisposition that seeks some kind of outlet and that will take violent forms if nonviolent channels are
closed off.[4]

THE OUTCOMES OF DEPRIVATION:
A RECENT CROSS-NATIONAL STUDY

Despite these qualifications, one can accept for operational
purposes the thesis that relative deprivation is a necessary
condition for and a major source of civil strife. It also is evident that even when deprivation is intense and common to a
large group, it is not a sufficient condition for strife. Patterns
of social control and of facilitation have a great deal to do
with the outcomes. Whereas psychological variables such as
expectation levels are difficult to study directly on any large
scale, the intervening social conditions that determine their
outcomes are more amenable to comparative, cross-national
study. The remainder of this article summarizes some results
of one such study, especially as they relate to the problems
of contemporary urban disorder.

THE STUDY DESIGN: MEASURES OF DEPRIVATION
AND CIVIL STRIFE

The first operational task was to assess the relative extent
and intensity of some common kinds of deprivation among
the populations of 114 nations. The propositions described
above suggested many conditions that are frequently associated
with widespread discontent, and a variety of these conditions
were indexed. Short-term and persisting deprivation were separately assessed. The extent of short-term economic and
political deprivation in the late 1950's and early 1960's was inferred from the magnitude of short-term fluctuations in economic conditions and from politically repressive activities and
value-depriving policies of governments, as determined by
applying systematic coding procedures to news information.
Persisting deprivation was indexed by constructing and com-

bining measures of the extent of such conditions as economic and political discrimination, political separatism, religious cleavages, and lack of educational opportunity.[5]

Data on the types and characteristics of civil strife in all countries for the years from 1961 through 1965 were separately gathered from news sources. During these five years, incidentally, we found that only 10 of the world's 114 largest polities appeared completely free of civil strife.[6] From these data we determined the approximate proportion of each nation's population that participated in strife; the total duration of strife; and its casualties proportional to total population. These measures were combined to obtain measures of "magnitude of strife," which was divided into separate measures of magnitude of turmoil, conspiracy, and internal war. Although the United States was not among the 24 polities that experienced what we defined as internal war, it ranked 15th among the 95 polities that experienced turmoil. (Turmoil was defined as relatively spontaneous, unstructured, mass strife, including demonstrations, political strikes, riots, political clashes, and localized rebellions.) The United States ranked 42nd among the 114 polities in total magnitude of strife. Although not all turmoil is urban, by far the larger part of it is, and the generalizations the study suggests about turmoil can be applied with some confidence to urban disorder.

THE INTERVENING VARIABLES

Theories about aggression, and about the conditions of revolution, suggested a number of variables that affect the outcome of deprivation. I decided to index, and determine the relative effects of, four kinds of conditions: the legitimacy of the political regime in which strife occurs, coercive potential, institutionalization, and social facilitation. I suggested above that if men feel that deprivation is legitimate they are likely to accept it with less anger. The legitimacy of a political regime is likely to have a comparable effect: insofar as people regard their political system as a proper one, they are likely to tolerate some kinds of deprivation — in particular those for which the government is held responsible — without taking violent action against it. Two characteristics were taken into account in devising an indirect measure of legitimacy: the length of time each political system had persisted without substantial, abrupt reformation; and the extent to which it was indigenously developed rather than borrowed or imposed from abroad.

Conventional wisdom and studies of strife both emphasize the importance of actual or threatened coercion on the outcome of deprivation. If men are sufficiently afraid of the consequences, the argument goes, they will not riot. But comparative studies of civil strife, and psychological theory, both suggest that the relationship is not so simple. Some kinds of coercion are more likely to increase than deter violence, specifically "medium" or sporadically-applied coercion, which seems to be associated with greater strife than either low or high coercion. Two measures were used to estimate some of these effects: one a measure of the relative size of military and police forces, proportional to population, the other a measure which took account both of the size of these forces and their past and concurrent (1961-65) loyalty to the regime.

Many social scientists have argued that strife is minimized to the extent that organizations like labor unions, political parties, and the government itself are broad in scope, command large resources, and are stable and persisting. The existence of such organizations may have several essentially psychological effects for the discontented. Men may see in them greater opportunities for satisfying their expectations. At the same time, they may be able to express their discontent through them in routinized and typically nonviolent ways; a strike vote or an anti-government political rally can provide a safe outlet for considerable anger. To assess the effects of "institutionalization," a measure was devised that takes account of the relative strength of unions, the stability of the political party system, and the fiscal resources of government.

Legitimacy, coercion, and institutionalization are social variables that hypothetically serve in various ways to minimize or redirect the destructive consequences of deprivation. A great many conditions can have the opposite effect, to stimulate men to take violent action; for example, the presence of revolutionary organization and leadership, the availability of sanctuary or outside assistance, and beliefs in the rightness of violence. I devised measures for two different kinds of "facilitation." First, there are good historical reasons for assuming that in countries with a history of civil strife violence is likely to become an accepted way of responding to deprivation, i.e., that traditions of civil violence tend to develop and persist. To assess this effect, a measure was constructed of the relative levels of strife among the 114 nations for the years from 1946 through 1959. Second, a composite measure was developed of three kinds of conditions that seemed likely to facilitate strife directly in the early 1960's: the relative size and activity of Communist parties (excluding countries in which they were in power), the extent to which each country had rugged terrain that rebels could use for sanctuary, and the extent to which foreign countries gave direct assistance to rebels.

SOME RESULTS AND THEIR IMPLICATIONS
FOR URBAN DISORDER

The measures of deprivation, of magnitudes of strife, and of the postulated intervening variables were correlated, with the results shown in Table 1.[7] All but one of the independent variables are significally related to the measures of strife. The one exception is the size of coercive force, whose effects were anyway expected to be complex. This variable, and the measure combining size and loyalty, are plotted against the magnitude of turmoil in Figure 1. Large military and police forces alone appear to have no consistent deterrent effect on turmoil, at least at this very general level of comparison, and in some as yet undetermined circumstances turmoil tends to increase as the coercive forces increase in size.

The next step in analysis was to determine how much of the magnitude of turmoil is explained — in the statistical sense — by all the variables acting together.[8] Total magnitude of civil strife is remarkably well accounted for by the eight variables, as shown in Table 2: the multiple correlation coefficient is .806. Turmoil is less well accounted for, with an R of .533. The difference may be partly the result of the way in which turmoil was categorized. Equally important, a great many temporary and local conditions of kinds not represented in the summary measures of deprivation are probably responsible for much turmoil. The results are nonetheless significant enough to indicate that the variables represent some of the major sources of civil strife.

TABLE 1

CORRELATES OF CIVIL STRIFE

VARIABLES	1	2	3	4	5	6	7	8	9
1 Short-term deprivation* (+)									
2 Persisting deprivation (+)	04								
3 Legitimacy (-)	-20	-04							
4 Coercive force size** (±)	-07	-21	25						
5 Coercive force loyalty (-)	-42	-14	48	53					
6 Institutionalization (-)	-17	-37	02	27	41				
7 Past strife levels (+)	34	-04	-05	31	-14	-19			
8 Facilitation (+)	34	17	-15	04	-37	-40	41		
9 Magnitude of turmoil	32	27	-29	-01	-35	-26	30	30	
10 Total magnitude of strife	48	36	-37	-14	-51	-33	30	67	61

NOTE: The proposed relationships between the independent variables and the strife measures are shown in parentheses, + or —. The correlation coefficients are product-moment r's, multiplied by 100. Underlined r's are significant, for n = 114, at the .01 level. Correlations between 18 and 14, inclusive, are significant at the .05 level.

*In regression analyses, separate measures of short-term political and short-term economic deprivation were used. Their separate correlations with other variables are not significantly different from those of the summary short-term deprivation measure.

**This measure was not used in multiple regression analyses.

The next and most important step was to analyze the causal interrelationships among the variables. A basic supposition for evaluating proposed causal relationships is that if X is a cause of Z whose effects are mediated by an intervening variable Y, then if Y is controlled, the resulting partial correlation between X and Z should be approximately zero. The

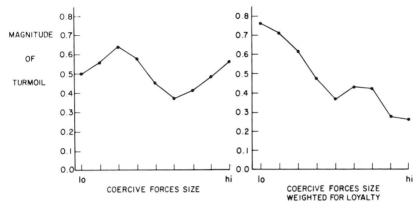

FIG. 1   COERCIVE FORCES AND MAGNITUDE OF TURMOIL

NOTE: The vertical axis gives the average magnitude of turmoil score for (left) deciles of countries with coercive forces of increasing size and (right) deciles of countries with increasingly large coercive forces relative to their loyalty. The range of turmoil scores for the 114 countries is 0.0 to 1.57, their mean is 0.52, their standard deviation 0.40. Units on the horizontal axis represent numbers of cases, not proportional increases in force size/loyalty. Only nine points are shown because the decile score curves were smoothed by averaging overlapping pairs of decile scores.

results of this analysis for magnitude of turmoil provide some surprises.

Remember that I suggested above that all the social conditions intervened between deprivation and strife. The causal model obtained by analyzing a series of partial correlation coefficients is summarized in Figure 2, and suggests a somewhat different sequence of events.[9]

Among nations generally — the qualification that applies to all these results — only three variables are direct and important causes of turmoil: long-term deprivation, a history of strife, and the legitimacy of the political system. These three variables control or mediate the effects of all others. It is important to recognize that this does not mean that short-term deprivation and institutionalization, for example, do not have anything to do with turmoil, but that "something happens" to them en route.

TABLE 2

DETERMINANTS OF MAGNITUDE OF STRIFE:
MULTIPLE REGRESSION RESULTS

Independent Variables	DEPENDENT VARIABLES AND PARTIAL CORRELATIONS	
	Magnitude Turmoil	Total Magnitude of Strife
Short-term deprivation:		
Economic	(07)	24
Political	(08)	(09)
Persisting deprivation	23	39
Legitimacy	-19	-26
Coercive forces loyalty	(-09)	-17
Institutionalization	(-05)	(07)
Past strife levels	21	(04)
Facilitation	(04)	55
Multiple R	.533	.806
Multiple R^2	.284	.650

NOTE: Partial correlations represent the relative contribution of each independent variable to the explanation of the dependent variable when all the other variables are controlled. Those in parentheses are significant at less than the .05 level. Since these analyses are concerned with what is in effect the entire universe of polities, all the correlations are in one sense "significant," but those in parentheses are of considerably less consequence than the others.

The "something" that happens to short-term deprivation is that its outcome is effectively controlled by three mediating conditions. If coercive forces are large and loyal, if few facilitative conditions are present, and most important if there is no tradition of civil violence, short-term deprivation is unlikely to lead to violence. If these intervening conditions are substantially different, however, even mild deprivation is likely to result in turmoil.

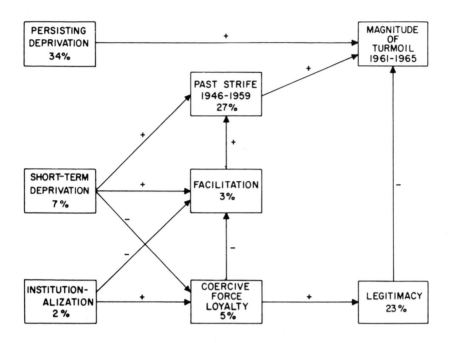

FIG. 2   SEQUENTIAL CAUSES OF TURMOIL

NOTE: The percentages are the proportions of **explained** variance accounted for by each variable when the effects of all others are controlled. Also see note 9.

These general conclusions provide some basis for speculation about the United States. The economic and political measures show relatively little short-term deprivation in the United States during the late 1950's and early 1960's: economic conditions were generally good and improving, and there were few government actions that were likely to antagonize large numbers of citizens. The position of American Negroes may not be reflected in these aggregate measures, however. As a group they did not experience any absolute economic decline, but civil rights activism in the early 1960's began to meet increasingly hostile and often brutal responses at the local level, especially but not solely in the South. We have considerable evidence, comparative and specific, that police or military repression has effects similar to deprivation:

it infuriates its victims, the more so to the extent that they believe their acts are legitimate, and their fury is likely to be contained only by the strongest of internal or external controls.[10] How strong external controls must be for containment can be suggested by examining the policies of the Soviet Union in the thirty years after the October Revolution, and of South Africa since the passage of apartheid legislation. The interpretation I am suggesting with specific reference to the United States is that police clubs falling on marchers' heads in Selma for a national television audience might as well have fallen on the heads of every American Negro, so far as their effect on Negroes' anger was concerned. The consequence, in the language of the theoretical model, was a sharp increase in short-term deprivation.

The three mediating conditions that affect the outcome of short-term deprivation seem at first to point, in the United States, to a minimization of turmoil. The military and police are numerous and unlikely to join in rioting. Extremist organizations were inconsequential in the early 1960's, and material foreign support for urban violence was and is nonexistent. But Americans are not, historically, as peaceful a people as they like to believe. Ethnic, religious, and labor violence are chronic in American history, and with reference to the late 1960's there is little doubt that a new pattern of racial violence as a response to deprivation has emerged, too new to be called a "tradition" but potent in its facilitative effects on deprivation in the future. When this new pattern is taken into account, in conjunction with the growth of extremist organizations that advocate covert and violent protest, the inference is that turmoil will be chronic in the near future. This interpretation — and it should be emphasized that it is only that — has some policy implications. Of the three intervening conditions that are generally relevant, it may be possible to minimize one kind of facilitation, that provided by extremist organizations, but "traditions" are unalterable in the short run. Moreover, coercion seems to have no consistent effects on angry men: it may in fact impel them to violent resistance. One general policy based on these findings would be to minimize coercion and to eliminate the specific short-term deprivations that contribute to violence.

Institutionalization, as it was measured in the cross-national research design, does not generally have a direct controlling effect on short-term deprivation nor any separate effect on turmoil. Instead it seems to be an underlying cause of the other mediating conditions. Countries with large, stable unions, parties, and governments tend to have high coercive potential and few of the immediate conditions that facilitate strife. Extending these generalizations to the United States, institutionalization is relatively high — on this index the United States ranks 36th among all 114 nations in the study — but such summary statistics conceal major internal variations. Negroes in many regions tend to be underorganized and underserved by local government. Moreover, the first major riot in the 1960's, that of Watts, occurred in a community in which by contemporary accounts associational activity and poverty programs had been less effective than those of almost any other large Negro community. Whether degrees of institutionalization are a major factor in American urban disorders is one of the many questions that needs to be studied on a city-by-city, or better, neighborhood-by-neighborhood, basis.

Legitimacy seems to be causally related to strife in the 114 polities, independent of either deprivation or the other intervening variables. About half of the initial correlation between legitimacy and strife is the result of the apparent causal relation between legitimacy and coercive force loyalty: not surprisingly, large and, most importantly, loyal military and police establishments are characteristic of legitimate regimes. Separately from this, however, high legitimacy is an important and independent source of low turmoil. The implication is that political legitimacy is itself a desired value, one so consequential that its absence constitutes a deprivation that incites men to violent political action. One implication for urban racial violence in the United States can be drawn from this. Although most Americans regard their political system as a legitimate one, a growing minority of Negroes say they do not because of what they regard as its deliberately dilatory progress toward racial equality. Whatever the merit of their judgment, it seems likely to persist; and insofar as the general principle applies to the United States, higher levels of turmoil can be expected as a consequence.

The most striking result of the cross-national analysis is that the extent of persisting deprivation has a major, direct effect on the magnitude of turmoil. There seems to be a certain inevitability about the association of such conditions as systematic discrimination, political separatism, religious cleavages, and lack of educational facilities, on one hand, and the extent of strife. No patterns of societal arrangements nor coercive response seem to have any consistent effect on its impact. (The same relationship holds between persisting deprivation and the magnitudes of conspiracy, internal war, and total civil strife.) [11] If the general relationship holds for the United States, then the country is likely to be afflicted by racial turmoil so long as racial discrimination persists. The potential for turmoil has existed since the founding of the Republic. Why so much of it has exploded in *this* decade is suggested by some of the preceding interpretations, to the effect that deprivation has intensified and that social conditions increasingly facilitate its violent manifestation. But the results of the research I have summarized give no general theoretical or empirical reason to believe that turmoil based on pervasive inequality can be permanently deterred or diverted. The only effective and enduring solution seems to be to remove its causes.

NOTES

1. For a more detailed and systematic discussion of the psychological sources of civil strife, see Ted Gurr, "Psychological Factors in Civil Violence," *World Politics,* XX (Jan., 1968). The propositions in this article provide the theoretical basis for the cross-national research discussed in this paper, as well as for the proposed interpretations of the sources and consequences of discontent among American Negroes.
2. William Brink and Louis Harris, *Black and White* (N. Y.: Simon and Schuster, 1967), esp. p. 258.
3. *Loc. cit.*
4. Franz Fanon, *The Wretched of the Earth* (N. Y.: Grove Press, 1963, 1966); Konrad Lorenz, *On Aggression* (N. Y.: Harcourt, Brace and World, 1966). See Gurr, "Psychological Factors in Civil Violence," *op. cit.,* for a critique of some alternative views of the sources of human aggression.
5. No comprehensive report on this research has yet been published. Some preliminary data and analyses are reported in Ted Gurr, *New Error-Composed Measures for Comparing Nations: Some Correlates of Civil Violence;* and in Ted Gurr

with Charles Ruttenberg, *The Conditions of Civil Violence: First Tests of a Causal Model* (Center of International Studies, Princeton Univ., Research Monographs No. 25, 1966, and No. 28, 1967). A summary description of all the measures referred to in this article will appear in Ted Gurr, "A Causal Model of Civil Strife: A Comparative Analysis Using New Indices," *American Political Science Review,* December, 1968 (forthcoming).

6. All polities — nations and colonies — with more than one million people were included, except for five that were excluded because of insufficient data: North Korea, North Vietnam, Mongolia, Laos, and Albania.

7. Correlations with magnitudes of conspiracy and of internal war are not reported here, but are of similar order and significance. Separate measures of short-term economic and political deprivation were also used; the short-term deprivation measure whose correlations are shown in Table 1 is the sum of these two measures and is reported here to simplify comparisons.

8. The "coercive force size" variable was eliminated from the regression analyses; the separate short-term political and economic deprivation measures were used rather than the summary measure.

9. The fundamental arguments on which causal inference analysis is based are summarized in Hubert M. Blalock, Jr., *Causal Inferences in Nonexperimental Research* (Chapel Hill: Univ. of North Carolina Press, 1964), chaps. 2 and 3. The brevity of this article precludes technical description of the calculations which suggested the relationships in Figure 2, but those who wish to reconstruct the analysis and to test the fit of alternative models to the data can do so on the basis of the correlation matrix in Table 1 and the data in Table 2.

10. See Gurr, "Psychological Factors in Civil Violence," *op. cit.,* for some evidence. Two recent papers that make this point with reference to specific cases include Gil C. AlRoy, "The Peasantry in the Cuban Revolution," *Rev. Politics,* XXIX (Jan., 1967), 87–99, and Bryant Wedge, "Student Participation in Revolutionary Violence: Brazil, 1964, and Dominican Republic, 1965," paper read at the 1967 Annual Meeting of the American Political Science Association, Chicago, Sept. 5–9.

11. The inconsequential correlation shown in Table 1 between persisting deprivation and past civil strife is a consequence, first, of the fact that in many countries "persisting" depriving conditions have changed — typically decreased — between the 1940's and the 1960's; and second, that during the 1940's and 1950's reported strife in most traditional and colonial polities was minimal, although objectively they had many of the conditions that led to violence, once their inhabitants were caught up in the "revolution of rising expectations."

A MODEL OF
DISCRIMINATION
AND TENSIONS

Marilyn Gittell
Sherman Krupp

■ The concepts of equilibrating systems can be used to construct a simplified model of race relations. Myrdal used such a framework in his classic study. Our system might be considered an extension of his logic.[1] The basic variables of our model are: (1) the *conventions of discrimination* habitual in the white community; (2) *expectations* by the Negro community; (3) *action* by Negroes to alter discrimination; and (4) a *set of conditions* that describe the contexts of the behavior. These factors, taken together, constrain the workings of the system and provide an explanation of racial tensions.

INTRODUCTION

The *conventions of discrimination* are understood as the accepted patterns of belief and behavior which preserve inequality between Negroes and whites.[2] Each context will reveal its own pattern of conventions. A belief that Negroes cannot easily learn the higher skills, practices that reduce economic opportunities, regulations that reduce Negro suffrage, are examples of such conventions.

These conventions can be ranked in order of intensity and range. The specific intensity and range form an outer margin, or frontier, which is the region in dispute—the convention concerning which whites are indifferent and barely willing to change. A pattern of discrimination, in relation to its context, describes a "discrimination function," or what we will call an "acceptance function." Any change in context conditions will introduce a shift in the rank order, generating a family of functions that vary over the range of that change. In this manner, a new training program to teach Negroes higher skills, or a broadening of voting rights, alter the range and intensity of discrimination—that is, they shift the function and its frontier.

Expectations map the general ideals of Negroes into a stated context, focusing on the immediate possibilities for change in the conventions of discrimination. The achievement of equal education is an ideal; the integration of a local school district is an expectation. Although usually greater than the possibilities for immediate improvement, expectations can be compared with the actual conventions and their changes to provide a measure of the degree to which Negro expectations are realized.

Expectations vary among members of the Negro community, by class, education, institutional affiliation, or as individuals. But insofar as expectations are also norms, they directly affect Negro actions to alter the scope and intensity of discrimination. This means that the actual level of discrimination is not explained by the discrimination function alone, but by its relation to an action function.

The conditions of the model set the context of its application and introduce those factors which act on the system. It is never possible to include all influencing conditions. The selection of conditions is an evaluation of the relevance and importance of the factors that act on the system and shape the functions. If a condition is relevant, a change in its magnitude will affect the outcome through the functions that constitute the system's internal relations. Often, conditions are not included as relevant in the initial formulation of an equilibrating system are demonstrated to be important later on. Such factors can frequently be comprehended by examining their effect on the relevant conditions or by reshaping the functions according to their estimated influence. Policy, for example, can conveniently be analyzed as a change in conditions.

Racial tensions are the main output of our system. Tensions arise because Negro expectations usually exceed realized improvements. An index of tensions, or the potential for tensions, is $\frac{\text{expectations - actual discrimination.}}{\text{expectations}}$ This measure varies from 0 to 1. However, a distinction between the long-term potential for tensions and shorter-run tension-making situations can be important for many contexts. The measure, $\frac{\text{expectations - actual discrimination,}}{\text{expectations}}$ suggests a *level* of tensions determined by the *levels* (magnitudes) of expectations and discrimination. However, the *rate* of change in expectations relative to the *rate* of change in discrimination suggests an important short-run factor. The evaluation of tensions requires a weighting of these two kinds of considerations: the determination of levels and the relative rates of change.

The field which our model represents—conditions, relations, and outcomes—describes change in three kinds of situations: first, situations which are stable over relatively wide changes in initial conditions; second, circumstances where changing conditions perceptibly affect the outcomes of the model; and lastly, situations where changes in initial conditions are so modified by interactions and feedbacks among the relationships that the direction of change can be sharply affected or the model alter its basic boundary conditions. A conflict model is described in the language of equilibrium analysis. The setting conditions and relations represent what we believe to be reasonable empirical approximations.

DESCRIPTION OF THE SYSTEM

THE DISCRIMINATION FUNCTION

The discrimination function derives its characteristic shape from three basic sets of conditions. These are: (1) power relations, (2) the distribution of wealth and income, and (3) the racial composition of population. These circumstances may be viewed in the context of the larger society or within the local community. It is also assumed that industrialization, which for many purposes could be ranked as a distinct condition, affects the system by operating on our three conditions.

Power Relations

The distribution of power is the determinant of the shape of the discrimination function. Its distribution can be considered in three general categories based on the sharing of power between Negroes and whites. These categories facilitate the measure and scaling of power relations.

(1) In some situations, Negroes have little or no political power. Here the unorganized Negro community exists as an alienated mass within the society. There are few mechanisms for the expression of group views and interests. Negro leadership is passive to white control and often represents the choice of whites. This kind of leadership, described as "Uncle Tom" by Thompson,[3] is more loyal to the white community and tends to reinforce the existing high level of discrimination. This category describes the power relations in most Southern communities and some Northern cities prior to 1960. Clearly, the Negro community has little political influence under these circumstances, and the scope and intensity of discrimination are maintained at high levels.

(2) Negroes only have limited political power as an organized group within society. Their power is then manifested as political pressure rather than by direct participation in formal power positions. Negro organizations will develop their own leadership, but compromise with white power holders is the means of influence. The threat of a Negro vote or recognition of Negro support may stimulate white concessions which move the frontiers of discrimination. Frequently the concessions are insignificant and the frontiers are maintained with little change. The shift of frontier, however, will usually depend on the already existing level of discrimination. But even where the level of discrimination is already low, only minimal reduction may occur. A recent study of school desegregation policy in several large cities by the National Opinion Research Center indicated that the pressure by Negro groups to achieve school integration was readily satisfied by public statements by the Board of Education and the Superintendent, although no real change in the school policy was experienced.[4] This situation, where Negroes have little direct political power, is typical of the largest Northern cities, although in New York City,

Chicago, and Detroit transitions to direct political power are occurring.[5] Sometimes Negroes elect one or two legislators, but this is usually a trade-off for support in other elections. Although discrimination is usually less than in situations where Negroes have little political power, the leaders will not press the white establishment for change. In fact, the power of the Negro so much depends upon the white community that the Negro leaders gain little by reducing discrimination and are frequently satisfied with minimal concessions from the whites. Typical of this was the role of the Urban League in most cities throughout the country. Prior to 1965, the Urban League in New York City, probably the largest in the United States, was satisfied to find jobs for 18 Negroes each year.[6]

(3) Negro power is greatest where a segment of the Negro community are active participants in decision-making. This would be reflected in the election of Negroes to public office, the active participation of Negroes in prestigious interest groups, the development of influential Negro interest groups, and, possibly, the development of Negro bloc voting. The Negro community can under these circumstances function as an active social and political force, pressuring whites to give them decision-making roles. Negro organizations take on greater significance, seek added political strength, and actively engage in reducing the existing levels of discrimination.[7] Negro leaders do not enhance their power by minor concessions from the white establishment; they draw their power from direct participation. The Negro vote is itself important to the power relationship. It may become the pivotal vote, as it was in the national election in 1960 or the New York City vote in 1965, and greater concessions are necessary to assure support.[8]

But where discrimination is low, and Negro expectations and action especially high, there is always the possibility of "white backlash." White backlash is a reaction to the rapid emergence of Negro power or action. Blacklash, if it is effectual, will retard the rate of change in the frontiers of discrimination and may even result in a temporary increase in the level of discrimination. This is evident in the response of whites to the current straining for "Black Power" in certain Negro groups. Not only is there stiffening resistance among more conservative white groups to pressures for change, but there has been witharawal of some liberal groups from the integration movement.

Power relations serve as the major constraint to change within
the system. White restraints on Negro power act to maintain the
level of discrimination, increasing the stability of the system.
Where the level of discrimination is high, the power structure is
usually more unfavorable to Negroes. This unequal distribution of
power acts as an effectual brake on the reduction of discrimination; in a more equal balance of power, the same set of influences
would have greater impact. As inequalities are reduced, discrimination becomes more responsive to changes in conditions. Where the
power of Negroes is greatest, the frontier of discrimination is also
most shiftable, but here there is also the possibility of white backlash. Backlash can retard the forces making for racial equality.

The Distribution of Wealth and Income

The Negro share of national wealth and income is substantially
smaller than their relative population. Redistribution, creating
greater equality with whites, increases their market power, forcing
the business sector to reduce discrimination in order to reach and
control that market. There are some indications of the effect of
even slight changes in recent years in advertising campaigns which
use Negro models to appeal to Negro consumers. Where the Negro
represents a significant market, practices which excluded Negroes
from stores or recreation areas have been relaxed. It is possible,
also, that resistance to integration and educational equality by the
business community is reduced because improvement in education
increases the supply of scarce skills in the labor market. The
unequal share of income received by the Negroes sharply reduces
their capability to improve their skills.

Improvement in the distribution of skills among Negroes, which
is closely related to the distribution of income, would increase
Negro productivity, change the occupational structure among
Negroes, and shift the discrimination frontier. There has, in fact,
been an increase in the number of Negro white collar workers,
although concentrated in the area of government services. This
group has emerged as a small but growing middle-class population.
They have created a demand for housing and recreation, encourag-
ing a reduction in discrimination to satisfy their new economic
position. In relatively low discrimination areas of the North, they
have begun to make headway in decreasing discrimination in
employment, housing, and education. In the high discrimination

areas of the South, similar groups have been the source of recent efforts to reduce discrimination by active support of sit-ins in restaurants and in transportation. Obviously, the more mobile, middle-class Negro is especially concerned with the reduction of discrimination. Middle-class efforts to reduce discrimination tend to recognize the role of political power, the use of the ballot box, the elimination of gross social inequalities, and the necessity for gradual change in the system. They rely less on violence as a form of action; indeed, the non-violence technique is a middle-class phenomenon.

However, the redistribution of wealth is especially difficult. It is more difficult, the higher the initial level of discrimination. Clearly, the Negroes receive relatively larger shares of wealth in the North than in the South, and migration and educational improvement tend to improve their position. But in a market economy, the redistribution of wealth requires, short of radical revision of the institutional structure, large reduction in discrimination and improvement in Negro productivity.

The Distribution of Population

The most important change in the character of the American population in the last three decades has been the migration of Negroes to cities, particularly Northern cities. The migration pattern was first a movement from rural to urban areas in the South and then a movement to the Northern city. (Over 70% of the Negroes in America now live in cities.) This change in the distribution of Negro population has influenced discrimination in several ways. In several large Northern cities, the Negro population exceeds 25% of the total city population.[9] The segregation pattern in these cities has intensified; black belts are concentrated in the core city, and suburban exclusion prevails. Grodzin notes that the added factor of white middle-class movement to the suburbs (generally people of greater tolerance), combined with the increased frustrations of ghetto life, produces increased interracial tensions.[10]

Although there has been a slight decrease in the overall number of Negroes in some Southern cities, the percentages have changed very little, and in the "black belt" Negroes still make up a majority or close to a majority of the population. In areas where there are very few Negroes, the discrimination level is generally low. As the percentage of the Negro population approaches a "tipping level," discrimination may increase.[11] Key demonstrated the political implications of large Negro population in his classic study, *Southern Politics*.[12] In those communities in the South where Negro population was highest, the political support of discrimination was greatest. In an empirical test of Key's hypothesis, a later study found the strongest white protest to school integration in the South in the areas with the largest Negro population.[13] In part this is due to the character of the migrants, predominantly lower-class, poorly educated Negroes with few of the skills actively demanded by the industrial North. A shift in population may also affect political power in a given community. However, in the South the increase in urban population will produce different results than in the North, the failure of the Northern cities to raise economic and social conditions rapidly enough has undoubtedly contributed to the tensions and volatile responses in the poorer Negro communities of the North. The riots in Watts and other Northern cities are symptomatic of this response.

In sum, the discrimination function has the following shape: Where discrimination is already very high, the function tends to be unresponsive to changes in conditions and to Negro action. It is "boundary maintenance." Where discrimination is less pronounced, as in the North, it is more sensitive to changing conditions and to action. Where discrimination is already very low and Negro action high, the function changes rapidly, but uncertainly, relative to changes in conditions.

THE EXPECTATIONS FUNCTION AND NEGRO ACTION

Expectations in the Negro community provide the second major function which will determine the output of racial tensions. The conditions which influence expectations can be grouped into three main categories: (1) the educational level of the Negro, (2) his economic status, and (3) the ecological setting. A Negro action function is directly derived from the behavior of expectations.

The Education Level of the Negro

As the education level of the Negro increases, there is greater exposure to the styles of life of the dominant culture, and his horizons are expanded. Educated Negroes will want better jobs, improved housing, and better education for their children. They will see few differences between themselves and whites that can account for the great gap in economic and social condition. Although the goals of less educated Negroes may be great in relation to what is possible, their concrete expectations are always more limited. In the South, the education level is low, and expectations are also low. The Southern Negro has been taught to live with his condition. The same Negro who accepts poverty and discrimination in the South, however, is less likely to be content with only slightly better conditions in the North. It is not that the less educated Negro expects a great deal. Urban poverty is more devastating because of the continual confrontation with higher standards which produce unrealizable hopes. The less educated Negro does not normally choose organizational or political action when his expectations are not fulfilled. The short-run failure of expectations to be filled, in the context of the looser social and political controls in the North, permits racial tensions to be expressed in forms of social disorganization such as riots.

One would normally think that the level of education would directly affect discrimination. This is not necessarily so. The discrimination function is mainly affected by the power relations, and the effect of education is indirect. It acts through the power relationship. To the degree that more educated Negroes are concerned with political power, they can use political influence as an instrument for change in discrimination. Perhaps in the very long run, education can shift the discrimination function by decreasing the skill differential between Negroes and whites. However, education is tied to relatively stable patterns of income distribution. Also, most data suggests that increased education does not result in a commensurate increase in economic status for the Negro.[14]

Relative Economic Status Within the Negro Community

The level of expectations of segments of the Negro community varies according to their internal class status. Although middle- and upper-class Negroes may withdraw from the larger lower-class Negro community, their expectations will probably increase because of their increased interaction with the whites.[15] They are especially sensitive to the conventions of discrimination which prevent their achievement of social status in the white as well as the Negro community. To this extent, their expectations may be less constrained than among the less successful. The important place of the middle-class Negro in reducing discrimination in housing, education, and recreation is an indication of their concern and their willingness to act. The frontier of discrimination for the middle-and upper-class Negro need not be the same as for the lower-class Negro.

Lower-class Negroes are more concerned with minimal economic goals. Protest leaders who articulate lower-class needs focus on employment, welfare, and basic housing and education facilities. The relative deprivation of lower-class Negroes, particularly in the Northern ghetto areas, may foster greater frustrations as a result of relatively high (as compared to the lower-class Southern rural Negro) expectations. When the discrimination level is high, expectations are constrained by the boundaries of the system. The caste mentality of the Southern lower-class Negro will limit the range of his hopes. The Northern lower-class Negro anticipates more from his environment because of the Northern (unfulfilled) ideology of equality.

There are also apparent differences in the forms of action by different classes. Middle-class Negroes are more likely to use long-term conventional political and social pressures. Lower-class Negroes suffer more immediate frustrations and tend toward more volatile short-run reactions.

Ecological Setting

Ecological setting embodies the historical conditions of a local culture. The social structure of a community outlines accepted practices for Negroes and thus sets limits on their expectations. In the South, the caste system influences all facets of Negro behavior as well as the discrimination level of the white community.[16] An increase in the educational level achieved by Negroes, or an increase in economic position, will increase expectations, but within the limits set by the ecological environment. Rohrer and Edmundson clearly indicated only small differences in the expectations of the educated middle-class Negroes, in contrast to lower-class Negroes in New Orleans.[17] The distinctive pattern of the Southern Negro protest movement which is directed at removing Jim Crow practices is another indication of the importance of the ecological setting.[18] In urban industrial areas in the North and the South, exposure of the Negro to the white culture is greater, expectations will therefore increase, and action will be directed at different conventions.[19]

In general, the expectations function will show the following behavior: Where discrimination is high and the Negro is poor and uneducated, his expectations will be low and unresponsive to changes in conditions. Moreover, improvement in his expectations does not result in proportionate increases in action. In contexts where discrimination is low and education levels and economic status high, expectations will be high and upwardly responsive to improvements in condition, possibly more than proportionately. For every change in expectations, action will increase at least proportionately. As compared to the discrimination function, the expectations and action function is generally more volatile.

Tensions result from the interaction of expectations and actual discrimination. The intensity of tensions may be measured operationally and quantitatively. Tensions are also measurable less directly by aggregate acts of deviant behavior. An indicator of tension is the extremes of behavior. Deviant behavior reflects alienation from and disagreement with the norms of society; riots therefore occur mainly in lower-class neighborhoods, as a form of protest. But any form of collective behavior by Negroes would suggest frustration and tension. Increase in membership in the Black Muslim movement, for example, is a sign of increased frustration among elements within the Negro community.[20] Black nationalism in any form—e.g., black power, or what might well be called "black backlash,"—is an important indication of tensions;[21] the same may be said for "white backlash." Tensions can be measured by behavior as well as by attitudes. In some cases tensions can take the form of withdrawal as well as counterattack. Negro parents removing their children from public schools in large numbers would be indicative of withdrawal, yet symptomatic of tensions. Increases in the membership and number of Negro protest organizations, and action such as strikes and boycotts of schools, reflect the counterattack manifestations of tensions. Negro attitudes towards whites, readily ascertained in public opinion surveys, could measure changing tensions.

Although tensions are an output in our system, they are also fed back into the system to influence the relations and conditions. In low discrimination areas, for instance, increases in the membership and activity of Negro protest groups could re 't in an increase in expectations and in an increase in the poli._ ,al power of Negro groups. In turn, this could effect a reduction in the level of discrimination. Feedback from tensions in high discrimination areas could operate to encourage white backlash. Rioting by Negroes intensifies the response of hostile whites, increases tensions further, and encourages black backlash.

Increased identification with black culture and organizations is one of the overt signs of tension and frustration with the dominant white culture. Black power is one of the more extreme recent manifestations of this kind of movement. The feedback from these tensions may have varying results in high and low discrimination areas. The form that white backlash may take will vary with the degree of discrimination. Where discrimination is already high, and power relations extremely unequal, white backlash, as a special form of resistance to Negro action, is likely to be more successful.

All of the major conditions determining the character of the two functions are measurable in principle, and often they can be measured by data now available or readily available in comparative form. In most cases, tabulation of the data will even permit the development of indices and rank order classification of variables by an ordinal scale. In many instances, historical treatment is also possible. Geographic areas can be compared to each other in measuring high and low levels of expectations and discrimination; within an area, historical comparison is possible as well.

Changes in the level of discrimination can be measured not only by attitudes (i.e., the Guttman scale) but by changes in policy (i.e., voting laws, school integration, housing legislation, etc.) and public practices. Changes in the levels of expectations are also measurable by attitudes. They may be measured as well by observable increases in the number and kind of Negro organizations, their stated policy goals, and their action programs. This macrosystem is oriented to the utilization of operational research techniques in which the relevant properies can be isolated for the purpose of prediction of change and the evaluation of policy alternatives.

A QUALIFYING CONDITION

The general thesis that almost no change in functions is possible in high discrimination areas is also borne out by the historical resistence to change in the black belt areas in the South. In comparing the communities in the South to each other, a pattern becomes clear. The relatively lower discrimination sectors have been more responsive to action than the areas of high discrimination. Changes in federal policy have been more influential in lower discrimination communities. Federal policy has had only nominal effect in the higher discrimination areas, and it has produced violent white backlash in the most extreme areas of discrimination.

However, radical change in a condition in society which is generally imposed externally can usually shift responses in local communities, regardless of the other conditions. Extremely rapid industrialization, for instance, may directly affect the stability of the system by shifting the basic functions and by altering their shapes. Similarly, a major upheaval in the American economic system could produce the kind of redistribution of wealth which significantly might alter the patterns of response.

NOTES

1. Gunnar Myrdal, *The American Dilemma* (N.Y.: Harper, 1944, 1962). Myrdal juxtaposed ideological commitment of the society with performance; the gap is reflected in frustration.
2. Peter I. Rose, *They And We* (N.Y.: Random House, 1964), p.79.
3. Daniel Thompson, *The Negro Leadership Class* (Englewood Cliffs, N.J.: Prentice-Hall, Spectrum, 1963). Thompson discusses the evolution of Negro political leadership in the South. In describing "Uncle Tom" leaders, he notes: "The Uncle Tom is fundamentally a preserver of a biracial system which perpetuates white paternalistic men of power in their status as hosts, and Negroes as parasites," (p.63). Wilson in describing Negro politics in the North suggests that strong political machines among Negroes in Northern cities are unrelated to or may hinder the ability of Negroes to secure important appointive offices in state and local government. James Wilson, *Negro Politics* (N.Y.: Free Press, 1960), p. 39.
4. Robert L. Crain and David Street, "School Desegregation and School Decision-Making," *Urban Affairs Q.*, Sept., 1966.
5. Wilson, *op. cit.* note 3, p. 99. Wilson suggests that the differences are in the cities themselves. Where there are smaller Negro communities, as in Boston, Minneapolis, and Denver, revenues for support are lacking. Also see Robert Dahl, *Who Governs?* (New Haven:

Yale Univ. Press, 1961). The development of Negro power in New Haven is described by Dahl as evolutionary, and he projects that the Negro will gradually acquire direct political power.

6. Marc Wallman, *Negro Organizations in New York City*, (unpub. honors paper, Queens College, N.Y., 1964).

7. Blumer strongly suggests the importance of organizational pressure and influencing decision-makers. Herbert Blumer, "Social Science and the Desegregation Process," *Annals*, March, 1956, p. 142.

8. Several Republican party leaders have indicated the need of the party to appeal to the Northern urban Negro voters in national politics. See also Brink and Harris' discussion: William Brink and Louis Harris, *The Negro Revolution in America* (N.Y.: Simon and Shuster, 1964), pp. 78-95.

9. Karl E. Taeuber and Alma F. Taeuber, *Negroes In Cities* (Chicago: Aldine Publishing Co., 1965), pp. 11-14.

10. Morton Grodzins, "The Metropolitan Area As a Racial Problem," in Earl Raab, *American Race Relations Today* (N.Y.: Doubleday, Anchor Original, 1962), pp. 97-98.

11. *Ibid.,* p. 91. Grodzins notes the unwillingness of whites in Northern cities to live in proximity to *large* numbers of Negroes. He suggests limits on Negro migration as a possible solution.

12. V. O. Key, Jr., *Southern Politics in State and Nation* (N.Y.: Knopf, 1949), pp. 513-519. Similarly, in a study of Jews, Dean hypothesized that "the larger the proportion of Jews in the community, the more Jews are excluded from socially elite organizations and residential areas." John P. Dean, "Patterns of Socialization and Association Between Jews and Non-Jews," *Jewish Social Studies,* July, 1955, pp. 249-251. Rose notes: "Comparable research on the relations of other minority groups to the larger community suggests that...this hypothesis states a valid generalization about majority-minority relations in the United States." Rose, *op. cit.,* note 2, p. 59.

13, Allan Gross, *A Reevaluation of Key's Hypothesis on School Integration.* (unpub. M.A. thesis, Queens College, N.Y., 1964). Wilson points out that the Northern states with the most liberal legislation from the standpoint of Negroes are often states whose largest cities have only a small proportion of Negroes. He attributes this to the need to satisfy the interests of other liberal groups. Wilson, in fact, generally gives greater credit to non-Negro political leadership for waging the battle for anti-discriminatory legislation. Wilson,*op. cit.,* note 2, pp. 99-100.

14. Thomas F. Pettigrew, *A Profile of Negro America* (Princeton, N.J.: Van Nostrand, 1964), pp. 188-189. V. W. Henderson, *The Economic Status of Negroes: In the Nation and in the South* (Atlanta: Southern Regional Council, 1963).

15. Pettigrew, *op. cit.,* note 14, pp. 181-183.

16. Herbert Blumer has noted: "Race prejudice becomes entrenched and tenacious to the extent the prevailing social order is rooted in the sense of social position. This has been true of the historic South in our country. In such a social order, race prejudice tends to become chronic and inpermeable to change." Blumer, *op. cit.,* note 7, pp. 6-7.

17. John Rohrer and M. Edmundson, *The Eighth Generation Grows Up* (N.Y.: Harper Torchbooks, 1960).

18. This is supported in a recent study by one of the authors in which Southern middle-class Negroes stated a preference for Negro colleges, while Northern middle-class Negroes preferred Ivy League

colleges and state universities. Marilyn Gittell, "A Pilot Study of Negro Middle-Class Attitudes Towards Higher Education in New York," *J. Negro Education,* Fall, 1965 pp. 385-394. See also Pettigrew, *op. cit.,* note 14, p. 49.

19. Drake and Cayton suggest that Negro migrants from the South gradually lose their "caste mentality." This would suggest that everyday expectations of Northern Negroes (lower-class) are greater than those of Southern lower-class Negroes. St. Clair Drake and Horace R. Cayton, *Black Metropolis: A Study of Negro Life in a Northern City* (N.Y.: Harcourt, Brace, 1945), pp. 174, 759.

20. E. U. Essien-Udom, *Black Nationalism: A Search for an Identity in America.* (Chicago: Univ. of Chicago Press, 1963).

21. Arnold Rose and Caroline Rose, *America Divided* (N.Y.: Knopf, 1948), p. 218.

POWER, CONFLICT, AND SOCIAL CHANGE

James H. Laue

■ Viewing *power* as the *control over decisions,* and *social change* as the continuous process of *redistribution of power* within social systems, this paper examines the role of conflict in the extensive changes in American racial patterns in the last ten years. It develops a framework to help explain what has happened, and raises questions about the nature and directions of racial change today and in the next few years. The model[1] is intended to have general applicability to processes of change through conflict.[2]

I have seen communities go through a common sequence of stages in working out desegregation and elimination of certain discriminatory practices in response to challenges from minority groups. The phase-structure of social change presented here was developed in response to the data from my research on the direct action movement of the early 1960's in more than one hundred southern communities, with particular emphasis on the sit-ins and Freedom Rides.[3] In every community, the pattern seemed to be the same: challenge by the minority group, a period of overt community conflict cresting in a crisis, the drawing of hitherto uninvolved elements of white power into negotiation, and the working out of some change.

MINORITIES AND POWER

The model is called the 7 C's. It is an attempt to systematize the process which I saw occurring in city after city as desegregation was achieved. It starts with the assumption implied at the beginning of the paper, that significant social change takes place when new combinations of community power groups emerge to force an alteration in the perceived self-interest of the powerful and, therefore, a change in their priorities for actions. Direct action achieved success as a technique for forcing certain concessions because it developed a new source of power within groups which did not have sufficient amounts of the normal sources of power-for-change in a democratic system. Lacking sufficient political and economic power, Negroes used the only form remaining to them—their bodies. Through the testing and refining of direct action techniques (literally "putting your body on the line"), the movement developed negotiable forms of power. In turn, legislation and other action stimulated by the civil rights movement have begun to equip Negroes more adequately with the standard forms of political and economic power.

In explaining how redistribution of community power occurs through the 7 C's phase-structure, I begin with an eighth C: Competition. Competition is a constant in all social interaction. All the stages of the model are varying forms of the competitive process. Competition is the process whereby persons quietly oppose one another in seeking the scarce rewards they have learned to want. It is, in this statement, close to what Sumner meant by "antagonistic cooperation."[4] It is a natural, ever-present condition of all social interaction, and the starting point for any analysis of social change. The holders of power in all social systems try to suppress the notion that certain types of competition exist. The American race-related version, once restricted to the South but now heard increasingly in northern cities in only slightly altered language, is "our colored people are (or were) happy."

THE 7 C's

The 7 C's—the stages communities have transacted in working out change in racial patterns—are:

1. Challenge
2. Conflict
3. Crisis

4. Confrontation

5. Communication

6. Compromise

7. Change

1. *Challenge* is the open and dramatic presentation of demands and grievances by the minority, including the range from single specific grievances to an attack on the whole pattern of systematic discrimination in a community. The challenge is usually a last resort after less public approaches (educational and legal, for instance) have failed to bring significant communication or change. The classic example is the sit-in. Others are marches, picketing, or other well publicized challenges to the status quo, such as lawsuits or boycotts.

2. *Conflict* is intensified competition of which a substantial proportion of the community is now aware. An accumulation of challenges brings ever-present competition to the surface, and it breaks through into overt conflict. Traditional mechanisms of social control are no longer able to manage the increasing frequency of challenges.

3. *Crisis* exists when elites with the power to change the discriminatory patterns challenged *define* the situation as severe enough to demand immediate action and rapid resolution. They have, that is, changed their perception of their own self-interest and what it will take to maintain and expand it. When the powerful decide to act, the crisis has been achieved. Crisis thus is a subjective power-term, determined by the plans and actions of the powerful, rather than by the objective conditions of the situation. Some of the conditions which have persuaded the powerful to make the crisis-definition are demonstrations, boycotts, economic loss, damage to the city's image, a court decision on desegregation, violence—or the seriously anticipated threat of any of the above. Once the crisis-definition is made, for whatever reasons, the meeting of at least some of the demands presented in the Challenge is inevitable, and the process usually hurries through the final stages to Change.

4. *Confrontation* takes place when the decision-makers being challenged recognize that the minority group has legitimate

demands which can no longer be explained away, but which must be dealt with. If there is a stage in the process in which normative awareness develops on the part of the powerful, that is it. Awareness of the moral legitimacy of demands results from the harder realities of economic and political awareness learned in the previous stage. At least it may be said that the challenge-targets are more suggestible and more receptive to previously rejected change proposals and persons.

5. *Communication* is direct, face-to-face negotiation between the challengers and the dominant group(s), each now bargaining from a position of power. Post-crisis communication is always more frank and goal-directed than pre-crisis communication, for the powerful have been forced to drop defenses and take positive steps to remedy problems of which everyone is now consciously aware. The newfound power of the challengers consists of threats (to create further crises if certain demands are not met) and promises (*not* to create crises if demands *are* met).

6. *Compromise* is a result of the bargaining which takes place in stage 5, and usually requires the enlistment (often covert) of community power resources beyond the parties in the Communication process. There are victories and concessions for all communicating parties, with each faction usually asking for more than it expects to gain, in preparation for the anticipated giving-in of the Compromise stage.

7. *Change* is, by definition, the achievement by the protesters of at least some of the goals set forth in the Challenge.

OTHER DIMENSIONS OF THE MODEL

There are several qualifications and subsidiary processes regarding the operation of this system. First, the model as outlined here recognizes that there are many centers and sources of power in communities and nations, and that the 7 C's process therefore operates simultaneously at many levels. Community power structure is not seen as monolithic, but rather as a series of interacting and interlocking systems—which means that the conflict process may operate within and between systems as well as in relation to the total community. It is macro as well as micro.

Second, the system *cycles*—that is, it runs its course on different issues over and over again in communities, with the stage 7 resolution of one conflict situation becoming the plateau upon which new challenges arise. A good example is Nashville, which in the years from 1960 through 1963 went through virtually the same process of appeal, Negro demonstration, white violence, boycott, arrests, community concern, and change on four different desegregation issues: lunch counters, theaters, retail employment, and the better restaurants. The emergence of the early stages does not always mean the cycle will run its course, however. It may get bogged down at any stage, then remain latent until sufficient challenges are taken up again. This was the case with Albany, Georgia, in 1961 and 1962, when the process never got past the conflict stage despite white retail losses of more than sixty percent. Local white influentials refused to make the crisis-definition, and eventually the movement lost momentum.

Lateral communication operates when non-conflict cities move to make change because of conflict and crisis in other communities. An avoidance model quickly develops. In Carolina and Virginia cities, significant changes which had been the object of protest for many months came about in a matter of days in mid-1963, because Birmingham was on. A typical northern example came from Philadelphia in August, 1967, largely as a result of the message from Detroit. The city's business leaders discovered 1,200 jobs "for the idle poor" in response to the Mayor's "call for help in easing ghetto tensions" and his statement that such hiring is "not now a matter of charity, but an investment in our community."[5]

Short-circuiting is the process whereby communities move directly from stages 1 or 2 to stage 5 or even 6. They have recognized, based on previous local experience or contemporary lateral communication, that the change being demanded by the challengers is inevitable, and that it is in the self-interest of the powerful to move directly to Communication (stage 5) or to the sixth stage of active re-ordering of priorities and enlistment of other community resources for change. A dramatic southern example is the immediate compliance of Albany, Georgia, officials with the 1964 Civil Rights Act, telling newsmen, "Of course we're going to obey the law. We don't want Martin Luther King coming down here again."[6] Short-circuiting of the change process also will be one result, I believe, of the current public and private sector activities responsive to the summer's urban racial violence.

The other side of the short-circuiting process, however, is the *neutralization of crisis.* Crisis tolerances change as communities learn to combat direct action and other forms of challenges. In most cities in the early 1960's, sit-ins were enough to stimulate a crisis-definition, but today they are dealt with as a matter of course and are generally not effective as a change technique. A surprising exception took place in early June, 1967, in Boston, however, when policemen tried to disperse a group of welfare mothers sitting and lying-in overnight in a public building, and triggered two days of violent conflict between Negroes and the police.

From the viewpoint of systemic integration, it can be added that the operation of the 7 C's is, in the long run, a healthy process. Although conflict is disruptive in the short run, it serves to bring systemic stresses to the surface and forces communities to confront them and deal with them forthrightly, as Lewis Coser and others have pointed out.[7] Conflict is not the cause of social system stresses—rather it is a symptom, a process whereby the stresses may be transformed for remedial action. Andrew Young of the Southern Christian Leadership Conference has summed it up best in lay terms. "The movement did not 'cause' problems in Selma, as Sheriff Jim Clark and others claimed," Young says. "It just brought them to the surface where they could be dealt with. Sheriff Clark has been beating black heads in the back of the jail on Saturday night for years, and we're only saying to him that if he still wants to beat heads he'll have to do it on Main Street at noon in front of CBS, NBC, and ABC television cameras."

CASES IN POINT, SOUTH AND NORTH

There are other examples from hundreds of communities of how conflict-produced crises have forced White Power to enable changes in racial patterns, a few of which may be mentioned here. Desegregation of lunch counters took place in more than two hundred southern cities within a year after the sit-in movement began in February, 1960. The Freedom Rides took place in May of 1961, Alabama whites provided the violent counterpoint, the federal government intervened—and by November a Federal Communications Commissions order was issued (and enforced) banning discrimination in interstate transportation facilities.

Atlanta and Birmingham offer other instructive empirical cases. The different crisis-tolerances in these cities is directly reflected in the ease with which change takes place, and the extent of the change. In Atlanta, a few sit-ins and arrests, SNCC picketing and Klan counter-picketing were enough to mobilize the necessary power centers for change. But in Birmingham it took Bull Connor, police dogs, fire hoses, 3,500 arrests, and pictures transmitted all over the nation and world before the "Committee of 100" got itself organized to make changes it had been capable of making all along.

It has worked in the North, too. In 1967 in Cleveland, Sealtest short-circuited the 7 C's by responding to the threat of a boycott with a $300,000 program for job training and community development in the Hough ghetto—a program it had been capable of delivering for some time. In Chicago, the Daley machine responded to SCLC's challenges in 1966 by giving minor concessions at the first push on housing inspection and enforcement and other targets. The Chicago Metropolitan Council for Open Cities, which is now working actively with the endorsement of corporate, labor, religious, and political elites, came into being after a summer of Negro open-housing demonstrations and violent white responses. The Council was, in effect, a trade for King's agreeing not to lead a massive march into Cicero.

At a national level, the Civil Rights Act of 1964 and the Voting Rights Act of 1965 are direct results of the operation of the 7 C's process. After three years of direct action throughout the South, Birmingham produced a crisis-defintion in the national administration in May of 1963. President Kennedy proposed the Act within a month, communicated on nationwide television using ever-present but previously unused Biblical and Constitutional value statements, and the Administration fought the bill through the compromise process of the Congress to enactment and the beginning of the change process within a year. The phase-structure of passage of the 1965 Voting Rights Act was much the same: the beating at the Selma bridge in March of 1965, a dramatic film clip on national television the next night, mobilization of national support by the movement, President Johnson going before the

Congress and national prime-time television several weeks later to request legislation and say "We shall overcome," the passage of the Act four months later in July, and the registration of more than 1.5 million new southern Negro voters within the next two years.

BUT DOES IT STILL WORK?

But a look at Congress, the urban violence, and the poll data on white attitudes today makes me seriously ask, "Do the 7 C's still work?" The last part of the paper is directed toward that question, based on an analysis of the changing conditions of racial conflict we have experienced and may expect in the next few years. This concluding section offers few answers but raises many questions bearing on the viability of the 7 C's model in the North in 1968 and beyond.

I question whether the model has long-range applicability as the racial focus moves urban and North because of two conditions which must be met if a social system is to achieve change through conflict:

1. The 7 C's can only operate in a social system in which the superordinate power *permits* conflict as a way of doing change. The 7 C's would not have worked in Nazi Germany; rather a kind of short-circuiting from Challenge (1) to repressive Change (7) would have taken place.

2. It appears that the superordinate power must not only permit conflict, but must, in fact, approve of the goals of the protesters.

What operated in the South, then, was a concessions model in which Negroes challenged politically expendable cities, counties, and states, while the federal government permitted the conflict process to run its course through to change, and tacitly (and often openly) supported the goals of the protesters as being consistent with basic American values. The permissive superordinate power was the federal government, supported by the mass media and the white liberal-labor coalition.

But now it is the superordinate power itself which is under attack. The targets now are the federal government, the large northern cities (with their nonexpendable Democratic machines), "white liberals," the mass media, judges who do not enforce housing codes, the real estate industry—in short, the whole middle-class establishment, including people who write articles like this and people who read them. It was easy for the superordinate power in its various forms to support conflict in Alabama and Mississippi: the moral issues were seemingly clear, the devils were readily available in brutal sheriffs and unattractive politicians, and it was far away. But the same persons, newspapers, private organizations, and government officials who endorsed the movement's implicit use of the 7 C's process in the South, now confuse demonstrations with riots and label southern protest techniques turned northward as ill-considered and irresponsible.

An important theoretical question, then, is whether the American black movement needs and can find a higher superordinate power to legitimate and support its activities. Contacts of civil right groups with revolutionaries and established governments in Africa and Latin America have been increasing in the last year. SCLC is beginning to convert its program to an international base, as presaged by the late Dr. Martin Luther King's anti-war activity and the continuing work of some SCLC staff members in Africa. And the movement has long considered soliciting United Nations support in the struggle against American racism.

In the light of these recent shifts in targets and allies—and of the urban racial violence of the past four summers—an even more important question is "Under what conditions will the crisis-definition become repressive?" In the current statement of the 7 C's, the crisis-definition is essentially oriented to democratizing change. It assumes, within the context of a sympathetic superordinate power, that crisis always moves the targets of the challenge toward actions supporting desegregation and equal opportunity.

In addition to specifying the conditions under which the crisis-definition becomes repressive, we need to develop an alternate version of the stages following crisis.

Political and military reactions to the 1967 urban summer violence lead some to speculate that political leadership at both local and national levels is closer to a repressive crisis-definition in relation to an American minority group than at any time since the Japanese "relocation" in World War II.

"Don't reward the rioters" sentiment is high among whites, and black ghettos are full of rumors that the concentration camps are ready—and the wry comment that ghettos are too ecologically strategic for the Man to bomb the people.

The rise of black nationalism confirms the hostility of white leaders: "Look at all we have done for them in the last few years, and now they're saying maybe they don't even want integration!" To the consternation of whites, the slain Malcolm X is coming to be the same kind of hero for many black ghetto people that the slain John Kennedy is for many white liberals.

Despite the intensifying confrontation between white sentiment for law-and-order and black sentiment for change, there are some tentative signs that the 7 C's are operating for change on schedule, at least at the national level:

¶ Vice-President Humphrey called for an urban Marshall Plan in a Detroit speech in August, 1967—a concept developed and popularized by Whitney Young, Bayard Rustin, Martin Luther King, Roy Wilkins, and others more than two years ago, but rejected by the Administration.

¶ The President made strong value statements about poverty, discrimination, and other structural conditions of urban ghettos in his televised address to the nation in response to the riots, in addition to strong remarks about law and order.

¶ The Urban Coalition was formed by New York Mayor Lindsay, David Rockefeller, the U.S. Conference of Mayors, and others on July 31, 1967. It has begun a national political mobilization to get Congress moving on urban legislation, urged the nation not to "penalize the majority because of resentment of the criminal acts of a tiny fraction," and has "called upon the nation and the Congress to reorder our national priorities, with a commitment of national resources equal to the dimensions of the problems we face."[8]

¶ Both the late Senator Robert Kennedy and Vice-President Humphrey have in recent movements put the issue in basic value terms in speeches about Negro alienation and identity, Kennedy saying that the ghetto poor have been denied "the most fundamental of human needs—the need for identity, for recognition as a citizen and as a man."

¶ The New Detroit Committee formed after 1967 disorder and composed of board chairmen and presidents of the automobile industry and the most powerful local retail businesses, has mobilized significant political and economic power for change in a short time. They have actively and personally lobbied for statewide open-housing legislation with the Michigan Legislature and, from August, 1967, to February, 1968, opened up 50,000 jobs in Detroit—half of them going to Negroes. Corporations like Chrysler and Michigan Bell have placed employment offices in Detroit ghettos, and are exercising new flexibility regarding educational backgrounds and arrest records of prospective employees.[9]

But in the face of these positive change-oriented responses, we must raise the question of how long the essentially expressive violence of the summer outbursts will be tolerated by the national political system. We can predict with some degree of confidence that the system will soon develop repressive crisis-responses if the violence moves from the expressive level to the strategic or political. And with somewhat less certainty, I suggest that city and national administration leaders will not long permit the kind of massive applications of nonviolent direct action techniques to northern urban systems of government, transportation, and commerce that SCLS led in Washington, D.C., in the spring and summer of 1968.

THE THEORETICAL TASKS AHEAD

I look forward to the unfolding of events in the next few years as they inform the unfolding of this particular change model. In summary, I believe that the next tasks in the theory's development are:

1. Develop a typology of the antecedent conditions which can predict whether the crisis-definition will be *democratic* or *repressive,* and develop an alternate version of the stages that follow a repressive crisis-definition.

2. Determine the social structural prerequisites for movement of the process from phase to phase.[10]

3. Develop a model to sort out the interaction patterns and interchanges within and between systems in conflict, with special attention to repressive and counterrevolutionary variables.

4. Develop a hierarchy of negotiable issues—and of meaningful concessions available to urban white power—as a guide to prediction of racial conflict and change patterns in the next few years.

If negotiable issues between urban blacks and whites can be found, then the 7 C's process will continue to operate in the urban North as satisfying concessions are elicited. If not—if the super-ordinate powers under attack move steadily toward repressive crisis-definitions—then the kind of conflict theory discussed here will seem pale in the face of the realities of violent change to come.

NOTES

1. This paper's outline of the change-through-conflict process is con-sciously called a "model" rather than a "theory" because, at this level of its development, it provides only a set of sequential cate-gories for understanding the process—not structural determinants for explanation and prediction.
2. Other change processes involving lesser degrees of conflict include drift and rational means-ends chains (planned change through legis-lation or moderate private social reform measures).
3. Data throughout this paper are drawn largely from my doctoral dissertation, "Direct Action and Desegregation: Toward a Theory of the Rationalization of Protest" (Harvard University, 1966) and from written and personal sources available to me in the course of my work in the Community Relations Service since 1965.
4. William Graham Sumner, *Folkways* (N.Y.: Dover, 1959; originally published in 1906), pp. 16-18.
5. United Press International, August 10, 1967.
6. Associated Press, July 3, 1964.
7. Lewis Coser, *The Functions of Social Change* (N.Y.: Free Press, 1965), and *Continuities in the Study of Social Conflict* (N.Y.: Free Press, 1967); Kenneth E. Boulding, *Conflict and Defense* (N.Y.: Harper and Row, 1962); and Thomas Schelling, *The Strategy of Conflict* (Cambridge: Harvard Univ. Press, 1960).
8. *New York Times*, August 1, 1967.
9. Report of the New Detroit Committee, January 29, 1968.
10. This task, begun in my dissertation, is being expanded in *Black Protest: Toward a Theory of Movements*, with Martin Oppen-heimer, in preparation for Blaisdell and Company.

Part II

PERSPECTIVES ON CIVIL DISORDER

Introduction

The riots, disorders, and outbreaks of civil violence which have shaken American urban centers in recent years, and have occurred from time to time throughout the nation's history, are complex phenomena. Although usually referred to as uni-dimensional events, riots are in fact multi-dimensional phenomena, composed of various behaviors on the part of different individuals, groups and institutions in different situations over time. As with other complex social phenomena, where one stands on the riots depends on where one sits in the social structure. This is as true for social scientists as it is for public officials and the public. The analysis and interpretation one makes of civil violence will be in part at least a function of the perspective one has or is willing to adopt. In the case of the public and the public official, interpretation will be a function of the attitudinal and normative stance of the individual or the community. (And how the riots are labelled may have serious consequences for how they are handled.) For the social scientist, analysis will be a function of his intellectual-theoretical orientation.

Thus the interpretation or analysis of riots made by the participant, observer or analyst will be influenced by at least two factors—one related to personal or intellectual bias, i.e., the overall orientation to the phenomena, and the second related to the particular phase of the violence and/or the behavior(s) one chooses to focus on. In Part II, five perspectives on civil violence are offered which, although admittedly not exhaustive, (e.g., there is no exponent of the Communist "conspiracy," nor anyone to speak for the Black revolutionaries themselves), we feel they are indicative of the various explanations and interpretations of the phenomenon under consideration.

Allen Grimshaw picks up the theme of multiple riot perspectives in the opening essay, where he identifies three different interpretations of the riots of the 1960's (civil disturbance, racial revolt, and class assult) and the "labeling process" employed by the proponents of each. Racial rightists and elected officials have tended to label the events as "civil disturbances" or "insurrec-tions" as if there were no racial overtones. Grimshaw reasons that rightists see the riots as symptomatic of a general breakdown of the normative order, and further, they are politically sensitive to the Negroes' plight; elected officials take this view because many are sympathetic to the Negro situation and are fearful of endanger-ing programs designed to alleviate ghetto conditions. The second label–"racial revolt"–is accepted by militant blacks (as a rallying cry), by traditional Negro leaders (as a mode of organizational or personal survival), and by whites (in order to maintain access to the militant movement). "Class assault" is a definition of riots attributed by Grimshaw to committed leftists (Socialists and Com-munists), poverty workers (influenced by the obvious economic problems of the ghetto), and some social scientists and journalists, who see marked differences in the patterns of the 1960's and earlier urban riots. Professor Grimshaw concludes that because the current riots are complex phenomena there is probably supporting evidence for each of the three perspectives from which individuals and groups, located differently in the social structure, interpret civil violence.

Sociologists Kurt and Gladys Lang agree with Grimshaw that riots are complex phenomena but take a somewhat different tack. Arguing that the riots cannot be adequately explained simply as pathology or a symptom of social change (but on this latter point see Quarantelli and Dynes immediately below), they examine the developmental stages in the dynamics of civil disorders from face-to-face confrontations through the epidemic spread of disruptive behavior and the acceptance of violence as "a technique of pro-test." The Langs focus on riots as a form of collective political protest, the evidence for which they find in the pattern of riots throughout the nation: "However spontaneous the elements that underlie any incident and its particular pattern of expansion, the riots reflect at the same time the stirrings of a major social-political movement." The resort to violence, they feel, is indicative of social, and not individual, pathology. The need is to develop organizational alternatives to "collective bargaining by riot."

A third sociological perspective is provided by Quarantelli and

Dynes in "Looting in Civil Disorders: An Index of Social Change." Their background as experts in the study of natural disasters in general and looting in particular, leads them to an analysis of the looting phenomenon in civil disorders. Eschewing the common interpretation of looting as a manifestation of man's irrationality in periods of social disorganization, the authors find that the spiraling pattern of looting can be interpreted as an indicator of change in the accommodative relationship of blacks and whites. Like Grimshaw's "class assault" model in some respects, and the Langs' "collective protest," the view of riot looting as a "more institutionalized system of articulating demands and responses" about the distribution of property ("a shared understanding about who can do what with the valued resources of the community"), posits riot behavior as rational and instrumental rather than meaningless and senseless. The implications of the Quarantelli and Dynes analysis is for the development of institutionalized nonviolent means for redistributing certain property rights. Lacking that, they conclude that looting may become established as a major structural device for change in the social system.

As this introduction is being written, members of two Chicago youth gangs (the Blackstone Rangers and the Devil's Disciples) are under investigation by a Congressional committee, for alleged misuse of poverty grant funds and not, interestingly enough, for participation in the Chicago riots. Urban youths, however, do receive a great deal of blame for the violence and destruction that occurs in the riots. In the fourth essay, focusing on the role of youth gangs in riot situations, Irving Spergel offers the perspective of an expert on the study of juveniles in urban society. He takes the position that youth gangs, especially fighting gangs, are necessary to stabilize a community system of limited access to social and economic opportunities. His observations of youth gangs in Chicago over a three-year period suggest that the gang, as an organized unit, neither starts nor participates in urban riots, and, indeed, gang leadership is often called upon by officials and agencies to prevent or control riot behavior among peripheral or low status gang members and even adults.

Spergel feels the reason for this is quite simple, although not commonly understood. The gang is a relatively stable set of roles and rewards which permits a certain range of deviant behavior but excludes others (e.g., riots) which may upset the equilibrium or viability of the "system" of relationships between gangs and legitimate agencies of the community, who provide status and economic rewards in exchange for the curtailment of violence (agency "success").

Threats to the stabilization of this system come from unaffiliated deviant youths and alienated gang members, significant shifts in policy and programs of relevant agencies (police, social agencies, funding organizations), which may create status frustrations, and the emergence of a new opportunity structure for status satisfaction —civil rights, Black Power, and revolutionary groups. The latter has been most successful where gang structures are weak and youth are "on the make," but even then, the difference between the pragmatic, "goodie-oriented" gang youth and ideological, "cause-oriented" group may preclude affiliation. Spergel concludes that if the community goal is to preserve stability, and not summarily to destroy fighting gangs, it may be more important "to manage, control, and if possible redirect energies and interests of [gang] members into both community and sub-culturally relevant enterprises."

The concluding paper in Part II is an example of the differences that exist in the intellectual-theoretical orientations of social scientists referred to at the beginning of this introduction. John G. White, a political scientist, describes a pilot study in the use of aggregate data and statistical analysis in the development of a model of urban riots. His goal is to place the explanation of the objective clauses of riots into the broader context of conflict theory. He focuses on the social structural approach to conflict, rather than on the study of individual and group behavior (e.g., the frustration-aggression model discussed by Berkowitz in Part I). The structural approach emphasizes objective social conditions as underlying causes of conflict. In his essay, White selects a set of variables which index such things as social distance between blacks and whites, structural conduciveness for conflict, and the structural strain and anomie involved in urbanism, and statistically relates them to the presence or absence of conflict in 262 cities with populations over 100,000. His findings in this pilot study are not conclusive (he can explain only 22 percent of the variance with the eight variables used) which suggest that some significant sources of variation have been excluded. Nonetheless the approach used is judged by White to be promising in the search for a macro-model of urban violence.

—L. H. M. and D. R. B.

THREE VIEWS OF URBAN VIOLENCE
Civil Disturbance, Racial Revolt, Class Assault

Allen D. Grimshaw

■ Most large and heterogeneous societies experience domestic violence during periods of rapid social change; this country is no exception. Easily identifiable ethnic differences have frequently served to define boundaries of conflict groups; again the past history of the United States provides many examples. In societies where the boundaries of ethnic membership are co-terminous with those of socially enforced patterns of subordination and dominance, the intensity of social violence is likely to be greater. This has been true in the United States not only for Negroes and whites but for a variety of other groups including European immigrants, Asiatics, American Indians, and Mexicans.[1] As long as subordinate groups — whether or not they be ethnically identifiable — are willing to accept their lower status, an accommodative relationship obtains which helps minimize acts of outright violence. When, however, the subordinated group actually assaults the accommodative structure, or when the dominant group perceives that such assault is occurring or threatened, social violence is

AUTHOR'S NOTE: *This is a slightly revised version of a paper delivered at the annual meeting of the American Association for the Advancement of Science, New York City, December 28, 1967.*

likely to follow. The historical pattern of Negro-white relations in the United States has been one in which the dominant group has, from time to time, responded to such threatened or real assault by direct attacks on the minority. There is little evidence that a conscious policy of violence has been frequently pursued by leaders of the dominant group; there is little evidence that leaders of the subordinated group have, historically, consciously followed a policy of direct and violent assault upon the system.

In the period before Emancipation, there were occasional rebellions, and individual assaults and murders, by slaves. After the Civil War, there were vigilante-style groups which undertook to enforce Negro subordination by terrorism, and there were many individual lynchings. The period of "classic" race rioting in the United States, however, which dates from about the time of the First World War, was one in which whites responded to Negro "insubordination" and "pushiness" by direct assault upon the minority — direct assault in which mobs of white civilians took part.[2] In these riots, a difference from earlier "pogroms" emerged, in that Negroes fought back, and in some instances racial mobs attacked smaller groups or individuals of the other group.

The Detroit "race riot" of 1943 was such a case, one in which large mobs of whites and Negroes directly confronted members of the opposite group in a pattern of racial warfare.[3] The most immediately obvious difference between the Detroit disturbances in 1943 and those in 1967 is that in the latter there were no significant cases in which black and white civilians directly attacked members of the other race. Indeed, while the significance of such activity has been overestimated, whites and Negroes occasionally cooperated in attacks upon the police and upon commercial establishments. Moreover, there were not in 1967, as there had been in the earlier violence, widely circulated rumors in each of the groups about cross-racial assaults upon women and children.

The pattern in Detroit in 1967 closely parallels those other urban disturbances of the Sixties which involved the Negro minority: Philadelphia, Rochester, Bedford-Stuyvestant, Watts, and Newark, to name some major instances. In their lack of direct confrontation between civilian whites and Negroes, they also parallel the Harlem disturbances of 1935 and 1943, although in the latter instances there had been rumors of heinous cross-racial assault (upon Negroes). The two Harlem riots and the riots of the Sixties may also differ from earlier riots in that, while there was improper behavior by police and other control agencies, it never compared to that of earlier riots — for example, in East St. Louis in 1917, formal control agencies actively participated in large-scale assault upon the minority group.[4] Furthermore, while much of the mass media has been critical of the black community for being insufficiently grateful for changes which have already taken place and for endangering future improvements by "hoodlumism," media treatment — both in news reporting and in editorial posture — has generally been far more sympathetic than was true in the past.[5] In this greater sympathy, the media have either led or followed a greater sympathy and concern in substantial sections of the dominant white community.

The large-scale urban violence of the first half of the century clearly had economic overtones. The rhetoric, however, was racist, and racial identity was the prime factor in determining attitudes and behavior alike. Disputes over housing and recreational facilities in the decade following World War Two and, in the latter part of that decade, disputes over educational desegregation, were clearly racial.[6] In contrast to the earlier riots, the events of the 1960's have a complexity of motivation and of relations to the larger social structure which eludes any easy interpretation. Again in contrast to earlier violence, events of the last few years have been the focus of a large variety of formal studies and of interpretations from within and without the affected communities in which it is difficult to find a common thread of explanation.[7] In this paper I will identify three main sets of interpretations of the occurrences of the last four years, identify principal proponents of the several perspectives, and attempt to relate the various explanations to the locations of their proponents within the social structure. The three interpretations can be most easily identified by using the labels for the violence given by their adherents: civil disturbance, racial revolt, class assault.

THE LABELING PROCESS

A number of social characteristics and several ill-defined variables are involved in the process by which individuals and groups label the violent events of recent years. Some of these, such as race and involvement (whether as rioter or as official), have a more obvious bearing than others. The interplay of motivation and structural constraints which culminates in a labeling decision is, however, no less complex and difficult to unravel than the "causation" of the urban eruptions themselves. Moreover, if it is not at all clear what the long-range consequences of the disturbances may be, it is certainly clear that the process of labeling, and the emergence of one or another set of labels as predominant, will have consequences for future events. To oversimplify, if society at large (or significant and powerful segments of the society) agrees with linguistic labeling of events as "criminal" and "rebellious," then an atmosphere will be created in which pleas for the strengthening of police (and other agencies with legal monopolies of force) and for "stricter law enforcement" will strike a responsive chord. If, on the other hand, identification of the same events as "a legitimate revolt against impossible conditions" is accepted, then people will be predisposed to accept solutions which attack sources of the behavior rather than solely problems of control. Similarly, certain critics have claimed that characterization of the events as an expression of "class assault" will have the result of arousing fears of "Communism" and related threats, with the consequence of producing still another responce.

Race and involvement were suggested above as social characteristics with obvious influence on labeling perspectives. Related to involvement is the question of the official position of the labeler: is he an elected or appointed official; is his constituency formal or informal; is his primary responsibility for social control or for welfare; and so on. These questions lead in turn to a consideration of perceptions: is the threat seen as immediate or remote; is the activity seen, for example, as legitimate or criminal; and so on. Reporters on the scene may witness "criminal behavior," while editorial writers may have in mind statistics about unemployment and poor housing. On the other hand, reporters may witness "unnecessary use of force" by police, while editorial writers may have in mind the passage of civil rights legislation and the changing pattern of court decisions. Middle-class Negroes in the area of violence may be subjected to police insult or threats to their own property, while middle-class Negroes who live away from the ghetto may, in the first instance, see legal improvements and ameliorative programs and, in the second, have in mind the same statistics about unemployment and poor housing. White liberals living in insulated small towns and protected college communities will respond differently from those in urban areas who can see the flames and hear the shots and sirens. Ecological and social distance from the actual events will both have an influence on perceptions.

Perspectives in labeling are also influenced by ideological postures and, given ideological positions, by tactical considerations. Some few observers have been ideologically neutral and have simply chronicled events.[8] Most, however, fall somewhere on a "right–left" continuum — or rather on continuua, since there are different meanings to "right–left" within the white and Negro communities. Right–left categorizations are made more difficult by the presence of different strategic orientations within the several groups and by preferences for legalistic, as contrasted to amelioristic, strategies, and, in the left groups, by disagreements over the primacy of political and social, as contrasted to welfare, goals.

Tactical perspectives are, of course, related to ideology. A wish to deny the race and/or class aspects of the events or to minimize the magnitude of the importance of such characteristics may lead to relatively neutral labeling of them as disturbances or disorders. Someone sharing the same general ideological perspective, however, but wishing to maximize the legal aspects — whether or not concerned with race and class aspects — may label the same events as lawlessness or insurrection. More generally, labeling may represent threatening as contrasted to conciliatory tactics. Thus "tough" military men and "hard" policemen may join not only with political rightists but also with black militants in labeling disturbances as "revolt," "rebellion," or "warfare." Similarly, elected officials oriented to the status quo, as well as sections of the mass media, may join with more moderate Negro leaders (the old-line "Negro" leaders, as contrasted to "black" leaders) and with white liberals in assigning more neutral and conciliatory labels such as "disturbance" or "disorder." In these instances, those who use the same labels may have very different purposes in mind: "enforcement" types want rigorous suppression of the "revolt," "black power" advocates are seeking recruits for the overthrow of the current social structure; white liberals and moderate Negroes want ameliorative social changes; some officials who talk about disorders want, immediately, to "cool" the situation, although they may also be sincerely interested in improvements in the conditions of minority group members.

The several variables suggested above are all of importance in influencing perspectives in labeling. Whatever their interaction, however, and whatever the weights of their mutual influence, there is another variable which in many instances may outweigh even ideological posture and immediacy of threat. This variable, which is of importance not only in the selection of an original position but also in its maintenance or rejection, is that of social supports and relevant reference groups. It is not yet clear what the boundaries will be of the new conflict groups emerging in American society. It *is* clear that processes of boundary definition are in operation. Insofar as the boundaries become more clearly defined and rigid and as the society becomes more polarized, there will be strong pressures on individuals to choose "for" or "against" ideological and tactical positions.

Thus, apparently, white "liberals" attending last summer's National Conference for a New Politics were constrained to accept more and more "extreme" positions and labeling in order to maintain an even grudging acceptance from their black colleagues (there are some extremely complex and interesting issues of psychological motivation involved here, which I have neither the space nor the competence to fully examine). Thus, before a large public audience Dick Gregory can castigate the whites (much to their apparent pleasure) and can then ask Negro students to rise, following this with the directive, "All of you who don't think there wasn't enough burning last summer, sit down" — and can thereby manipulate social structure so as to coerce a public acceptance or rejection of a "militant" stance. Thus, "white, liberal, so-called intellectuals" who have long identified with the aspirations of the Negro revolution are constrained to give up their status as objective observers and to accept the interpretations and verbalizations of black militants uncritically, or else to face complete rejection — a quandary which leaves some of them immobilized, others emasculated, and others schizophrenic, while still others are driven into retreat from the situation.

Two further points may be made. First, labeling perspectives may be somewhat less stable than is the case with some other emotionally and politically important attitudes. There is evidence that there have been shifts in perception as a consequence of peer pressures, of superordinate policy shifts, of information on the scope and magnitude of events. Thus, some elected officials may in anger initially condemn rioters as "hoodlums" engaged in criminal violence. Their characterization may shift (perhaps in response to a review of the full political implications of their position) to one which labels the events as civil disturbances generated by impossible conditions. On the other hand, the New York *Times*, which initially emphasized the conditions which "caused" the riots, shifted as the summer of 1967 wore painfully by to a position where they stated editorially (July 25, 1967), "the arsonists and looters have to be dealt with as the criminals they are (whatever the root causes)."

The second point to be made now is that while the variables discussed operate in complex ways, they have influences on some more mundane and measurable characteristics of persons who ultimately do the labeling. Negroes with different sex, age, occupational, class, and educational characteristics do respond differently to queries about the meanings and reasons of the riots and about their possible consequences for the Negro "cause." Middle-, lower-, and working-class Negroes do have different sets of complaints and different perspectives on goals as well as on tactics. They also have differential access to the opportunity structure and differential exposure to social slight and insult. Differences in responses by age categories can clearly be linked to differences in the experiences of different generations. Other papers will document differences on the basis of these more traditional socio-economic variables; I simply want to underline the fact that behind the distribution of responses, there is a complex inter-play of structural features of the social system with individual attributes.

CIVIL DISTURBANCE AND/OR INSURRECTION

> . . . open rebellion . . . criminal insurrection . . . an atrocity
> . . . plain and simple crime and not a civil rights protest.
> New Jersey Governor Richard J.
> Hughes on the July, 1967,
> "disturbances" in Newark.

> Victims of civil disorders report here . . . .
> Notice outside of office of
> municipal social service
> agency, Detroit, 1967.

Both radical rightists and elected moderates have chosen to label events of recent summers as if there were no racial overtones, although in detailed exposition both sets of ob-servers have referred to the fact that most of those involved have, indeed, been Negroes. The rightists have chosen to keep their labels racially neutral for two reasons. First, they have chosen to depict the disturbances as resulting from leftist agitation and from a general breakdown of the norma-tive order, and they see the agitation and breakdown as characterizing the entire society, with Negroes being only somewhat more susceptible because of the fact that they can't or won't "make it" in American society; and, secondly, be-cause in spite of their ideological predispositions rightists are politically sensitive to growing sympathies which exist, at least in the abstract, for the Negro plight. Thus, while in many instances they have referred to "ingratitude" and to the fact that "appeasement" will only lead to more violence and to further inflated demands, they are cautious about alienating possible sources of support in the larger community.

Elected officials like Governor Hughes have responded viscerally during the actual eruption of violence and have generally moderated their characterizations in the post-riot period. Many of them, and I single Hughes out only because of the widespread attention given to his pronouncements, are fundamentally sympathetic to the situation of the Negro, but are simply unable to understand why Negroes are not aware of "what is being done for them." They *do* want peace and order, but at the same time they want to avoid racial labeling because of the dangers of either white or black "backlash," or both. They *are* concerned about conditions in the ghetto, and may in some instances use racially neutral euphemisms because they do not want to endanger programs directed to improving those conditions. Some officials, moreover, may choose to use racially neutral terms because of the implications which admission of racial meaning of the disturbances might have for the conduct of American foreign policy.

Individual police officers and many enlisted military personnel doubtless see the disturbances as race riots, "pure and simple." Law enforcement officials and many military officers, on the other hand, have responded to disorder in the abstract and have seen the events simply as problems of law enforcement and peace maintenance, and have seen their duty simply as that of restoring law and order. While many are doubtless sympathetic, and while others may be strongly prejudiced, they have been preoccupied with questions of logistics and tactics and only after disturbances have been quelled (or have simply run down out of inertia) have they moved from control problems to interpretation of causes. Thus, the reflections of at least some officials have been directed to the relative merits of different patterns of the commitment of police officers and/or troops, the advantages of containment as contrasted to dispersal, and the effectiveness of tear gas as compared to that of night stick or bullet.

RACIAL REVOLT

... If you know the culture and gain access to the heart of the community, you come up with one astounding pattern and that is, they hate Whitey — they literally hate Whitey, all of them. And even with the middle-class Negroes — you're not going to get them to say, "Let's go and kill Whitey!" or something like that, you're not going to get that — but I'll tell you what. Try talking to them about their jobs. Where the highest level among many of them is to get to be some kind of bullshit supervisor, and they know damn well they're smarter than the honky who's over them. Get them talking about that sometime.... Everybody, well, not everybody, but particularly the liberals do not want to face the aura of hate that is inside the community. They don't want to deal with it. They don't want to deal with the tremendous racial aspect of what has happened. It's just too ugly.

> From an interview with a Negro intel-
> lectual, Detroit, August, 1967.

Particularly since the events of the summer of 1967, Negroes of every political persuasion and of every ideological hue have increasingly identified the current activities as "the Negro revolt" or, in some instances, "the black revolt." These terms have superseded the earlier label of "Negro Revolution" which had, in spite of its implications for a complete restructuring of society, come to be identified as a peaceful revolution which would use the courts and the ballot as well as non-violent confrontation as tactics. The term "revolt" is used in its dictionary meaning as "a renunciation of allegiance and subjection to a government; rebellion; insurrection."

In the case of the militant blacks (and they by no means constitute an ideologically homogeneous bloc), the label is used descriptively and also as a rallying cry and a coercive linguistic weapon in the definition of the boundaries of a conflict group. The threat, however, is not directed against whites or the white Establishment or its mercenaries — for there is an attitude that there is little to be gained from Whitey, that his institutions will respond only to forceful change, and that white reaction will be the same no matter what labels are used or what tactics adopted. The threat is, rather, directed to moderate Negroes and more traditional leaders — "Join with us or see your organization wiped out — and we can't promise safety for you!"

For some traditional-style leaders, this threat has been enough; they choose to adopt the militant "black" rhetoric as a mode of organizational and, perhaps, even personal survival. Others, however, accept the labeling primarily in order to use it as a threat to the white Establishment. Thus, national leaders have stated that the labels are descriptively correct, that the reasons for the emergence of militant revolt lie in a failure of the Establishment to fulfill the more moderate requests that they have been making over the years. They state, "Meet the kinds of demands that we have been making or you will have to deal with wild-eyed radicals and guerilla warfare in the streets, rather than with intelligent and reasonable men like ourselves." These leaders are in a far more difficult situation than the black militants. They are dealing with multiple constituencies and must, while publicly calling for reason and for peaceful solutions, privately mobilize the entire battery of threats which the militants imply in their labels. They must also, somehow, adopt enough of the militant rhetoric in dealing with dissidents within their own organizational structures to disarm them while not losing credibility with the white Establishment.

It has been suggested above that some whites have, in order to maintain access to the militant movement, also accepted its rhetoric (as at the NCNP). There are other whites, however, who have insisted on the continuing importance of racial factors in the etiology of violence. This has been true even of scholars who have found significant class aspects in the violence. Thus, for example, the Murphy–Watson study of Watts found that middle-class Negroes were even more hostile to whites than the very poor, who were primarily angered about "welfare" issues.[9] The same hostility is suggested in the quotation which introduced this section, and which is accompanied by the question, "Just what the hell do we have to do in order to be accepted? We've done everything that has been demanded of us in terms of obtaining education and acting like middle-class people — but we're still subjected to continuing insult and social exclusion."

Thus, in this case and in that of those labeling the disturbances in racially neutral terms, a wide variety of perspectives, motives, and tactical outlooks is involved.

CLASS ASSAULT

Initially, I like to term this thing as an economic revolt. Initially, it had no racial overtones at all. It was just the looting, etc., and as you know there was integrated looting in the 12th Street area. I saw integrated couples over there. No one said anything. Whites and colored were standing on the corner together. It started out, as I said, like an economic revolt. . . .

Really, on this racial overtones, I think the police and the National Guard brought this in. . . . Some of us said (at that time) if you bring in the National Guard, you'll bring in the racial connotation that heretofore there has not been. As you know, the National Guard and the police with their brutality, etc., have done this, people who were not angry before are — because of what happened — because of this hotel–motel thing.

. . .

It's being played up by some people that there's this schism between lower- and middle-class Negroes and there is this class type thing. . . . I know that there is a lot of feeling in the community now — not only in Detroit but all over the country — that middle-class folk have not done as much as they can for the brethren. I'm sure that you've heard the expression many times that "When he gets into the system, he becomes whiter than Whitey."

> From an interview with a Negro
> activist, Detroit, August, 1967.

This is one of the major problems on reporting. Everybody from the newspaper reporters to these so-called intellectuals, they come on with their preconceived notions. The newspapers have their side and their specific interests within the framework of the entire control process, so they report it in a certain way. Intellectuals and social scientists usually have theoretical fancies which they use phenomena to support — they have certain ways of looking at things. . . . A reporter, if he is to have veracity, cannot have any preconceived notions.

> From an interview with a Negro
> intellectual, Detroit, August, 1967.

Three sets of commentators have emphasized the class and economic aspects of the summer violence. Committed leftists, including theoretically oriented socialists as well as activist Communists, are influenced by ideological concerns in their search for understanding of the events. Poverty workers are influenced by the obvious economic disadvantages of ghetto residents and may be more likely to notice the "economically rational" behavior which accompanied simple cathartic or more punitive destructiveness. Some social scientists and journalists in their emphasis on class aspects of the riots may be influenced by the sharp differences in behavior patterns which have distinguished the disturbances of the Sixties from those of 1943 and earlier (with the aforementioned exception of the two major Harlem disturbances in 1935 and 1943).

Some of these differences were mentioned above, particularly that in which, in contrast to earlier riots, those of the Sixties were characterized by an absence of direct confrontation between large groups of civilians of the two communities. But there have been other, more subtle differences as well. Williams, in his studies in smaller communities done in the Fifties, reported that militancy (as measured by fairly routine types of civil rights goals) was higher among the educated, young, and middle-class Negroes — but that prejudice toward whites was also lower in this group.[10] It seemed likely that as militancy became redefined — and it clearly has — the middle classes might lose their role as militant leaders to new leaders with greater demands, but that at the same time they would remain lower in prejudice toward whites. There was evidence that although progress was slow, an increasing number of Negroes could be characterized as middle-class. The overall gap in education between whites and Negroes narrowed; and although the income gap increased, this was more a function of larger proportions of Negroes who had in some sense dropped completely out of the economic structure than it was of continuing discrimination on a large scale against Negro professionals and others with middle-class occupations. Indeed, as was suggested in the interview quoted above, some middle-class Negroes were being coopted out of the Negro community. If anything, then, it seemed likely that as middle-class Negroes became more successful and lower-class Negroes less, it could be anticipated that there would be a growing estrangement between the two class groups within the minority community.

Moreover, there had been some evidence that, as ties between class groups within the community became attenuated, new linkages might grow up between the underclasses of each of the two communities — in other words, that lower-class whites and lower-class Negroes, both victims of economic exploitation and of diminishing opportunities in a social world demanding, for example, increasing education, might act together in common cause. Thus, while middle-class Negroes continued to be victims of discrimination (as suggested in the quote at the beginning of the section on racial revolt) and subject to police indignities and social affront, at the same time they would increasingly identify with the White Establishment, while their less successful brethren would begin to see the identity of their own interests with those of unsuccessful lower-class whites. This set of events did occur; and Bayard Rustin, among others, began to suggest that there were identities of interest amongst all the very poor. It began to look as if there might be processes in motion which would establish new group boundaries and new conflict alignments in American society.[11]

These trends, if they existed, were essentially cut off by the course of actual events. It is clear that while the initiating incidents in the disturbances of the Sixties were frequently if not always racial in character, nonetheless, the events that initially followed showed a reaction to economic conditions as well as to discrimination. Moreover, at least in Detroit, there were cases of cross-racial solidarity. However, as the disturbances were drawn out, the role of the police and of the National Guard was such that middle-class Negroes, whatever their initial feelings about the rioting, were sharply reminded of their racial identity and of their common cause and fortune with their brethren.

Murphy and Watson in their careful survey of Watts in the aftermath of that catastrophe were somewhat surprised to find that middle-class Negroes were more rather than less hostile and prejudiced toward whites than their less successful fellow community members.[12] Lower-class Negroes were preoccupied with "welfare" problems, poor housing, jobs, education, high prices, and bad food. Middle-class Negroes reported anger and hostility toward whites. It would be interesting to know what kinds of responses these same middle-class Negroes would have given to the same sets of questions prior to the riot, and what role the behavior of police and the National Guard had in redefining their attitudes. Quite clearly, such redefinitions did take place during the course of the rioting in Detroit; it can be assumed that similar redefinition took place elsewhere.

SUMMARY AND CONCLUSIONS

As is the case with every pattern of social behavior, there are no simple explanations of the terror we have witnessed through the last four summers. This is clearly *not* a case of conflict between well-bounded and homogeneous groups. Lower-class Negroes and lower-class whites have been brutalized by the police and have been victims of an exploitive or indifferent economic system. More militant blacks, however, have little sympathy for lower-class whites because, no matter what their difficulties, they have the advantage of being white and yet can't "cut it" in a society where skin color is the most important characteristic a man has. Middle-class Negroes, on the other hand — who, as some felt, were slipping into white society — have had the importance of their color driven sharply home not only by recent events, but also by a continuing pattern in which they have not moved successfully within the white Establishment and where they have frequently witnessed more rapid advancement of whites whom they feel are substantially less qualified. The situation is further complicated by the fact that while all Negroes are angry at "Whitey," some are more concerned about social slights and some more concerned about welfare issues, the nitty-gritty of jobs, housing, bad food, and wretched schools. Even among those who agree on goals, there are sharp differences on tactics. Perhaps one of the things which has prevented greater success of the Negro Revolution is the multiplicity of factions within the Black Community. There are more complicated dimensions to this issue than to any other I have ever examined in my role as a sociologist.

The situation is not simple. The events *are* disorders and they *have* involved criminal elements. There *are* clear elements of revolt against the economic power structure which can be seen in the pattern of attacks upon merchants in the ghetto and upon those identified as mercenaries of that structure, namely, the police and the National Guard. However, as a consequence of questionable practices by the police and by the National Guard there is, at least in Detroit and probably in other major cities as well, a growing increase in the strength of the previously attenuated solidarity between the Negro middle and lower classes. There is probably, at the same time, a decrease in whatever bonds may have been growing up between the Negro proletariat and the lower-class white "honkies." As a consequence, we may conclude that there is some accuracy in each of the three perspectives from which people, located differently in the social structure, see urban disorder. As has already been suggested, selection of one or another of these labels by policy-makers in our society will have major consequences both for the immediate possibilities of improvement and for the likelihood of recurrence or non-recurrence of major urban violence.

NOTES

1. For a review of this inter-ethnic violence and of periods in Negro-white violence in the United States, see Allen D. Grimshaw, "Lawlessness and Violence in the United States and their Special Manifestations in Changing Negro-White Relationships," *J. Negro History,* XLIV, 1 (Jan., 1959), 52–72.
2. See, e.g., Allen D. Grimshaw, "Three Major Cases of Color Violence in the United States," *Race,* V, 1 (July, 1963), 76–86.
3. Allen D. Grimshaw, "Urban Racial Violence in the United States: Changing Ecological Considerations," *Am. J. Sociol.,* LXVI, 2 (Sept., 1960), 109–119.
4. Allen D. Grimshaw, "Actions of Police and the Military in American Race Riots," *Phylon,* XXIV, 3 (Fall, 1963), 271–289.
5. For some notion of these changes, compare current editorials in the New York *Times* with these comments on the riots in Washington in 1919 (July 23, 1919):

   The majority of the negroes [sic!] in Washington before the great war were well behaved. . . . Most of them admitted the superiority of the white race, and troubles between the two races were undreamed of. Now and then a negro intent on enforcing the civil rights law, would force his way into a saloon or a theatre and demand to be treated the same as whites were, but if the manager objected he usually gave in without more than a protest.

   Nevertheless, there was a criminal element among the negroes, and as a matter of fact nearly all the crimes of violence in Washington were committed by negroes. Had it not been for this fact, the police force might well have been disbanded, or at least reduced to very small proportions.

6. See, e.g., Allen D. Grimshaw, "Negro-White Relations in the Urban North: Two Areas of High Conflict Potential," *J. Intergroup Relations,* III, 2 (Spring, 1962), 146–158; and Allen D. Grimshaw, "Factors Contributing to Color Violence in the United States and Great Britain," *Race,* III, 2 (May, 1963), 3–19.

7. See, especially, the publications of the Los Angeles Riot Study undertaken by the Institute of Government and Public Affairs, U.C.L.A. Probably the most useful of these studies for purposes of this paper is Raymond J. Murphy and James M. Watson, *The Structure of Discontent: The Relationship Between Social Structure, Grievance, and Support for the Los Angeles Riot,* published by the Institute in 1967. Two other major studies, still in the data analysis state, are those of Dr. John Spiegel of Brandeis (the Six Cities Study) and the ongoing study sponsored by the Detroit Urban League and Michigan's Survey Research Center.

8. An excellent and generally "neutral" study done by a journalist is Robert Conot, *Rivers of Blood, Years of Darkness* (N. Y.: Bantam, 1967).

9. *Op. cit.*

10. Robin M. Williams, Jr., et al., *Strangers Next Door: Ethnic Relations in American Communities* (Englewood Cliffs, N. J.: Prentice-Hall, 1964). See, more recently, Gary T. Marx, *Protest and Prejudice: A Study of Belief in the Black Community,* (N. Y.: Harper and Row, 1967).

11. For a fuller review of some of these changes, a review now in need of substantial revision, see Allen D. Grimshaw, "Changing Patterns of Racial Violence in the United States," *Notre Dame Lawyer,* LX, 5 (1965), 534–548.

12. *Op. cit.*

RACIAL DISTURBANCES AS COLLECTIVE PROTEST

Kurt Lang

Gladys Engel Lang

■ To overlook the purposive meaning of acts which comprise a collective disturbance implies an acceptance of the official perspective of the law enforcement agency, whose judgment of what is or is not a riot is simply a matter of the degree to which it is felt to be a menace to public order, a judgment apt to depend on the time and place it occurs. A relatively minor disturbance in closed quarters such as a dance, a rally, or a sporting event becomes a riot if it attracts the attention of police. Or in a tense racial situation, even a small incident will be suppressed as if it were a riot, as in some gang friction in East New York in the summer of 1966. Similarly, many diverse incidents — vandalism, looting, and brawling — are together declared a riot when they are so concentrated in time and space that to cope with them requires an unusual show of force. Thus, the term *riot* — especially in the present political climate — is often used indiscriminately to refer to rather different events which constitute a single category only because they evoke a similar official response.

In other words, the kind of disturbance that has become almost commonplace in the United States cannot be adequately explained or dealt with simply as a pathological manifestation or as an inevitable product (and symptom) of social

AUTHORS' NOTE: *This is a slightly revised version of a paper presented before a joint session of the American Sociological Association and the Society for the Study of Social Problems, San Francisco, August, 1967.*

change. In what follows, we present a brief outline of what we see as the underlying dynamics of these disturbances. This includes, first of all, the face-to-face confrontations that precipitate the polarization of a collectivity to a point where violence functions as a spontaneously shared defense against anxieties that individuals experience; second, the epidemic spread of disruptive behavior to nearby areas, and mutations in the pattern of rioting; and third, the ways in which direct action in the form of violence or other illegal acts becomes accepted as a technique of protest, so that the pattern is repeated in other cities even without deliberate organization, the movement being carried along by the myth of the violent uprising.

THE PRECIPITATING EVENT

The collective disturbances discussed here can be fruitfully viewed as a spontaneously shared collective defense, i.e., a collectively sanctioned defense against demoralization through the spontaneous coalescence of individual reactions in a distressing situation. In such a situation, the members of an aggrieved population act directly and coercively to assert certain norms against established authority, or to impose their conception of justice against deviants defined as a threat. Though such action may involve a deliberate defiance of authorities, the willful violation of laws, and savage acts of intimidation, violence, and destructiveness, it nevertheless represents at the same time a method of social control, no matter how unconventional.

The basic postulate here is that the standard practices by which any society or group defends itself against demoralizing tendencies are in some sense analogous to the characterological defenses of individuals. They mobilize sentiment and affect to support and maintain social solidarity. This mobilization of sentiments to uphold a threatened norm is evident in loyalty parades or propaganda rallies to counteract "heresies."

The probability that violence in some form will emerge as the spontaneously sanctioned collective form of defense is increased when institutionalized channels for the expression of grievances are ineffective or when they are, or seem to be, lacking. The potential for violence furthermore depends on the degree to which (1) there is a threat and (2) the threat touches on common moral sentiments. The greater the perceived threat, the stronger will be the response and the greater the need for action with visible consequences, to allay any sense of outrage. Accordingly, any population is capable of violent reactions.

Wherein lies the special potential of arrests (which in most instances appeared to be a legitimate exercise of police functions) to evoke moral outrage? First of all, there is a disposition *common to all segments of the population* to view the use of force by police as provocative and offensive, especially when the suspect's protests and policeman's reactions are highly visible, while the reasons for the arrest are obscure.

Second, this reaction is all the more likely among a population where many persons have suffered severe damage to their sense of self-esteem at the hands of the impersonal authorities; welfare, police, and other agents of what is often seen as an alien "occupying force."

Third, the residents of the Negro ghettoes constitute in a very special sense an "isolated mass," segregated together by color and with little sense of participation in the larger community. This facilitates the generalization of sentiment, even if the grievance is, to begin with, minor.

Finally, there are the latent effects of civil disobedience as a political tactic to force compliance with nationally proclaimed politics for eliminating discriminatory practices. Often such action has been met by one-sided law enforcement clearly directed against the demonstrators. Publicity has centered as much on police against demonstrator as on demonstrator against police. Thus the image of its struggle for equality that does disseminate into the Negro community can be invoked to justify acts of overt resistance, even in the instance of legitimate arrests.

The disposition of groups of slum-dwellers to view the use of force in any given situation as threatening, accounts for the special potential of certain incidents to evoke a spontaneous collective defense. Sentiment quickly rallies behind the victim of an apparent outrage, as some of the bystanders begin to assert a non-debatable demand. *The polarization with the victim and against the police hinges on the presence of a "critical mass" of susceptible individuals ready to go into action at what they perceive to be provocation.* How many susceptible people must be on the scene for a collective outburst to occur depends on the level of existing grievances and on the amount of counter-force at hand. The time and place at which these incidents have typically occurred contributes to the likelihood that people are ready to act against authority from whatever mixture of reasons would be present in large numbers of people.[1] The volatile among them help polarize action against the police, and even though the mass of bystanders may be passive and themselves indisposed to commit any violent acts, their mere presence lends tacit support to those who initiate action. Very few dare to intervene decisively in the interest of law and order.

EPIDEMIOLOGY: THE RIOTING SPREADS

During the first phase, hostile action has usually been directed primarily against the police equipment immediately involved in the incident. Thereafter, as reinforcements arrive, "rioting" quickly extends to all law enforcement personnel. Then, in a third phase, any visible representatives of the "white power structure" – the press, autoists, curiosity seekers – and their property become targets of attack. This pattern may change and is, in fact, beginning to change already. However, if the initial eruption of violence depends on the formation of a physical-contact group, the spread of a local disturbance no longer does. Once order breaks down, people begin to experience a sense of their power. For a time at least, there is general immunity from punishment. Destructiveness then becomes less discriminating, and looting for personal gain begins to proliferate.

From all evidence, only a minority of the residents of Negro ghettoes have participated in the various kinds of "riotous" behavior, and in all likelihood most will continue to remain immune to its contagious appeal. However, given the widespread sense of grievance, it has been extremely difficult to organize an active opposition once the rampage has begun. Civil rights leaders who have attempted to intervene on the side of order have not been able to make themselves heard.

Press reports on the predominance of unemployed youths among the "rioters" are not substantiated by the few statistics available. Thus, the *typical* Negro male rioter arrested in Rochester in 1964 and in Watts in 1965 was in his upper twenties; only one-fourth under twenty-one. Likewise, the majority were employed — albeit usually in an unskilled job.[2] Similarly, self-reports on riot activity obtained by Murphy and Watson in their post-riot survey, while revealing a slight concentration of activity among the 15- to 24-year-old group and among the unemployed, still show these two categories to have been in a minority.[3] All indications are that those active in the riots approximate in certain respects a cross section of the younger male population in the ghetto areas.

It is nevertheless most unlikely that any incident can expand into a pattern of rioting without the prior existence of groups that become the nuclei of trouble from which other incidents develop. These groups are of two kinds: sectarian agitators ready to foment trouble and/or to exploit any incident for their own purpose, and those who normally participate in all sorts of illegal activities and are therefore prepared to take advantage of any disorder as a cover for their usual pursuits.

The part played by agitators is undeniable when it comes to creating a climate and spreading the initial incident. However, the extent of their responsibility for keeping the rioting going, once it becomes widespread, is difficult to fix.

As regards the presence of so-called criminal elements, police and court statistics on arrests, despite their inherent biases, continue to be the major source of evidence. Of adults arrested in the Watts riot, one-third had no previous arrest records, another third had "minor" and the remainder "major" criminal records. Given the slim chances young, poor Negroes have to avoid arrest for 20-odd years, the following statement by the California Bureau of Criminal Identification and Investigation is significant: "A review of their prior criminal history fails to show a record as serious as that generally present in many non-riot felony bookings usually handled by urban police and courts."[4]

One cannot conclude, nevertheless, that looters are "amateurs" who cannot resist taking things that are there to be had. Ponder the fact that more of those arrested for burglary and theft had records of previous arrests or imprisonment than those charged with assault and homicide. The degree to which conflict with law enforcement agencies becomes generalized and leads to arrests may be gleaned from the case of the Rochester woman who, after having saved the life of the police chief, was herself arrested while helping conduct people out of the riot area. The tendency on the part of the police to overreact and to be undiscriminating in their counter-violence certainly fans the fires of conflict until they are put out.

THE REPLICATION OF A RIOT PATTERN

It is usually considered axiomatic that racial outbreaks of the kind that have rocked our major cities can be prevented by dealing with the basic structural deficiencies at the root of unrest in the slums. Direct action has typically been an avenue by which suppressed groups gain recognition and lend force to their demands. In the United States, for example, violence in the labor field has generally accompanied those disputes in which management refused to treat the union as a responsible bargaining agent. The machine-wrecking actions of Luddites had elements of what Hobsbawn has called "collective bargaining by riot."[5] Historical research has shown that attacks on mills and places of storage likewise served to reduce prices and levy money from the wealthy.

To what extent, then, are we dealing in these recent riots with the use of violence as an instrument of politics? The evidence bearing on this question is to be found largely in the pattern of replication throughout the country. That there should have been so many similar incidents of large-scale rioting within a short time span implies that, however spontaneous the elements that underlie any incident and its particular pattern of expansion, the rioting reflects at the same time the stirrings of a major social-political movement.

Two observations seem at first glance to contradict this depiction of the activity as a means to articulate some particular interest. First, most of the recent outbreaks have coincided with unusually hot weather and not with politically significant events. (Chicago 1966 and Newark 1967 are among the exceptions.) The diffuse nature of the outbreak gives it the appearance of an occasion for collective license, i.e., an occasion where normal restraints are no longer binding. Yet the political climate lends implicit sanction to this disposition to cast off certain restraints while channeling it into intergroup conflict. The evidence suggests that Negro rioters, far from going on a binge, were in fact highly discriminating with regard to their targets, venting their destructiveness primarily on stores of white property owners, while Negro-owned stores went largely unscathed. One finds that the majority of stores ransacked and burned were indeed owned by whites, and that some store owners were able to secure a measure of protection by exhibiting in thin show windows a declaration of their Negro ownership. But then it is generally true that most supermarkets, appliance, liquor, clothing and other stores favored by Negro rioters are owned by whites. In spite of the fact that they brought some personal gain and satisfied some feelings of revenge, one cannot view these attacks as a deliberately planned tactic of intergroup conflict. They also abolished many jobs held by Negroes, without offering alternative job opportunities. Even where they closed down white businesses, they made it possible only in a few instances for Negroes to come in. They brought about neither an extension of credit and/or a reduction of retail prices. The protest was primitive and anomic because most participants lacked any clear perception of how to advance the collective interests of the Negro community.

Our second observation refers to the militant rank and file of the civil rights movement. Except for a small number of persons on its extreme wing, they have dissociated themselves from the violence. Nevertheless, increasing militancy reflects the growing expectation among Negroes of all walks of life that they should be enjoying full equality and their fear that these claims are being denied. The appeal of mass civil disobedience was certainly the product of such a pairing of frustration with hope. Yet, mass civil disobedience was more than a tactic for dramatizing especially irritating Negro grievances. Participation in sit-ins and demonstrations carried with it the mystique of direct action. It often attracted recruits who may not have grasped the significance of the new tactic but nevertheless caught the new spirit. This then sets the context within which militant collective protests can spill over to become collective license.

The white power structure often responds to outbursts, or to threats of violence, by focusing on troublemakers while avoiding serious negotiations over an issue that touches the entire Negro community. Negro accommodationist leaders are then caught on the horns of a clear-cut dilemma. On the one side, they need dramatic successes in order to maintain their tenuous hold over their following. On the other side, their ability to gain a hearing within the power structure depends on their ability to restrain their constituencies. The prospect of violence reinforces their claim that legitimate Negro demands be met, and they can therefore exploit disorder to present themselves as the better of two alternatives.

The ambivalence of many Negro leaders about civil disobedience and extremes of militancy as tactics, coupled with a tendency on the part of the mass media to give a disproportionate amount of attention to violent incidents and to would-be leaders advocating violence, however limited their following, causes such acts to attain some legitimacy as a means of conflict. It comes to be expected more and more.

A society with a mature civic culture is not prone to counter violence with physical repression. In such cultures, the use of force to suppress rioting is apt to come in for very strong criticism. The forces for law and order are committed only with the utmost hesitancy, and any misapplication of force by any of its members undermines the legitimacy of its use.

The emergence in these circumstances of personalities bent on stirring up the potential for disorder — always so close to the surface in depressed communities — expresses only the ambivalences and structural maladjustment of the society at large.

IMPLICATIONS

Recent racial disturbances, despite the presence of irrational factors, do not develop as automatic responses to frustration. They are complex social phenomena involving many different kinds of actions by different kinds of participants. They are part and parcel of a new pattern of Negro militancy developing among depressed masses alienated together, whose political skills are as yet poorly developed and whose level of political organization is low. Among the main carriers of movements of radical protest in their early phase are persons who, because of théir poor adjustment, exhibit many facets of pathology. Such persons were, undoubtedly, among the arrestees in these recent riots. But as efforts at reasonable negotiations are frustrated, and alternative collective solutions to problems disappear, even psychologically adjusted persons will be prone to tactics that include or invite violence. The resort to violence is indicative of social, and not an individual, pathology. The riots come more and more to fit the model of the anomic movement.[6]

Explanations of the origin of the recent disturbances must take the following into account:

(1) the prevalence within the urban ghettoes of a *subculture* that sanctions illegitimate means, including violence; (2) the presence there of large numbers of youths to whom *limited opportunity structures* for achieving status are available; (3) the *remoteness and impersonality of the "power structure"* and its apparent inertia when it comes to improving conditions in the ghettoes; (4) the *low level of political skill and of organization* among poor people which reduces the capacity for effective negotiations and stymies many self-help programs; (5) the high visibility of police, storeowners, and other "privileged" groups, with the inevitability of *frictions on the interpersonal level;* (6) the *apparent effectiveness of a disturbance* in forcing official cognizance of conditions in the urban slums; (7) *tacit legitimation* by other extralegal forms of civil rights protest; (8) the *sanctioning "by default"* of collective license due to the reluctance in the present climate to employ counter-force; (9) the *presence of core groups* whose agitation provides counter-norm sanction.

It may seem obvious that an integrated society is the best deterrent to rioting. If our analysis is valid, then rioting will continue for a time, and perhaps even increase, despite the crash efforts to upgrade the conditions of life in depressed areas of our cities. Rioting evolves as a form of collective pressure or protest where large numbers of people are crowded and alienated together, sharing a common fate that they no longer accept as necessary, though it may seem inevitable to them.

Even small incidents are likely to precipitate larger disturbances. Whatever their underlying cause, civil society cannot tolerate physical violence and destruction as a means of pressure without changing its character, but neither can it suppress them by force alone. Conflict needs to be rechanneled into more effective day-to-day negotiations with visible results. The main result of efforts toward the political organization of slum dwellers is to provide organizational alternatives to "collective bargaining by riot." Until this happens, the "isolated mass" within the Negro community is apt to continue to produce its own forms of anomic protest that can, at best, be contained.

NOTES

1. Raymond J. Murphy and James M. Watson, *The Structure of Discontent: The Relationship Between Social Structure, Grievance, and Support for the Los Angeles Riot* (Institute of Government and Public Affairs, U.C.L.A., 1967).
2. By "typical," understand "median."
3. Murphy and Watson, *op. cit.*
4. Statistical report on the Watts riot supplied to the authors by the State of California Department of Justice.
5. E. J. Hobsbawn, *Primitive Rebels* (New York: W. W. Norton, 1965).
6. G. A. Almond and J. S. Coleman (eds.), *The Politics of the Developing Areas* (Princeton Univ. Press, 1960); George Rudé, *The Crowd in History* (N. Y.: Wiley, 1964).

LOOTING IN CIVIL DISORDERS
An Index of Social Change

E. L. Quarantelli
Russell Dynes

■ Outbreaks of looting have increasingly become one of the core concerns of communities which have undergone large-scale civil disorders in America within the past several years. Most current press reports of such outbreaks have as one of their central themes the occurrence of looting, and frequently depict looters in action. Even after-accounts of the civil disturbances or editorial polemics often emphasize stories of plunder to illustrate the "breakdown of law and order."

Part of the intensified popular attention to looting undoubtedly stems from actual increases of incidents. In one of the very first large-scale disturbances, that in Harlem in 1964, 112 stores were looted.[1] However, about 600 establishments were plundered or burned during the 1965 Watts outbreak.[2] A peak was reached in Detroit in July, 1967, when, according to unofficial accounts, around 2,700 stores were raided by looters.

The explanation commonly given for such "anti-social" behavior is that, in periods of social stress, the thin veneer of civilization is stripped off the human animal, revealing man's basest nature.[3] Under more normal circumstances, these base tendencies are somehow held in check. However, under the pressure of crisis situations, man is revealed not as Rousseau's "noble savage," but as Hobbes' "creature," at war with all. Anticipating that certain kinds of large-scale emergencies activate this depravity, community officials often request additional law enforcement officers. The National Guard is alerted or mobilized, and a wide variety of supplementary security measures are undertaken.

Such steps are frequently initiated on first reports of the beginnings of a civil disturbance. Often, expressions of concern that looting will occur, and the steps being taken to prevent it, are among the first stories circulated by radio and television after reporting the event itself. In the absence of any actual information about what is occurring, mass media outlets often report that which is expected to happen.

As a consequence of this common interpretation of looting as being a manifestation of man's irrationality in periods of social disorganization, punitive control measures are most frequently advocated as befitting the situation. In addition, since at least current civil disturbances have a racial dimension, such behavior tends to reinforce both manifest and latent conceptions which many whites have of Negroes – i.e., looting is a manifestation of the bestial nature of the Negro, or at least his inherent anti-social nature. Such views tend to reinforce calls for action which are repressive in nature.

While there is no doubt that much behavior in current urban civil disorders is illegal, we suggest that the spiraling outbreaks of looting are also indicative of the end of a particular era of accommodation between American Negroes and whites. In effect, the plundering and looting increasingly signal the end of a period of time when existing "rights" in a community will be automatically accepted by a significant proportion of Negroes therein as being given. These signals, of course, can be read as an invitation to institute strong repressive measures, as they seemingly have been in most recent civil disturbances. (That the potential for highly repressive actions lies not far below the surface of American society is suggested by the herding of most Japanese-Americans into detention camps at the start of World War II.)[4] However, looting can also be seen as a rather violent beginning to a new process of "collective bargaining" concerning rights and responsibilities of various groups in most American communities. The behavior, defined as anti-social by the larger community and unlawful according to legal norms, actually marks the end of one era and the beginning of a new one in racial intergroup relations in American society. In short, looting is an index of social change. (From another perspective it is also an *instrument* for societal change, but we will not develop that point in this article.)[5]

The reasons for seeing looting as the end of one era and the start of another are perhaps not self-evident. The same difficulty probably applies also to the meaning of looting and its implication. An understanding of both requires an analysis of existing definitions of property within a community.

As Kingsley Davis notes: "So ingrained in human thought is the fallacy of misplaced concreteness that property is often regarded as the thing owned rather than the rights which constitute the ownership."[6] In popular parlance, property is generally equated with material goods or physical objects. Even the United States Supreme Court did not recognize that property refers to rights, rather than a tangible object, until the end of the nineteenth century.[7] Rights and obligations are not tangible in a physical sense, nor is the tangibility or intangibility of what is owned of great consequence. What is important are the rights and obligations with respect to something scarce but valuable.

Property thus is a set of cultural norms that regulates the relations of persons to items with economic value. "It consists of the rights and duties of one person or group (the owner) as against all other persons and groups with respect to some scarce good. It is thus exclusive, for it sets off what is mine from thine; but it is also social, being rooted in custom and protected by law."[8] In effect, property is a shared understanding about who can do what with the valued resources within a community.

The norms or rules, the legal ones in particular, specify the legitimate forms of use, control, and disposal of economically valued objects. These norms, besides defining the rights and responsibilities of owners, also delineate social relationships among other individuals, because the "right" of any person in relation to an object entails at the very least the "obligations" of others to respect that right. There is obviously considerable variation in what the norms specify in different time periods and different societies, but at any given point they are normally widely shared and accepted in a community.

In contrast, civil disturbances such as American communities have recently witnessed are *situations of temporary and localized redefinitions of property rights*. The urban disorders we are discussing represent conflict on community goals and manifest differences of opinion in the community regarding economically valued objects. In these situations, rights to the use of existing resources become problematical, and in many instances there are open challenges to prior ownership.[9] If property is thought of as the shared understanding of who can do what with the valued resources within a community, in civil disorders there occurs a breakdown in this understanding. What was previously taken for granted now becomes a matter of open dispute, expressed concretely in a redefinition of existing property rights.

The problematic nature of property in urban disorders can be seen by noting the pattern of looting in such situations. Two aspects of the pattern are particularly important. First, the looting is highly selective, focusing almost exclusively on certain kinds of goods or possessions. Second, instead of being negatively sanctioned, looters receive strong although localized social support for their actions.

The degree of selectivity can be seen in the fact that particular types of stores have been the prime focus of looting. In Detroit, 47 grocery stores were attacked, more than in any other category.[10] Furniture, apparel, and liquor stores are also frequent objects of looters, with more than a million dollars' worth of stocks of each being plundered during the Newark disorder.[11] In contrast, banks, schools, plants, and private residences are generally ignored, although some of the latter have been inadvertently damaged as a result of being close to burned business establishments. Looting, contrary to many initial press reports of such situations, has not been indiscriminate; in fact, certain kinds of consumer goods have been the only foci of attention.

In addition to the selective pattern it assumes, looting at its peak is almost always if not exclusively engaged in by local residents who receive support from segments of their local community. This appearance of normative support can be seen in the almost spiraling pattern that occurs in situations of civil disorder and which reveals cumulative shifts in redefinitions of property rights. The pattern appears to proceed roughly through three stages: (1) A primarily symbolic looting stage, where destruction rather than plunder appears to be the intent. It often seems initiated by alienated adolescents or ideologically motivated agitators in an area. (2) A stage of conscious and deliberate looting, in which the taking of goods is organized and systematic. It frequently appears spurred by the involvement of omnipresent delinquent gangs and theft groups operating on pragmatic rather than ideological considerations. (3) A stage of widespread and nonsystematic seizing and taking of goods. At this point, plundering becomes the normative, the socially supported thing to do. Property rights become so redefined that it becomes permissible if not mandatory to transfer to different private ownership the possession of certain material goods. The legal right does not change, but the group consensus supporting the prerogative to appropriate valued resources in the community does shift, among a segment of the population.

In the first phase, little looting, if by that is meant the taking of goods, occurs. Instead, destructive attacks are most often directed at objects symbolic of the underlying sources of conflict. Police cars and stores operated by white merchants are attacked. These attacks signal the start of the redefinitions of property rights. Illegal use is made of possessions normally and generally accepted as being under the control of formal community representatives (e.g., police and fire department equipment) or "extra-community" agents (e.g., stores in urban black ghetto areas owned by whites). In actual fact, many outbreaks of civil disorders up to the present have not progressed beyond this initial phase of window breaking, car burning, tossing of isolated fire bombs, and the like.

In the second stage, there is a definite change. Looting of goods rather than destruction of equipment or facilities becomes the mode. White merchants dealing with consumer goods particularly become the object of attack. However, that the white merchants have goods which are readily moved probably makes them the focus of looters as much as the fact that the owners are white. Negro-owned stores of the same general type are not always spared by the marauding bands operating during this time period. There are some indications that a "soul brother" designation has become less and less of a protecting device as the disturbances have increased in intensity over the last several years. The racial dimension, while not absent, appears to be secondary to the economic factor in the behavior of the looters.

In the third stage there is a full redefinition of certain property rights. The "carnival spirit," particularly commented upon in the Newark and Detroit disturbances, does not represent anarchy. It is, instead, an overt manifestation of widespread localized social support for the new definition of the situation. The new consensus that emerges in such situations is suggested by the almost total absence of competition or conflict by looters over plundered goods. In fact, in contrast to looting in other situations such as disasters,[12] such behavior in civil disorders is quite open and often collective. Goods are openly taken, not by stealth. Looting is often undertaken by people working together in pairs, as family units or small groups; seldom is it carried out by solitary individuals. The availability of potential loot is frequently called to the attention of bystanders, and in some cases, strangers are handed goods by looters coming out of stores.

Not only is most looting in large-scale civil disorders by "insiders" (i.e., local community members) and not outsiders, but there is evidence suggesting that participants are from all segments of the population. Looters do not come only from the lowest socioeconomic levels or from neighborhood delinquent gangs. Arrested looters are, typically, employed persons, and roughly similar to persons generally participating in the disturbances. There is definite evidence that the latter are from all segments of the community. Thus, a statistically random sample revealed that all participants in the Detroit outbreak were, in about the same proportion, across all income brackets.[13] A U.C.L.A. survey in Watts discovered that those active in the disorders there — perhaps a fourth of the residents — along certain dimensions, represented a cross section of the younger male population in that ghetto area.[14]

This type of phenomenon is not new in history. Rudé has analyzed nineteenth-century demonstrating mobs in England and France.[15] He found that they were typically composed of local residents, respectable and employed persons, rather than the pauperized, the unemployed, or the "rabble" of the slums. As in the instance of current disturbances, the more privileged classes of those times defined these popular agitations as criminal, i.e., as fundamentally and unconditionally illegitimate.

Certainly most contemporary community authorities see looting as essentially a legal problem, and consequently as a matter largely of law enforcement. Many segments of American society, particularly middle-class persons with their almost sacred conception of private property, also tend to define the problem in the same way. Legislators, in response to pressures generated by such perceptions, move to strengthen "anti-riot" laws and other repressive measures.

There is, of course, no question that looting is criminal behavior, violating in various ways numerous statutes and ordinances. Viewed primarily in this context, looting, as well as the civil disorder, can be seen — as stated in FBI and other reports — as "meaningless" behavior.[16] However, such a view obscures something more fundamental.

The laws themselves are based on certain dominant conceptions of property rights. The legal framework is the residue of the past consensus regarding the distribution of property. It reflects an accommodation arrived at sometime before the present.

We suggest that the current civil disorders in American cities are communicating a message about the society. A time of social change, particularly with regard to the distribution of valued resources in communities, is at hand. The old accommodative order defining certain limits to property rights of American Negroes is being directly challenged to the point of collapse, although this seems presently more recognized by the subordinate rather than the superordinate group involved.

Perhaps the current situation has many parallels to the situation in the United States over a hundred years ago. The Civil War symbolized a period of time of disagreement about human beings as property, and the rights of their owners. The reluctance to redefine in a peaceful manner the legal structure which supported these property rights resulted in tremendous social costs to the society. Some of these costs were immediate, while others are still being collected today.

Viewed in this context, the attack against existing property rights is neither "irrational" nor "senseless." This is particularly so if it leads to a more institutionalized system of articulating demands and responses in which the rights and obligations of the contending parties become a matter of general community consensus. If this is the case, the current looting will mark the initial steps in the evolution of a social system in which certain heretofore urban segments of the society can nonviolently express their views, and in which the more favored groups and the elites will listen.

If more responsive and representative institutions cannot be established, certain groups in American urban communities will continue to engage in disorder and violence or, in our earlier terminology, to indicate their racial discontent and economic aspirations in periodic and increasingly costly redefinitions of property rights. There have been incidents of looting in earlier outbreaks in urban ghettos, some as early as two decades ago, as in Harlem in 1943.[17] However, the scope and intensity of current attacks indicate that increasingly larger number of persons no longer share the consensus about property rights held by the larger community. If property is seen not just as physical goods, but as a shared understanding about the allocation of valued resources within a society, a growing lack of consensus will progressively manifest itself in open conflict.

In actual fact, a point of no return may already have been reached. Lambert, in his study of communal violence in India,[18] found that a breakdown in the formal means of social control accompanied broad changes in the social organization of Indian society in the decades immediately preceding independence. Police officers there came to be viewed, not as impartial arbiters of social disputes and as operating within a system of legal redress for grievances; rather they were seen as armed representatives of their socio-ethnic groups. This interpretation of the policemen's role was accepted by members of the opposing group, by their own groups, and, increasingly, by the police officers themselves. "When this occurred the usefulness of the police in social control was sharply reduced and, in some cases, police activities contributed to further disruption of social organization."[19]

Much of this reads as if it were written of local police actions in American ghettos. A typical popular interpretation is to see all of this as a breakdown of "law and order." In one sense, it is that. However, in another more fundamental sense, as in Indian society, the failure or inability of the police in a community to prevent looting (apart from those instances where their own actions may initiate such behavior) can be seen as marking the end of an era. The psychological controls which really are the bases of police control in a community no longer suffice. The sheer power of National Guard or regular military units, when disorders reach a peak, is the only formal control left to communities.

Given any foreseeable combination of circumstances, military forces will prevail. However, it would seem that American society, if it wishes to insure domestic tranquility, should move to institutionalize nonviolent means for redistributing certain property rights. Looting can only be a temporary and localized redefinition of property rights. But if no other solution is found, the pattern itself may become routine across more and more American communities. If that is the case, instead of being an index of social change, the looting that has increasingly appeared in recent civil disorders may establish itself as a major structural device for change in the American social system.

Similar patterns of behavior have so established themselves in the past. Rudé, in the analysis mentioned earlier, notes that the disorderly demonstrations became a means of protest that in time enabled a segment of the urban population to communicate to the elite.[20] Hobsbawn, in his similar analysis of the pre-industrial "city mob," states the point even more strongly. He observes that the mobs did not just riot to protest, but because they expected to achieve something by their disorder. They assumed that the local authorities would be sensitive to the disturbances and make attempts to deal with the implicit demands of the mobs. According to Hobsbawn, "this mechanism was perfectly understood by both sides."[21]

A similar situation could develop in American communities. Some militant Negro ghetto leaders have almost been explicit about such a possibility. However, the cost to the society would be high and would not really settle the underlying bases of the conflict.

Furthermore, an even greater threat to the society may develop in such a direction. Signs of it have already appeared. The participation of poor white looters in the Detroit outbreak hints at the possibility that the broader middle-class–lower-class consensus about property rights may also become subject to attack, if the more immediate problem is not solved. The development of such an open class conflict would make the current racial conflict a highly desirable alternative state of affairs.

Thus, a failure to see looting in current disorders as something more than "meaningless" or "criminal" behavior may eventually fragment the social consensus far more than it has been up to the present. This perspective upon looting as an index of social change may suggest alternative ways of dealing with property rights.[22] In fact, if nonviolent ways are to be found, there may be no choice on how to think about the current disturbances sweeping American cities.

NOTES

1. James Jones and Linda Bailey, "A Report on Race Riots" (unpublished paper), p. 37.
2. This and all other information not later footnoted has been acquired in field work on civil disturbances by members of the Disaster Research Center at Ohio State University. Data have been obtained primarily through personal interviews with organizational officials, supplemented by systematic observations and analyses of unpublished agency reports.
3. See Anselm Strauss, *Mirrors and Masks* (Glencoe, Ill.: Free Press, 1959), for a criticism of this point of view.

4. Morton Grodzins, *The Loyal and Disloyal* (Univ. of Chicago Press, 1956).
5. This is a point of view also expressed in Kurt and Gladys Lang, "The Significance of Recent Racial Disturbances for Theories of Collective Behavior," in *Proceedings of the Seventh Annual Intergroup Relations Conference,* Houston, Texas, 1966, pp. 2–15.
6. Kingsley Davis, *Human Society* (N. Y.: Macmillan, 1949), p. 452.
7. John Commons, *Legal Foundations of Capitalism* (N. Y.: Macmillan, 1924), p. 14.
8. Davis, *op. cit.,* p. 452.
9. As Davis observes, there is an important distinction between ownership and possession, since property rights in an object do not necessarily imply actual use and enjoyment of the object by the owner. *Op. cit.,* p. 454.
10. *Detroit Free Press,* August 20, 1967, p. 4B.
11. These figures were given in an AP dispatch of August 17, 1967, citing a report issued by the mayor's office.
12. See Quarantelli and Dynes, "Looting in Civil Disturbances and Disasters" (unpublished paper).
13. This is from a study by University of Michigan social scientists, reported in the *Detroit Free Press,* August 20, 1967, p. 1B.
14. Raymond Murphy and James Watson, *The Structure of Discontent: The Relationship Between Social Structure, Grievance, and Support For The Los Angeles Riot* (Institute of Government and Public Affairs, U.C.L.A., 1967).
15. George Rudé, *The Crowd in History* (N. Y.: Wiley, 1964).
16. *Report on the Nature of the City Riots* (U. S. Dept. of Justice, 1964).
17. Allen D. Grimshaw, "Urban Racial Violence in the United States: Changing Ecological Considerations," *Am. J. Sociol.,* LXVI (Sept., 1960), p. 112.
18. Richard D. Lambert, "Hindu-Muslim Riots" (unpublished Ph.D. dissertation, Univ. of Pennsylvania, 1951).
19. Allen D. Grimshaw, "Actions of Police and the Military in American Race Riots," *Phylon,* XXIV (Fall, 1963), p. 271.
20. E. J. Hobsbawn, *Social Bandits and Primitive Rebels* (Glencoe, Ill.: Free Press, 1959).
21. *Ibid.,* p. 116.
22. The acceptance of the problem of the Negroes as basically a labor-market problem is set forth in Norbert Wiley, "America's Unique Class Politics: The Interplay of the Labor, Credit and Commodity Markets," *Am. Sociol. Rev.,* XXXII (August, 1967), 529–541.

YOUTH GANGS AND URBAN RIOTS

Irving A. Spergel

■ The thesis of this presentation is that urban youth gangs, particularly those which are highly aggressive, are a *stabilizing* influence
in a community characterized by severe lack of social and economic opportunities.[1] Our interest here is not so much with the
generation of delinquency and youth gangs as with the institutional conditions which permit their survival and development.
Various other conditions—ecological, psychological, organizational, and subcultural—must exist to account for the creation of
group delinquent phenomena. For example, a sufficient number of
adolescents must be present in a delimited area, sharing somewhat
similar problems of social adaptation; they must be defined and
define themselves as deviants; the gang must come to represent a
refuge for social failures as well as a place to achieve status and
reputation.

The important analytic consideration here is that the gang may
provide for its members not only a relatively stable set of social
and psychological satisfactions in their interactions with each
other and with members of opposing gangs, but may also produce
a "surprisingly" rewarding system of external interactions with
representatives of the official and dominant society, e.g., police,
school authorities, social agency personnel, grass-roots leaders. The
successful gang demands and obtains a great deal of respect and
prestige in the community, and this becomes a basis for negotiat-
ing and exchanging a variety of resources both with other gangs
and with the official representatives of the community. A relatively
stable set of roles and rewards is established, which serves to manage a range of deviant behavior but to exclude others—for example,
riotous behavior or any behavior which may upset the system.

Riotous behavior by gang youths or former gang youths may be a consequence of the dissolution or failure of the established pattern of relationships between the fighting gang and the community or its legitimate organizations. This occurs when the gang is destroyed, no longer obtains adequate satisfactions or rewards from offical organizations in the community, or can no longer provide satisfactions or even entree to all youths who seek to become members. The absence or weakness of the gang structure contributes to the search by alternate solutions to their problems of social and pyschological failure. These solutions may include drug use, riot, or participation in revolutionary movements. I am proposing that youth gangs, especially fighting gangs, are necessary to stabilize a community system with limited access to social and economic opportunities. When these opportunities are further reduced relative to culturally induced aspirations—when gangs are weakened by decreased access to resources—collective youth frustrations are heightened. The response—transitory, spasmodic, and disorganized though it may be—includes participation in riot and civil disorder.

YOUTH GANGS AND RIOTS

Observations over the past two or three years suggest that relatively well organized and established aggressive gangs, particularly their leadership segments, tend not to participate in urban riots and revolts.[2] Furthermore, the gang as an organized unit usually does not start a riot. Limited evidence suggests that it is the more peripheral or younger members of the gang who are the ready and active participants in urban riots. Higher status and leadership members of the group may be drawn in, if at all, only later, when riot-control efforts appear to have failed. Indeed, the established gang leader may be used by the police and social agencies as an intermediary, a communications link, and a control agent with peripheral or low-status gang members and even with adults participating in the riot. The gang leader probably has been used more often to prevent a riot than to control one once it has begun, although it is the latter type of activity which has received primary public attention.

For example, in Chicago in the summer of 1967, following a fatal shooting by a white liquor store owner of a Negro adult, a large, noisy, threatening crowd gathered on the street in front of the store. Leadership of two major youth gangs in the area were contacted by representatives of a grass-roots organization and a social agency to assist in cooling tempers and dispersing the crowd. According to newspaper accounts:

"The gangs, The Disciples, and the Blackstone Rangers helped restore calm in the neighborhood.. . .A major flare-up was threatened, said Earl Doty, director of the Youth Action job training project, after Nicholas J. Nickolaou, 34, operator of Big Jim's Food and Liquor Mart, killed Julius Woods, 16.

Nickolaou, who was charged with murder, said he killed Woods in self-defense after Woods threatened him.

Doty told reporters, "There's absolutely no doubt in my mind that had it not been for the efforts of the Disciples, the situation developing in Woodlawn would simply have exploded."[3]

On the west side of Chicago during the summer riots of 1966, gang leaders known to youth workers of the Chicago Youth Centers were given arm bands and identification cards, and in some cases were paid to assist in the control of further violence and looting. They worked hand-in-hand with the police to pacify the crowds and communicate accurate information to youths and adults in the community. Gang leaders in other cities have played police auxiliary roles, both officially and unofficially, in the prevention and containment of riots.

The reasons why established gang leadership has played a key role in riot control activities are not difficult to understand. Gang leaders, as perhaps all leaders, are particularly ambitious and opportunistic. They strive mightily to achieve status and, once they have it, work diligently to avoid threats to it. They do not possess or ordinarily develop a radical ideology of any sort. They are conservative and are especially concerned to maintain or even augment their positions of leadership during community crises which may threaten them.

GANG LEADER OPPORTUNITIES

Gang leaders, as most members of youth gangs, tend to be non-ideological. They accept the democratic society, the free enterprise system, and even ghetto conditions as the nature of their daily living experience. Their gang activities are directed to adapting and manipulating, and only in the most limited manner to refashioning, the institutional roles which they confront. This is not to deny that gang members may sympathize with the goals of civil rights and Black Power advocates, although often they do not. They may even participate in civil rights and Black Power rallies, act as guards on protest marches, and even electioneer for candidates (Republican as well as Democratic or independent) to public office. When gang members, and particularly their leaders, do this, they do it primarily out of considerations of personal and gang-relevant interests, rather than out of ideological consideration or to achieve directed social change. Promises of jobs, special favors, income, or publicity have been extracted by the gang. The gang, and its leadership in particular, will support a given program, if immediate or tangible rewards are to be supplied. "What is in it for me" is the chief criterion of participation in organized (or unorganized) community activities.

The opportunism of gang leaders is rather clearly demonstrated by their participation in projects sponsored by various community action programs, youth-serving and detached worker agencies, and churches. It is an opportunism which is usually responsive to the cooptive efforts of these agencies, who need gangs and their leadership to serve organizational purposes. Social agencies and other community groups constantly seek to establish and confirm their legitimacy and success as social-control and opportunity-providing agents to deprived and alienated youths. They receive funds and grants from private and public sources to develop a wide assortment of programs to prevent, control, and treat delinquency. Agency prestige, fiscal stability, and program expansion depend in large measure on the ability of the organization to provide an "effective" service. Agencies must therefore coopt delinquents into the program, i.e., capture gangs, especially gang leadership, and "redirect their energies into more constructive channels."

Agencies increasingly "reach out." They offer staff positions, special programs, and facilities in order to "serve" and "change" gangs. However, agencies have to meet the interests and needs of gangs in order to do this. They have to come to terms with them. A bargain more implicit than explicit is arranged, although at times formal negotiations are held and a written contract as to the conditions of relationship, particularly in regard to a given crisis between gangs, is produced. The critical situation is usually augmented by open conflict or severe competition between community agencies. This may occur when social agencies and even the police find themselves in competitition for the right to "serve" or control gang youths. Gangs, particularly leaders, increasingly find they can bargain or offer themselves as aides or clients to the highest bidder.

A *quid pro quo* is arranged. The gang leader and his group support the agency's purposes, participate in its program, abide by certain rules, and even agree to stop, or at least limit, gang fighting, if, for example, jobs, special facilities, economic rewards, honorific positions, avoidance of arrest, dismissal of court cases, etc., are provided. Gang leaders may seek to capture the agency to meet the needs of the gang as a corporate entity as well as to achieve particular individual social and economic advantages. For some relatively sophisticated gangs, the political bargain serves as a basis or "cover" for "safer," more criminalistic, delinquent activities, such as extortion from other gangs or unaffiliated youths, deriving protection money from local storekeepers, and recruiting more members to the "successful" gang. All of this is a rather subtle process, and the participants—including agency as well as gang personnel—are not always clear about what roles or functions they are playing.

The gang leader stands to gain most from negotiations with community agencies and authorities. If he is successful, not only does he stabilize and partially legitimize the gang's status in the community, but he solidifies and augments his own position. Because he is used as a communications link and even a key staff person or agency "representative," he can distribute favors in a manner which best serves his interests. He comes more effectively to control the perquisites required by members of his group. He is placed or places himself in the role of politician, mediating the gang structure and the world outside. Some risks are entailed in the position he plays. If he identifies too closely with the interests and wishes of the agency, he may alienate the more aggressive and alienated members of the group, especially if there are not enough "goodies" to go around. If he identifies too closely with the interests of his gang, particularly for conflict purposes, he is vulnerable to sanctions from the authorities. Agencies may attempt to isolate and displace him as a leader.

On the other hand, the successful gang leader is able to command loyalty and manipulate gang decisions and actions by virtue of his control of resources from agencies and community organizations. Furthermore, he continues to use the threat of gang violence (and even civil disorder or riot) to persuade agency representatives to provide additional perquisites to him and the group, and to retain him as its liaison and spokesman with the group. A major problem, however, is that his influence with the agencies depends upon both his threat of, and ability to control, gang violence. If the gang's violence is contained for too long, the gang may disintegrate. The central cohesive element of the fighting gang is lost, and the power of the gang leader may vanish with it. This problem is in part resolved if occasional skirmishes between gangs are permitted or even encouraged by the leader, to cohere the gang at critical moments. The value of the gang for agency purposes also depends to some degree on the gang's potential capacity to engage in aggressive action and validate the claim that the agency is performing an effective service and protecting the community.

An equilibrium or viable system of relations can be worked out between the gang, as deviant group, and the legitimate agencies in the community. The gang leader can succeed in entrenching himself. The gang is less threatening to the community, and in turn is less threatened by its official agencies. Benefits occur to all members of the subsystems immediately affected. Organized violence is measurably curtailed; gang members, particular leaders, are accorded new prestige, power, and economic rewards; and agencies have established successful programs. Whether basic causes of lower-class delinquency are modified, substantive social and economic opportunities provided, and delinquency rates in fact lowered, is open to question. Much depends, it would seem, on the relevance and scope of the resources provided by the intervening agency and the willingness of the community to develop alternate meaningful structures for the gang. Sufficient resources and positive alternatives to delinquent gangs have probably not yet been developed by any agency, community action program, or community.

In any case, the community system can be stabilized. More opportunities and success or prestige have been provided to key members of the various subsystems, although basic structures and goal orientations have not been significantly modified. Most important, the community's cultural criteria for success and the economic and social means for achieving them have not been substantively altered on behalf of delinquents. The old system has been slightly modified and more strongly reinforced.

THREATS TO STABILITY OF THE SYSTEM

The system, however, is subject to stress and strain. Threats to the equilibrium established between the community and its gangs arise from within the gang and from unaffiliated deviant youth, from the nature of the resources provided by and controls imposed upon gang youths by agencies, and from social change forces in the community.[4]

There is an inherent replacement dynamic in youth groups, including fighting gangs. Adolescent ganging is by definition a phenomenon of the teen-age years. The most active members of fighting gangs tend to be of early and middle teen years. At the age of 18 or 19, youths tend to obtain full-time jobs, get married, enter military service, receive "stiff" prison sentences, or enter an alternate deviant subculture, traditionally one committed to drug use, petty racketeering, or thieving. The society begins to "open up" for older adolescents and young adults, or at least to treat them legally, vocationally, and socially as adults. The norms of the fighting or conflict subculture, furthermore, tend to support the withdrawal of older teen-agers and young adults from fighting and gang activities. Important status and "room at the top" must be provided for "up-and-coming" gang fighters, if the vigor of the gang is to be sustained. Often an entire leadership or hard core group "withers away" naturally in this process. Sometimes younger, status-hungry, more aggressive members may oust existing leadership or break away to create their own rival group. Youths seeking access to the gang and denied entry may establish their own fighting group. Although particular gangs are modified, vanish or are replaced, the fighting gang system is sustained.

The competitive quest for status through gang fighting is sustained by the limitation of access to social and economic resources and the the delinquent subcultural tradition in the area or larger community. The fighting gang has, since World War II, provided a major systemic means for obtaining prestige and recognition for youths defined as failures by schools, youth agencies, families, and police. The gang has served as a mechanism for the realization of youthful energies, abilities, and aspirations which have been blocked, directly and indirectly, by established legitimate structures in the community.

However, certain basic community changes may be occurring. Older teen-agers or young adults no longer find that, with increasing age, access to desirable status is improved. Unskilled and semi-skilled jobs are increasingly difficult to obtain. Furthermore, aspirations have risen, and low-status adult positions may no longer be culturally or subculturally acceptable. On the other hand, rewards continue or are augmented for the "successful" gang leader. Consequently the older leader does not readily graduate to adult status. At the same time there is less opportunity for younger members to rise or take over the gang. The situation would not be untenable if new sources of legitimate (and possibly illegitimate) status were provided to these ambitious youths. But sufficient alternate resources within the gang or through agency programs may not be available. Also, although new fighting groups may be formed to threaten the established gang, the efficiency of gang prevention or control programs is enhanced by the cooperation of established gangs and their leadership. Social agencies, the police, and the established gang leadership attempt to deter threats to new relationships. Nevertheless, the existing system of pressures continues to build.

Threat to the established gang system may also arise from significant shifts in policy and program of socializing and authoritative agencies. The police are particularly vulnerable to pressures from within and outside their own organization to change policy. The police department may suddenly come under pressure, for any of a variety of reasons, to assert its authority and dominance in the law-enforcement area and abrogate a policy of *modus vivendi* with youth gangs. With sufficient resources of man-power, organizational efficiency, and the support of influential sectors of the community, it may completely destroy the fighting gang structure. A stable set of community relationships which the youth gang has established over a period of months and years may be smashed overnight by the action of the police.

Youth agencies and funding bodies also contribute to destabilization of the system. Programs devised for gang youths tend to be crisis-oriented. When the crisis or the "hot summer" is past, program funds are often withdrawn. Also, on occasion, a program does prove to be highly "successful" and gang fighting diminishes or ceases to exist. Gang fighting is no longer meaningful to youths. Gangs dissolve, but their membership is not fully or satisfactorily incorporated into existing legitimate programs. Deviant youths are still present in the community, but in more diffused and inchoate clique structures. With the dissolution of fighting gangs may come a shift in agency strategy. Increased attention is directed to serving disturbed individuals. Consequently the needs and drives of youths for corporate affiliation are further blunted, increasing the disorganized character of adolescent life in the slum community.

Furthermore, a new opportunity system which can serve the needs of these aspiring but unsuccessful youths and young adults may be developing. Certain organizations, civil rights, Black Power, and revolutionary groups do promise a measure of meaningful status. These youths and young adults are intensely frustrated, bound only by tenuous ties to organizations and peers. They are ready for any action which promises immediate reward and the most transitory improvement of their present unbearable status. They are prepared to topple any and all existing community arrangements, including that which has been worked out by the established gang system. They are prepared to participate in short-run disciplined or undisciplined activities or even destructive enterprises, such as riots.

With the destruction of gangs and the failure to provide alternate corporate youth-socializing structures, status frustrations are aggravated. Deviant youth become available for ephemeral, quixotic mob or riotous actions. While bonds with gang structures and peer culture have been weakened, bonds with a variety of deviant and change-oriented adult organizations may be developing. The destruction of gangs may serve, under certain conditions, to de-isolate delinquents from receptive adult sectors of the community. Gang leadership itself may be the first to sense the changed conditions existing and the non-viability of traditional gang structures.

They may make strong attempts to exploit receptive adult organizations to fulfill their own status needs and those of other youths.

COOPTATION OF GANG MEMBERS BY
NEW RADICAL ORGANIZATIONS

The presence of the fighting gang may be viewed as a function of the need of the adult community or its various subsystems to sustain institutional relationships and a pattern of restricted resource allocation. The fighting gang has generally been a response to the lack of available roles for youth within legitimate and illegitimate adult structures. A delinquent youth structure has evolved, partially integrated into the need structures of the police and social agencies. Functionally this has been less satisfactory than the older type of criminally-oriented youth groups—which is still, however, prevalent in many parts of the urban community. This latter type of youth group, concerned principally with theft and robbery, served more effectively to link youth and adult deviant roles. Delinquents could look forward to careers in professional and quasi-organized crime. In recent decades, increased bureaucratization and sophistication of criminal techniques has severed many of the links between unskilled delinquents and highly skilled and professionalized criminals. Routes to status in the criminal system have become extremely difficult to find for most slum-dwelling delinquents.

With the breakdown of the function of criminal youth groups and, more recently, fighting gangs, deviant youths are in the market for more satisfactory types of group delinquent or deviant adaptations. A new type of deviant adult system which has been developing in urban communities may constitute a means for securing such status satisfaction. The protest organization, the civil rights, Black Power, and even revolutionary groups are elements of this system. These are not deviant adult groups in the traditional criminal sense, except as defined so politically. They do, however, constitute a threat to the established patterns of institutional relations and come to be regarded, in varying degrees, as violating accepted norms and laws. The norms of the more radical of these adult organizations are at least partially congruent with those of the members or former members of fighting gangs. A new deviant opportunity-providing system has thus become potentially available for alienated youths.

The new deviant organizations have not only become accessible to delinquent youths but increasingly seek to attract them. The assumption of the radical organization is that youth gangs are the most powerful force in the ghetto and constitute a leverage point for organizing youths as well as adults against the established system. The argument of the radical organization is that gangs and delinquents are wasting their energies and uselessly attacking their peers or "brothers" over problems which basically emanate elsewhere. Gang fighting generally tends to be intra-ethnic, and delinquency is generally directed against other slum dwellers. The gangs and their members are urged to turn their anger and hostility against the established system through the program of the particular social-change or radical organization. Gang leaders, furthermore, are viewed as possessing the prestige, power, and needed contacts to make the radical organization successful in the ghetto. One Black Power organizer formulated his approach as follows: "If I could get fifty gangs to join together, we could take over the city."

These newer-type change-oriented organizations, however, have not yet been successful in recruiting sizable numbers of gang youths for extended periods of time and commitment. They have, nevertheless, succeeded in securing the support of some gang leaders and gang segments for a particular activity such as a protest march, a picket, or a rally.

Organizers from these adult groups suffer a number of handicaps in their efforts to recruit gang support. They tend to be middle-class and highly ideological in their commitments, and thus are often unable to effectively communicate with and persuade gang youths. Gang youths in the ghetto may not be terribly interested in open housing in the suburbs or access to skilled jobs in industry. They seek immediate and concrete rewards and relief from frustration—e.g., unskilled and semi-skilled jobs, improved recreation facilities, less stringent police practices. They are suspicious of grand schemes and are concerned less with changing the community and society than with getting a larger share of certain visible "goodies." They may see no glory in arrest for a Cause. They want guarantees of bail money and tangible recompense for efforts expended, which ordinarily are not forthcoming from radical organizations. Perhaps most important, involvement in radical-change organizations and programs may be viewed as a diversion from more exciting and meaningful activities by members and leaders of the stronger, more cohesive gangs. A problem arises which tends to defeat radical organizational efforts directed toward cohesive gangs and powerful gang leaders: they are generally less inclined than weaker gangs and leadership to participate in change-oriented activities.

Radical social and political organizations have thus been most successful in their youth recruitment attempts in those communities where gang structures are already weakened, and youths are seeking alternate status-providing organizations to affiliate with. The established gang structure is not readily brought to support a program of change. Mainly older teen-agers, former gang leaders, and peripheral junior members of gangs are attracted. The radical organization is also most appealing to these youths when its objectives and pattern of activities are similar to the norms and goals of the delinquent subculture, e.g., when there is a strong commitment to illegitimate or violent means.

CONDITIONS FOR INCREASED PARTICIPATION BY GANG OR FORMER GANG MEMBERS IN RIOTS

I have proposed that fighting gangs as corporate entities tend not to precipitate or sustain riots and revolutionary activities, but certain types of gang or former members do contribute to the ongoing development of civil disorders. It is important for purposes of social policy and social control to determine under which conditions gang members are likely to increase their participation in extremist activities and social disorders. The identification of certain conditions or variables—independent, intervening, and precipitant—is important for the creation of appropriate strategies of action, particularly by those concerned with minimizing the possibilities of extreme community disruption. These interrelated conditions which contribute to gang youth participation in riots are:

A. Independent Variables

1. Continued delimitation of social and economic opportunities for youth in the ghetto, relative to culturally induced high aspirations for achievement; consequently, a growing sense of failure to achieve desired status, more particularly a decreasing number of status-providing positions, whether at school or on the job.

2. At the same time, more youths are defined or stigmatized as deviant and subjected to degradation ceremonies—school failure, unemployment, arrest, imprisonment, etc.

3. Alternate meaningful illegitimate means to status achievement are further delimited; e.g., adult criminal structures are closed off and fighting gangs are suppressed or destroyed through the efforts of police and social agencies.

B. Intervening Variables

1. Increased adult organized effort is directed to radical social-change programs.

2. More effective efforts are made to involve deviant ghetto youths in these change programs, particularly through the provision of more meaningful status and inducements.

3. Mass communications, at formal and informal levels, more forcefully indicate the futility of the existing system of institutional relationships as a means of providing solutions to the status dilemma of youths.

C. Precipitating Variables

1. Control and socialization mechanisms, mediated especially by the police and the schools, become more ineffective; e.g., the activities of the police are increasingly defined as brutal and the attitudes of teachers as callous.

2. Social agency efforts in relation to problems of income maintenance, housing, recreation are decreased, and frustration on the part of clients is increased.

3. Crisis phenomena (natural and man-made) such as heat waves, cold spells, or communications or transportation breakdowns occur and serve to disrupt normal patterns of activity and produce frustrations.

These would appear to constitute necessary and sufficient conditions for the participation of ghetto youths, particularly gang youths, in civil disorder or riots. The presence of all these variables is probably essential for the involvement of youths in a riot. Thus, the reduction or elimination of one or more of these conditions would contribute to stabilizing a given community system.. While the independent variables are the more fundamental conditions contributing to youth involvement in civil disorders, strategies of immediate control need to direct primary attention to the precipitant variables, where this is possible. Modification of police, school, and social agency attitudes and practices with a view toward at least minimally reducing status and situational frustration of ghetto delinquents would appear most useful. (Precipitant and intervening variables are, however, not unrelated to fundamental determining conditions.)

In short, it may be more important—if the goal is to preserve community stability—not to summarily destroy fighting gangs, but to manage, control, and if possible redirect the energies and interests of their members into both community and subculturally relevant enterprises.

NOTES

1. See Robert K. Merton, *Social Theory and Social Structure* [rev. ed.] (Glencoe, Ill.: Free Press, 1957), esp. chaps. 4 and 5; Richard A. Cloward and Lloyde E. Ohlin, *Delinquency and Opportunity* (Glencoe, Ill.: Free Press, 1960); Irving Spergel, *Racketville, Slumtown, Haulburg* (Chicago: Univ. of Chicago Press, 1964).
2. Many of the observations on which the conclusions of this article are based were made in the course of a curriculum development project supported by the President's Committee on Juvenile Delinquency and Youth Development during 1966.
3. *Chicago Sun Times,* Aug. 4, 1967.
4. For an informative article which supports the notion *"that the sources of gang cohesiveness are primarily external to the group,"* see Malcolm W. Klein and Lois Y. Crawford, "Groups, Gangs and Cohesiveness," *J. Research in Crime and Delinquency,* Jan., 1967, pp. 63-75.

RIOTS AND
THEORY BUILDING

John G. White

The theories of riots and urban violence that abound in the mass media and the social science literature seem to present the following overriding "consensus": riots in American cities are the direct result of the cumulative frustrations experienced by Negroes confined to the urban ghetto. This frustration, many observers argue, emanates from the appalling conditions within the ghetto and the treatment accorded the Negro by the dominant white society. According to one recent study, for example, social scientists and ghetto residents alike agree that among the primary contributing factors to the occurrence of riots, the following seem most important: police brutality (this includes both physical acts and police attitudes), living conditions, unemployment, poverty, and anger with local businessmen.[1]

According to this thesis, the urban Negro feels hopelessly trapped in the ghetto, where he is abused and brutalized by the representatives of the white world with whom he comes in contact.

This "theory" refers to ghetto conditions as the underlying causes of racial riots. This is a plausible argument, but viewed as a theory, it is very difficult to test empirically. In order to approach the problem a bit more systematically, perhaps it would be wise to place this kind of "theory" into the broader context of conflict

AUTHOR'S NOTE: *The author would like to express his appreciation to R. J. Rummel, for his support and encouragement; Michael J. Shapiro, whose criticisms of the manuscript were most helpful; and Gary Tanaka and Charles Wall, who helped with the data collection and analysis.*

theory in general. In this way, it might be possible to suggest the relevancy of various micro- and macro-level theories of conflict to the analysis of urban racial violence.

Some theorists argue that conflict can be best explained in terms of theory which emphasizes the behavior of individuals or groups. Others feel that conflict should be viewed within its social setting and that the societal structures should be emphasized. Behavioral theories of conflict emphasize attitudes and their formation. This would include such things as "degrees of strain and anomie in societies, the processes by which tension and aggression are generated, and the processes by which human beings are 'socialized' into their communities."[2]

The classical frustration-aggression model as it relates to conflict would fall under the rubric of behavioral theory. Davies has suggested that Dollard's original frustration-aggression model might be useful in explaining conflict at the societal level.[3] Davies argues that "relative deprivation" should be related to the occurrence of conflict: when there is a gap between expectations of individuals and their abilities to meet those expectations, frustration and subsequent conflict (aggression) will ensue. Following this logic, one would expect to see conflict arise when earning power declines sharply (a depression) or when a group of individuals is deprived of earning power relative to the rest of society. A number of authors have taken Davies' suggestion and attempted, with varying degrees of success, to operationalize the frustration-aggression concept at the micro-level. Much of the research utilizes some variant of "deprivation" as a unifying concept. For example, Bruce Russett tries to show that land and economic inequality are related to instability and conflict.[4] Gurr, on the other hand, attempts to use indicators of economic and political deprivation of certain groups as correlates of civil strife in selected polities.[5] Feierabend and Feierabend attempt to operationalize the concepts of "want-need" and "want-satisfaction" and show that the gap between the two is an underlying cause of aggressive behavior within polities.[6]

Structural hypotheses, as opposed to behavioral ones, emphasize objective social conditions as the underlying causes of conflict. The focus is usually upon economic conditions, social stratification and mobility, or geographic and demographic factors. Smelser, for example, points out that conflicts (and in particular, riots) can best be examined in light of what he calls "structural conduciveness and structural strain" in the existing society.[7] By "structural conduciveness" he means the presence of religious, ethnic, racial, or class cleavages from which hostility can emerge. He points out, interestingly enough, that the potential for hostility increases when economic, political, and racial memberships coincide. Conduciveness also involves the presence of channels for expressing grievances. Smelser notes that "it is important to inquire into the possibility of expressing protest by means other than hostility. Are these other means permanently unavailable? If so, aggrieved people are likely to be driven into hostile outbursts."

Structural strain involves the social situations that drive people to conflict, when structural conduciveness is present. Many strains are institutionalized and follow the cleavages outlined above. The chronic tensions that exist between Negroes and whites in this country would be an example of an institutionalized strain. This strain, combined with other conditions such as high in-migration of Negroes into a city, increases the possibility of conflict. Open conflict, then, tends to focus on a particular issue such as police brutality, job discrimination, or open housing. According to Smelser, when strain and conduciveness are present, precipitating factors sharpen and exaggerate the effect of these conditions and can easily provoke a riot. That is, the precipitating factors overlap with the broader background conditions, but provide an igniting element. It is often this element that immediately precedes a riot.

The reader may have already reached the conclusion that to talk about two distinct kinds of conflict theory is perhaps setting up a false dichotomy. To a certain extent this is true, as the two theoretical orientations overlap both conceptually and methodologically. Much of the empirical research tends to implicitly or explicitly combine the two orientations in attempting to explain specific conflict events. It is difficult, for instance, to apply a frustration-aggression model at the micro-level without stressing the importance of structural conditions or the social setting. It is because of this, and because of the problems incurred in operationalization and testing, that we choose not to directly apply general conflict theory to the analysis of riots. A more fruitful strategy would be to take our "cues" from existing theory and combine our knowledge of general conflict with what we know about race relations. This combination, then, can form the basis for the selection of factors or variables to focus upon in the analysis of such disturbances.

Social distance theory is a case in point. Although not a theory of conflict as such, the concept of social distance is certainly relevant to the study of racial conflict. Whether the social distance between groups has class, religious, or racial determinants,.its presence creates the potential for conflict.[8] Knowing what we do about the very distinct social distance between whites and Negroes in this country, what kinds of consequences might this have for the confrontation of these two groups in large urban areas? It suggests, for one thing, that we select factors which can define this distance and attempt to relate those factors to racial riots.

The central concern of this paper is an attempt to propose a method for the scientific investigation of racial disturbances. We hope to subsume some of our knowledge concerning the position of the Negro in American life, together with theories of conflict, under an overall explanatory scheme. That is, to follow the logic outlined above by selecting variables which define, at the macro-level, the social distance between Negroes and whites, and which define the racial cleavages, the structural conduciveness, the structural strain and anomie inherent in modern urban life. In other words, we would select variables which seem to index some of the important features of urban areas, and relate these features to open conflict. What, for instance, might be the relationship between such variables as population density, percentage of population which is non-white, unemployment levels of minority groups, and migration level with the occurrence of riots in a selected group of American cities?

The advantage in using this kind of aggregate data is that we can focus upon macro-level variation through the use of ecological correlations. Menzel has pointed out, for example, that high correlations between arrests and divorces do not imply that individuals who are arrested tend also to be prone to divorce.[9] However, it may indicate the presence of an underlying cause which is not inherent in the individuals but in the properties of the area where the individuals are living. Let us, for example, determine the correlation between two characteristics (variables) of American cities. We find a high correlation; they tend to vary together when all cities are taken into account. Peraps this can be seen as a pattern of variation. If we correlate the variables with the occurrence of riots and find a high correlation, then common variation exists. We might argue that the variation in riots is explained by the variation in the two characteristics. If the characteristics were present first and then the riots occurred, we have the basis for inferring a causal connection between the variables and riot behavior. That is, if riots tend to occur in cities with high levels of population density and with large Negro populations, we can infer a causal relationship between the two variables and riots. The concept of variation is crucial here, and is central to the kind of research being proposed. If we can relate variation and theoretically relevant characteristics of cities to variation in riot behavior, we can begin to replace speculation with empirical description and a more scientific understanding of the conditions underlying civil violence.

The present paper includes the results of a pilot study which is intended as an important first step in the method under consideration. The study uses data taken from a factor analysis of selected characteristics (structural variables) of American cities. The factor analysis, using aggregate data on 65 demographic and economic variables, determined the major patterns of variation (groups of variables that vary together) among the cities included. That is, since some of the variables were interrelated, it was possible to "reduce" them through the use of ecological correlations to several underlying patterns or dimensions of variation. In the present study the variation of a selected group of cities along these dimensions was related to the variation in rioting through the use of regression analysis.

The analysis consists of regressing a riot measure upon variables which index or define eight structural dimensions of American cities. The dimensions used in this study were delineated in a previous factor analysis and serve as input in our analysis, so that the multiple regression is the focus of concern. The factor analysis, along with the characteristics of the independent and dependent variables, will be described first. The results of the regression analysis will then be given.

Data for the present analysis is taken from the factor analytic study of American cities by Hadden and Borgatta.[10] This is perhaps the most comprehensive study of its kind to date. The authors selected 65 structural variables for 644 cities, and factored out 13 primary dimensions which accounted for 90% of the variation in their data matrix. The factor structure of the cities was then converted to separate profiles, so that the individual cities could be identified and then compared on the basis of their characteristics. Single items (variables), rather than the more complex factor scores, were used to index or define the independent dimensions. (One exception to this will be noted later.) This was done because in most cases the correlation of a single item with the corresponding factor scores was essentially unity, so that the content of the analytic structure was well represented.

The authors selected twelve profile items (variables) which reflected the factorial content of eight primary dimensions. Since only eight dimensions are represented, we have selected eight items for the present analysis. These eight items, which are shown in Table 1, are independent variables in the regression and represent independent sources of structural variation for the cities. The authors choose to convert the variables to decile score which were used in the present analysis, so that all items vary from 0 to 9. The variables, for the most part, are self-explanatory. *Median income* is a measure of family income and is intended to index what the authors call a socioeconomic status factor. *Median age* is the median age of the population of the city, and defines the age composition dimension. *Educational center* reflects the presence of universities and colleges. The decile score for this item is actually a weighted index computed from the three variables which load highest on the factor.

It should be re-emphasized at this point that the present study utilizes variables only from the Hadden and Borgatta study. Additional variables which might be theoretically relevant for the explanation of riots were not included at this time, since it would require an additional factor analysis. The ideal procedure would be to factor analyze both the present variables and other variables deemed theoretically relevant. Since this was not done, the present study is limited to that extent.

The dependent variable used in the present analysis is a dichotomous one which indicates riot and non-riot cities. Data on riots was collected for the years 1963 to 1967 primarily from news magazines. Cities where riots occurred in those years were coded as (1) and cities where no riots occurred were coded as (0). Riots were operationally defined as spontaneous collective assaults by

Negroes on persons or property, involving one hundred or more persons. Data for this variable as well as the independent variables was collected for each of 262 cities which had populations of over 100,000 at the time of the 1960 census.

Table 1 gives the results of the regression of the riot measure on the eight independent variables. The regression analysis program used here (BMD - 03R) provides us with the multiple correlation coefficient (R), the product-moment correlation coefficient (r), and the partial correlation (r_p with the dependent variable holding other variables constant. The regression coefficients (B) are expressed as standardized "beta weights." The partial correlations and the "beta weights" which are statistically significant are indicated. The reader should be aware of the fact, however, that tests of significance are reliable only when the independent variable is normally distributed. Since the data used here does not fall into a normal distribution, statistical significance should be viewed in that context.

TABLE 1

REGRESSION OF RIOT DUMMY VARIABLES UPON
EIGHT STRUCTURAL CHARACTERISTICS OF CITIES
REGRESSION COEFFICIENTS (B), PARTIAL (r_p) SIMPLE
(r), AND MULTIPLE (R) CORRELATION COEFFICIENTS

VARIABLES	COEFFICIENTS			
	B	r_p	r	R
1. Total Population	.025**	.170**	.29	.47
2. Density	.015**	.164**	.28	
3. Family Income	.023*	.131*	.10	
4. %Non-White	.052**	.280**	.30	
5. %Foreign Born	.004	.027	.06	
6. Median Age	.020	.114	.13	
7. %Same House 1955-1960	.010	.052	.02	
8. Educational Center	.0005	.003	.01	

* Would be significant at .05 if normal distribution
** Would be significant at .01 if normal distribution

The (r_p) column and the (B) column give an indication of the relative importance of the respective independent variables. From the "partials" one can tell, of course, which variables account for the most variance when other independent variables are held constant. The regression coefficients or the "beta weight" indicate the extent to which riots are functionally related to each independent variable. Most of the variation in riots seemed to be accounted for by total population, percentage non-white, median family income, and population density.

The multiple (R) indicates the amount of variation in the dependent variable associated with the total variation among the independent variables. That is, it indicates how well the regression equation "fits" the data and shows the predictive capacity of the model. In this case, since (R) equals (.47), only about 22% of the variance in riots is accounted for (R^2). Some possible reasons for this relatively low (R) will be discussed in the next section.

Although the primary concern at this point is in assessing the relative importance of the independent variables, the predictive capacity of the regression model is important. When the multiple correlation coefficient is relatively low, this indicates that, for one thing, some important sources of variation have been excluded. That is, some theoretically relevant explanatory variables are not included in the model. In the present analysis, it will be recalled, we have used primarily structural or demographic variables which index patterns of variation. The unexplained variance might be considerably reduced by excluding, for example, measures of police harassment, Negro store ownership in the ghetto, and Negro representation on the police force in a second factor analysis. Then one could use the output (factor scores) from this analysis as predictive variables in a second multiple regression.

A second possible explanation of the low (R) in the present study is perhaps error in the measure used for the dependent variable. Of course, riots expressed in some sort of continuous measure would be preferable to the dichotomous variable used here. However, data on such measures are not readily available. Two alternatives seem feasible at this time. First, one might attempt to obtain data on the scope and intensity of riots such as: length of the riot, number killed, property damage, and number of persons involved. These variables could be combined in a weighted index which would be expressed as a continuous variable. Because of data collection problems and for reasons of conceptual clarity, these kinds of measures were not used. The second alternative has been alluded to earlier; namely, that one could factor analyze the above variables and express the independent variable as a factor score. The present analysis is intended to be a pilot study, but does represent an important first step in the theory building process discussed earlier. The regression indicates which of the predictor variables are the most important in explaining riots. In fact, we could reduce the number of components by almost one-half and not significantly affect the predictive capacity of our model. That is, the first four variables account for almost all of the explained variance in the model. The assigning of relative weights to predictor variables at this point is not going to give us the key to the understanding of the causes of riots or allow us to predict riots. It is, however, imperative that we do this kind of analysis at the outset, so that we have some basis for proceeding with further investigations. The present analysis indicates that we can relate variation in conditions from one city to another to the variation in riots, using aggregate data. This fact has bearing on subsequent factor and regression analyses, and eventually will bear on the kind of simulation model which can be constructed.

NOTES

1. P. Meyers, *The People Beyond Twelfth Street* (Detroit Urban League, 1967).
2. H. Eckstein, "On the Etiology of Internal Wars," *History and Theory,* IV (1965), 133-163.
3. J. Davies, "Toward a Theory of Revolution," *Am. Sociol. Rev.* XXVII (1962), 6.
4. B. Russett, "Inequality and Instability," *World Politics,* VI (1964), 446-454.
5. T. Gurr, *The Conditions of Civil Violence: First Test of a Causal Model* (Princeton: Center of International Studies, Princeton Univ., 1967).
6. I. Feierabend and R. Feierabend, "Aggressive Behavior Within Polities," *J. Conflict Resolution,* X (1966), 250-269.
7. N. Smelser, *Theory of Collective Behavior* (N.Y.: Free Press, 1962).
8. H. C. Triandis et al., "Race, Social Class, Religion, and Nationality as Determinants of Social Distance," J. Abnormal and Soc. Psychol. LXI (1960), 110-118.
9. H. Menzel, "Comment on Robinson's 'Ecological Correlations and the Behavior of Individuals,' " *Am. Sociol. Rev.,* XV (1950), 674.
10. J. Hadden and E. Borgatta, *American Cities: Their Social Characteristics* (Chicago: Rand McNally, 1965).

Part III
THE SETTING OF
URBAN VIOLENCE
Some Empirical Studies

Introduction

This section concentrates on reporting findings of researchers who have investigated empirical conditions which give rise either to actual instances of civil disorder or to attitudinal structures or clusters which are supportive of the employment of violence to gain social or political ends. The findings actually reported here are drawn variously from Houston, Oakland, Los Angeles, Cleveland, Omaha, Rochester, Buffalo, and in the case of Latin American specialist D. P. Bwy, from several cities in Central and South America. We suggest, however, that the results reported here could inferentially be extended beyond the specific sampling sites involved.

We make this argument because, as will become apparent in reading the reports, there is a remarkable similarity of themes running through these investigations even though they were all conducted independently at different times and places. The central theme implicit in virtually every one of the essays is that the propensity to civil violence arises among those who feel themselves deprived or dissatisfied. Dissatisfaction or deprivation is here measured in several ways. Some of the investigators see it as deprivation relative to the perceived achievements of some salient reference groups; for example, the urban Negro ghetto dwellers' perceptions that they are not achieving the same standards of living as Whites. Other investigators see deprivation as a gap between what the individual feels he now has and what he aspires to have or expects to get (three of the papers use the Self-Anchoring Striving Scale, a technique for exploring aspiration and expectation levels compared to past and present levels of satisfaction). These differing research emphases are not, of course, antithetical. And both rest solidly on the concept of felt deprivation.

Related to this is the second theme: that the dissatisfactions or deprivations which individuals feel themselves to suffer cannot be, or are not being corrected by existing social institutions. In very general terms we may say that the concepts invoked here are alienation and legitimacy. If individuals regard the authorities as responsive to them and as properly constituted they are likely to be willing to tolerate a greater amount of deprivation without resort to civil violence. But in urban North America it is no overstatement to say that many of the traditional agencies of social control, in particular the police, are losing their legitimacy. Another way of making this point is to say that every government has a bank of moral capital which must from time-to-time be spent in the making of unpopular decisions but must also be replenished by acting to ameliorate the conditions of the potentially disaffected. A government unwilling or unable to make this kind of reinvestment loses its legitimacy, for moral capital, like all other capital, is easily dissipated.

The third and last theme common to these essays turns on the willingness of individuals to employ violence as a means for forcing political and social change. In most social systems individuals have internalized norms which inhibit the use of violence to a greater or lesser degree. These norms are in turn reinforced by external agencies of social control. But in situations where chronic deprivation is felt and where the external controls are no longer perceived as legitimate the internal controls on the employment of violence are weakened. In such situations there develops what the authors represented here have variously termed a "protest factor" or a "riot orientation"—in fine, a receptivity to the use of violence as a means of effecting social and political change.

Turning now to the specific findings, we can see how the individual authors have woven these themes through their analyses. The first paper by sociologists William McCord and John Howard, compares Negro opinions in three cities in the United States. Here the degree of dissatisfaction or deprivation is assessed by the general question put to respondents concerning the pace of desegregation. This means much more than merely the striking down of legal barriers to racial integration, for it is clear that Negro demands encompass operative equality with Whites in such areas as jobs, housing and education.

Illustrative of the two remaining themes is the authors' concern with investigating Negro grievances with the police and the degree to which Negroes would employ violence and view rioting as efficacious. As one would expect there are differences among the Negro populations of the cities examined as well as differences within the Negro community of each city based on education, age, income, sex, and the like. But in the eyes of these analysts the differences among Negroes are more than overshadowed by the unanimities of opinion among urban Negroes that life for them in these United States is dissatisfying and that the employment of violence is emerging as a preferred technique of change among all segments of the Negro population.

The second essay in this group, by political scientist D. R. Bowen, E. R. Bowen, S. R. Gawiser and L. H. Masotti, focuses on the first and third of the themes by relating a sense of felt deprivation relative to personal aspirations to what the authors term, "protest orientation." Having established empirically that such various activities as picketing, public marches, protest demonstrations, and riots are perceived by ghetto respondents as very closely related kinds of activities, the authors raise the question concerning exactly what kind of felt deprivation is associated with support of protest activities? And the answer is that both perceived upward mobility (the expectation on the part of the individual that he will be much better off in the future than he is now) and perceived downward mobility (the converse expectation, that he will be much worse off) are associated with supportive attitudes about the employment of unconventional and even violent techniques of political change.

The third essay in this section takes us outside the United States as D. P. Bwy examines social conflict at the city and provincial level in four Latin American countries. Bwy's analysis leads him to argue that there are two basic dimensions of social conflict: the first, "turmoil," involves such events as riots, strikes and demonstrations; the second, "internal war," includes terror, revolutionary invasion and guerrilla warfare. The first is more or less spontaneous, random and unorganized whereas the second implies a high degree of conscious organized planning. Taking these two dimensions of conflict as separate dependent variables Bwy then asks how they can be explained.

His answers to that question are found in his three major

independent variables: system response, legitimacy and satisfaction. System response opens an area of investigation not by other authors in this section (although Parts IV and V of this book address themselves to that problem). What Bwy is pointing to is that the cohesiveness of a ruling elite and their ability to muster force against those in dissent strongly affect the degree to which violence can or cannot be controlled. In particular does this seem to be the case with respect to his turmoil dimension—the events most like those in the United States.

Beyond this, however, Bwy's remaining two variables, system legitimacy and satisfaction iterate two of the major themes we have been discussing. And his findings are consistent with reports of rioting in this country. The more satisfied the populations of these countries are, the less likely are turmoil events to occur; and the more legitimacy the political order enjoys the less likely these events are to occur.

In the fourth essay of this group political scientist Harry Reynolds examines the actual grievances as stated by rioters and their neighbors in Omaha's ghetto areas. Overwhelmingly, Reynolds finds the major source of felt deprivation is jobs. It is not that the rioters are totally unemployed but that they are underemployed or employed only intermittently. Even more to the point is the finding that rioters tend to perceive the jobs available to them as ones which have no opportunity for meaningful advancement, thus underscoring the theme of deprivation relative to other reference groups.

Reynolds also points out that there is a high degree of cynicism or skepticism concerning the abilities of existing agencies (including the power of the Negro middle-class politicians) to effect any change in this situation. Indeed this perception is quite accurate according to this analysis, for agencies which have the power to effect such change are not necessarily responsive to the demands of ghetto-dwellers and those which are at least minimally responsive, such as the Negro power structure, have little power in the larger society. Within this context of powerlessness, then, Reynolds argues that the riot was not such much an ideologically motivated, consciously employed technique for achieving social change as it was a desperate cry for help.

It should be obvious to social scientists that the Black ghettos in the United States are not one undifferentiated social mass, and it is this fact which sociologist Jay Schulman has chosen to remind us of in the fifth paper in this section. Schulman notes that Negro residents of Rochester divide both objectively and in their own perceptions between those who live in the "better off" ghetto and those who live in the "worse off" ghetto. Given this distinction Schulman then raises the question whether such residence makes any difference either in the degree to which individuals are politically alienated or in the degree to which they are prone to engage in civil violence.

In the course of this analysis the point is also made that what we have called relative deprivation must be carefully and subtly employed by the investigator, for it is not simply the case that those who live in better off areas (presumably, therefore, having much higher expectations) are more alienated or more prone to riot than those who live in worse off areas. For example, Schulman discovered that those of lower SES in the better off ghetto and those of higher SES in the worse off ghetto were both far more likely to be alienated than their opposites, thus suggesting that the simple notion of relative deprivation should be coupled with hypotheses concerning status discrepancies.

In terms of the relationships between three central themes mentioned earlier, deprivation, legitimacy or alienation, and willingness to engage in violence, Schulman's overall conclusions are that residents of the better off areas are more likely to be riot prone and the residents of the worse off areas to be politically alienated. He believes those who are better off to be more riot prone because they are likely to be motivated by "status" demands or values as opposed to the "survival" values of the worse off. The perceived inability of the better-offs to realize these "status" demands may, in Schulman's terms, create a "free-floating externalized aggressivity." At the same time these status values inhibit political alienation for the better off because existing political institutions are, to a large degree, the product and con-servator of the middle-class value system. Conversely, the survival values of the worse off make them less riot prone because they suffer the most from actual violence and because they have been made so dependent on existing agencies; but it also renders them more alienated because they do not share nor express the status values and demands embodied in current political institutions.

The final investigation reported in this section is that of three political scientists, Everett Cataldo, Richard M. Johnson, and Lyman Kellstadt. Their report is one of the few available where the investigators were able to obtain their data before instances of civil disorder actually occurred, and is therefore free of any contaminating influences the actual experience of civil violence may have had on the formation and expression of attitudes among their respondents. It is thus extremely important to note that the results summarized here are extraordinarily similar to those reported in other papers in this section.

The authors indicate that among Buffalo's Negroes there is a very high degree of hope for the future, both for the achievement of their own personal expectations and also for progress in their community and in the United States. But as is commonly accepted among social scientists such high expectations may be extremely de-stabilizing if the political and social systems are unwilling or unable to meet them. And in the perceptions of these respondents governmental agencies are not doing enough, fast enough, to fulfill the kinds of expectations they have. This finding underscores two of the themes we have seen running through this literature: felt deprivation arising from a perceived gap between current achievement and high expectation, and alienation or dissatisfaction with existing political and social institutions.

The third major theme is likewise the concern of these investigators; namely, does dissatisfaction lead to a proclivity for unconventional and possible violent forms of political expression? That answer to that question in Buffalo, as elsewhere, is yes. Those segments of the population which are the most deprived and the most dissatisfied are also those which are more likely to withdraw from conventional forms of political participation and to view unconventional forms including civil violence and disorder as more efficacious.

—L. H. M. and D. R. B.

NEGRO OPINIONS IN THREE RIOT CITIES

William McCord
John Howard

■ For the past several years, we have examined the predicament of urban Negroes in America, their "style of life," and their response to ghetto conditions. This research has taken various forms. McCord, for example, reported on the riots in Watts (1965)[1] and Houston (1967), while Howard observed "on the scene" the riots in Hunter's Point, Newark, and Detroit in 1967. We have been equally concerned with analyzing the reaction of the white "establishments" to urban conditions, the political future of American Negroes, and such social-psychological issues such as why people join the Black Muslims.[2]

In this paper, however, we address ourselves solely to exploring Negro opinions in three cities: Houston, Los Angeles (Watts), and Oakland. In a number of characteristics, these cities differ strikingly from each other. Houston is, of course, the largest of Southern cities, while the other urban centers exist in the presumably "freer" atmosphere of the West. Watts has suffered from an extraordinarily high rate of unemployment (in 1965, for example, about 41% of young men in Watts did not have a job), while Houston has benefited from

AUTHORS' NOTE: *This research was funded by the Texas Department of Mental Health and Mental Retardation and the Advanced Research Projects Agency under ARPA Order No. 738, monitored by the Office of Naval Research, Group Psychology Branch, under Contract No. N00014–67–A–0145–0001, NR 177–909.*

an expanding economy and a relatively low rate of unemployment. The Negroes of Oakland and Watts have been forced into compact ghettoes, while in Houston the ghetto has been dispersed geographically into approximately twenty areas. Watts has had a major racial explosion while Houston and Oakland have, up to this point, suffered only minor riots.

All of these differences reflect upon the political structure of the cities, the potentiality for mass violence, and Negro opinions about jobs, schools, housing, and the police. These opinions were derived from three sources:

1. Formal interviews with 572 randomly selected Houston Negroes, gathered in the summer of 1966 (before the violence which characterized the spring and summer of 1967).

2. Formal interviews with 187 randomly selected Oakland Negroes, gathered in the winter of 1967 (during the Hunter's Point riot and a minor disturbance over school integration in Oakland).

3. "Natural dialogue" interviews with 426 randomly selected Negroes in Watts, gathered in the winter and spring of 1967. In the "natural dialogue" situation (developed by Blair Justice), the interviewers held conversations with fellow Negroes in bars, barber shops, pool halls, etc. The person did not realize that he was being interviewed. His opinions were recorded immediately after the conversation on a card carried by the interviewer. Naturally, less information can be secured in this manner, but — at least in some cases — the person is presumably more open and honest.

All of the 1185 interviews were gathered by indigenous Negroes of approximately the same social class as the respondents. In some cases, different questions were asked in different samples. Consequently, we are not always in a position to make direct comparisons. Nonetheless, the surveys in Houston, Watts, and Oakland all touched on the same basic issues.

A COMPARISON OF AMERICAN CITIES

In order to put Houston, Watts, and Oakland in some national perspective, we have relied on three publications which outline Negro opinions in a number of other cities. One is the John F. Kraft report;[3] the second is the research carried out by Gary Marx;[4] and the third is a set of *Newsweek* polls conducted by William Brink and Louis Harris.[5] Some of the questions asked in these studies were practically the same as some of the questions used in our research, so that a direct comparison was possible.

First, concerning the speed of integration, it is clear that the Far West and the North are most disturbed by the slow pace of desegregation, while, expectedly, a Deep South city such as Birmingham appears most satisfied:

SPEED OF INTEGRATION

	Too Slow	About Right	Too Fast	Don't Know
Oakland (N:187)	68%	23%	4%	5%
New York (N:190)	51%	39%	2%	8%
Watts (N:426)	48%	50%	—	2%
Houston (N:572)	44%	51%	4%	1%
Chicago (N:133)	38%	55%	3%	4%
Atlanta (N:192)	31%	63%	2%	4%
Birmingham (N:200)	24%	72%	1%	4%

The striking features of these findings are that the residents of Houston, a presumably Southern city, were about as concerned with the speed of integration as residents of Northern cities. Indeed, the level of discontent was higher in Houston than in Chicago. The extraordinary degree of dissatisfaction in Oakland may be due to the fact that the interviews were done just after a crisis on school integration.

When questioned about the main problems which they faced, Negroes in the different cities predictably disagreed about their specific grievances:

MAIN PROBLEMS FACING NEGROES

	Houston (N:572)	Watts (N:426)	Oakland (N:187)
Jobs	55%	20%	62%
Bad schools	14%	*	18%
Housing	11%	15%	7%
Police/discrimination/ other problems with "power structure"	10%	23%	7%
"Troublemakers"	*	10%	*
No opinion/no problem	10%	32%	6%

*Where a blank occurs, it does not necessarily mean that people in the particular sample were unconcerned with schools or "trouble-makers," but rather that the question was posed in a slightly different fashion and thus could elicit different responses.

In Houston, jobs seemed a paramount concern (although public disputes in recent years have centered on school integration). In Watts, the relatively greater concern with the police (cum "power structure") seems a natural outcome of the 1965 riot. Oakland's overwhelming concern with jobs again reflects an objective situation where unemployment and underemployment of Negroes characterize the city. Perhaps the most curious figure is the large number of Watts residents who did not specify any particular grievance. Conceivably this indicates some degree of success of anti-poverty programs or of the indigenous movements which have emerged since the riot. One finds it difficult to accept this interpretation, however, since, as we will later note, most Watts Negroes do not believe that such programs have been notably successful. It seems more probable that the people in Watts have learned not to express grievances openly, even to a casual acquaintance (one of our interviewers) whom they have just met in a bar or some other setting.

Despite the fact that residents of Watts never mentioned integrated schooling as the primary issue, and the citizens of Oakland seldom cited bad schooling, it is quite clear that they disapprove of their "neighborhood" schools to an even higher degree than Houstonians:

OPINION OF SCHOOLS

	Houston (N:572)	Watts (N:426)	Oakland (N:187)
Disapprove current system	42%	53%	73%
Approve/neutral current system	55%	47%	20%
No opinion	3%	—	7%

Specific questions about the quality of their housing were not asked in Houston, but comparable opinions from Watts and Oakland were derived. On the whole, Watts residents disliked their housing conditions more than Oakland Negroes; yet few in either group rated their living conditions as excellent or good:

OPINION OF HOUSING

	Watts (N:426)	Oakland (N:188)
Excellent/good	8%	29%
Fair	32%	45%
Poor	40%	21%
No answer	20%	5%

On some grievances, the citizens of the three cities disagreed strikingly. When asked, "What do you think about the police in your neighborhood?" only a minority in each community expressed approval (or even "neutrality"), but more Houstonians said that they liked the police or were neutral toward them:

The higher degree of resentment found in Oakland is difficult to interpret. Perhaps the reforms initiated in Watts after the 1965 riot have, comparatively, lowered active hostility toward the police. The complacency of Houston Negroes toward the police force cannot be attributed to the greater tolerance, restraint, or training of the Houston police. The most reasonable interpretation is that Houstonians are overwhelmingly of Southern birth: 98% of the Houston sample

OPINION OF POLICE

	Houston (N:572)	Watts (N:426)	Oakland (N:187)
Excellent/fair/ neutral	46%	16%	14%
Some all right, others not	*	45%	26%
Abusive	31%	36%	56%
No contact	15%	*	*

*Once again, where a blank occurs, it does not necessarily indicate a total absence of opinion in that category, but rather that the question was stated in a slightly different fashion and thus could elicit different responses.

was born in Texas or elsewhere in the South, as compared to 51% of the Oakland group, and only 21% of the Watts sample. In other words, the great majority of Houston Negroes have experienced only the behavior of Southern policemen; perhaps, then, their expectations about "fair" treatment from the police are much lower than that of the other groups.

In spite of their condemnation of ghetto conditions, very few Negroes in any of the cities have actively participated in trying to change their cities. When asked if they had taken part in the civil rights movement — by protesting, picketing, boycotting stores, etc. — only 18% of Houstonians and 26% of Oakland citizens claimed to have engaged in civil rights activity. Indeed, even in Watts, only 32% said that they had protested, and 25% responded that they were simply not interested. Thus, while many are verbally discontented, it appears that the massive apathy and hopelessness of the slums prevent most people from trying to alter their situation in life through collective action. This conclusion was reflected in the fact that the urban ghettoes exhibited a high degree of political ignorance. In Houston, only 16% of the respondents could correctly identify the mayor, and even in Oakland, a presumably more sophisticated city, only a minority (43%) correctly recalled the name of the mayor.

When questioned about the usefulness of riots in aiding the Negro cause, no systematic differences appeared among American cities — although New York, surprisingly, condemns rioting more than other areas, for 57% of Negro New Yorkers (in the Marx research) said that "no good can ever come from riots"; 38% in Chicago agreed, as did 26% in Atlanta, 39% in Birmingham (all from the Marx study), 41% in Houston and 27% in Oakland.

From our own data, it was possible to probe more deeply into opinions concerning urban disturbances during the period of 1966–1967. Formal questionnaires utilizing the same questions were administered in Oakland, *after* it had experienced a minor riot, and in Houston, *before* the violent incidents of 1967. As the following figures indicate, Oakland clearly emerged as a more militant city than Houston:

DO YOU THINK THE RIOTS IN WATTS AND OTHER CITIES HAVE HELPED OR HURT?

	Houston (N:572)	Oakland (N:187)
Helped	30%	51%
Hurt	37%	27%
Both	10%	2%
No effect	3%	12%
Don't know	20%	8%

When asked specifically, "In what situations, if any, do you think violence on the part of Negroes is justified?" both Watts and Oakland appeared to be more militant cities than Houston. It should be noted, however, that even in a Southern city such as Houston, only a minority of people responded that they were *always* opposed to violence:

USE OF VIOLENCE

	Houston (N:572)	Oakland (N:187)	Watts (N:426)
Always opposed	45%	20%	23%
Justified in self-defense	24%	44%	47%
Justified to "gain attention"/ "only way"/other	26%	13%	29%
No answer	5%	23%	1%

Expectedly, Watts residents appeared most willing to use violence, and also the great majority have made up their mind on the issue, while in Houston almost half of the citizens say they would never use violence.

The reactions of the Watts residents to the 1965 riots were strangely mixed. Although apparently willing to resort to violence, only 10% of the Watts sample thought that still another riot would aid them. At the same time, very few people (11%) believed that conditions had improved in Watts since the riot. And, despite the massive funds which have been poured into Watts since 1965, only a minority (31%) believed that the War on Poverty was doing a "good job" in their neighborhood. Thus, more people in Houston and Oakland – areas which had not undergone a full-scale riot ordeal – believed that the Watts riot had been beneficial than did the residents of Watts themselves.

THE INFLUENCE OF OTHER FACTORS

In addition to our concern with differences among the cities, we were also interested in the relationship of occupation, education, age, sex, and religiosity to a person's expressed opinions. In this section we will concentrate upon Houston and Watts. We have chosen this approach because of the sheer size of the two samples, and because the interviews were based on different approaches – and yet generally produced similar results. In other words, when one possesses hundreds of interviews, it is sometimes possible to match subgroups of people – let us say of the same occupation, age, sex, and education – and analyze whether their religious attitudes as such have any discernible influence upon their opinions about community issues. It might appear at first glance, for example, that highly religious people are most likely to condemn the use of violence. Yet, this could easily be a spurious finding if, for example, highly religious people were predominantly old – and if it turned out that (when all other factors were held constant) age seemed the best predictive factor concerning one's opinion about violence. Our strategy, then, was to find out the particular level of unrest or concern in these cities and, specifically, to detect who were the most concerned.

In this short article, we cannot explore in depth the influence of other factors on Negro opinion, or present our evidence in detail. Nonetheless, the general picture emerged in this fashion:

— Young people in Watts (holding other factors constant) were more discontented in every way than those in Houston. Yet, regardless of geographical area, the young were *not* generally more dissatisfied with their life and they were not more militant than older people. It appears, then, that the "Negro revolution" is by no means confined to younger people. The young may be "where the action is," but they apparently have the tacit support of many gray-haired men and women who do not directly protest or riot.

— Both in Watts and in Houston (again holding other factors constant), those who claimed to be "very religious" were least dissatisfied with ghetto conditions and least likely to protest in any way against the status quo. For example, in Houston, 58% of people who said they were very religious opposed the use of violence, as opposed to 26% who claimed they were "somewhat" religious, and 20% of those who were nonreligious. Our evidence, therefore, tends to support those writers who have contended that religion has served to deflect Negro concerns from this world to the "next world," to instill an ethic of humility, and to divert potential revolutionary urges into harmless religious channels. Religion, however, seems to be losing its influence over the Negro community. In Houston, only 22% of people said they were "very religious," and in Watts (using a more indirect question), 30% of the respondents said they thought religion was "very important" to other people in their neighborhood.

— In a reputedly matri-focal society, it is particularly important to know what the women believe. When we compared matched groups in Houston and in Watts, it became clear that males and females held to similar views about the main problems which face Negroes, the speed of integration, the quality of jobs and schools, etc. Only one topic of disagreement split the men and the women: the utility of violence. Of the males interviewed, only 28% condemned the use of violence under any circumstances, as compared to 40% of a matched group of women. Clearly, however, a majority of both groups believed that violence could usefully be employed at certain times. It appears, then, that women in Negro society are not exerting the "pacifying" effect that one might expect.

— — When other variables — age, sex, and education — were held constant, we did not discover any major differences between occupational groups in Watts. Whether in a white-collar profession or unemployed, the Watts sample seemed equally militant. In Houston, however, a more complicated pattern emerged. As one might predict, the unemployed in Houston regarded jobs as their major problem more often than other occupational groups. Again, predictably, the unemployed evinced greater hostility toward the police: 44% of the unemployed as opposed to 17% of white-collar groups regarded the police as abusive. Further, only 2% of the unemployed, compared to 26% of the white-collar group, claimed never to have had contact with the police. Half of the white collar group (51%) believed that violence was never justified, while only a quarter of the unemployed (26%) always condemned violence. In Watts, therefore, we may have witnessed a unification of occupational groups on basic issues, but in Houston, predictable differences in opinion between various strata continue to exist. Unification may be one effect of a major riot upon urban Negroes.

— — While education is, of course, highly correlated with such other factors as occupation, the level of one's education, in itself, was significantly related to people's opinions. In general, the higher the individual's education, the more often he expressed approval about the immediate conditions of his own life, such as the quality of his job or his housing conditions. Nevertheless, in viewing the condition of the Negro community in its entirety, the best-educated group expressed greater concern about the speed of integration, claimed to have participated in civil rights activities more often, and condemned police behavior with greater vehemence. Only two differences separated Houston from Watts: first, in Watts, education had no relation to claimed participation in civil rights activity; secondly, in Watts, college-educated people were least opposed to the use of violence, while in Houston, no consistent differences appeared between the various educational groups. In Watts, 21% of college-educated people and 28% of individuals with an elementary school education always opposed violence. On the other hand, 49% of a middle group of people who had received some high school education completely eschewed the use of violence. One effect of the 1965 violence in Watts appears to be an alliance of opinion between the least and the most educated urban Negroes.

CONCLUSIONS

The results of our study of three American cities contradict some of the usual impressions and should concern American whites. In comparing American cities, we found an unexpected unanimity in the opinions expressed by urban Negroes. Dissatisfaction with the speed of integration is stirring many supposedly Southern cities, like Houston, at about the same level as the Northern cities, such as Chicago. Only a minority of urban Negroes totally condemn violence under all circumstances (although, expectedly, the Southern Negro is more quiescent on this issue than Western Negroes). Discontent is not limited to the young or just to men. The "Black Bourgeoisie," the white-collar class, is exhibiting overt signs of interest in the Negro community as a whole. Since it is this group which has previously "kept things quiet" in the ghetto, there are indications that white city fathers should pay increasing attention to the grievances articulated by the white-collar, well-educated groups. Religion does indeed seem to exert a "quieting" effect upon urban Negroes. Yet, only a minority of our sample exhibits great interest in traditional religion, and one might venture to predict that the secularizing effects of urbanization will swell the numbers of the non-religious group. All of these facts suggest that the rash of civil disorders which has spread in American cities during the last few years will not soon abate — without fundamental changes in the urban American social structure.

NOTES

1. See, for example, William McCord, " 'Burn, Baby, Burn': The Los Angeles Riot," *New Leader* (Aug., 1965).

2. See, for example, John Howard, "The Making of a Black Muslim," *Trans-action*, IV (Dec., 1966), 15–21.
3. *A Report of Attitudes of Negroes in Various Cities*, prepared for the U. S. Senate Subcommittee on Executive Reorganization (Ribicoff Committee) by John F. Kraft (1966).
4. Gary Marx, *Protest and Prejudice* (N. Y.: Harper & Row, 1967).
5. William Brink and Louis Harris, *Black and White: A Study of U. S. Racial Attitudes Today* (N. Y.: Simon and Schuster, 1967).

DEPRIVATION, MOBILITY, AND ORIENTATION TOWARD PROTEST OF THE URBAN POOR

Don R. Bowen
Elinor Bowen
Sheldon Gawiser
Louis H. Masotti

■ Dissatisfaction breeds unrest. Dissatisfaction among the urban poor in the United States has led to protest, picketing, demonstrating, marching, and rioting. These statements are widely accepted, albeit there is lack of agreement about the sources and nature of the lack of satisfaction which leads to unrest in general, or to the unrest of America's urban poor in particular.

Perhaps the simplest hypothesis in this area is that there is a linear relationship between actual economic deprivation and protest. This hypothesis is sometimes, and incorrectly, attributed to Marx, who, in denying the revolutionary potential of the lumpenproletariat, was in effect indicating that severe

AUTHORS' NOTE: *The authors wish to express their appreciation to Greater Cleveland Associated Foundation which funded the larger project of which the research reported in this paper is a part.*

deprivation did not lead to revolution, but rather to political quiescence. The lumpenproletariat were incapable of action because they could conceive of no alternative to their impoverishment. More recent theorists have emphasized the importance of felt deprivation rather than deprivation which was objectively defined. The common assumption here is that the deprivation which leads to unrest is that which is relative to some other standard. Sometimes dissatisfaction is thought to be the result of deprivation relative to increased aspirations which are part of a "revolution of rising expectations" or of a "demonstration effect." For other theorists, dissatisfaction is the result of deprivation relative to an individual's own experience — e.g., downward mobility. In this paper we utilize survey data from a sample of persons living in the poverty areas of a large American city to explore the relationship between felt deprivation, personal mobility, and their evaluations of several kinds of protest activities.

In April and May, 1967, an opinion survey was conducted among the residents of Cleveland's nine officially designated poverty areas. Interviews were conducted with 500 persons selected on the basis of a block quota design, who constituted a 1.4% sample of the neighborhoods in which they lived. Of the respondents, 78%, representative of the pattern of race and ethnicity in their neighborhoods, were Negro. Of the remainder, 10% were native white, 9% were whites of identifiable ethnicity, and 2% were Spanish-speaking. The interviews were conducted nine months after the Hough riot of the previous summer.

THE PROTEST ORIENTATION

Since we wanted to investigate the degree to which respondents living in conditions of deprivation evaluated protest activities, a series of questions were asked about such activities as picketing, demonstrations, public protests, marching, and the like. Like most segments of the American population, a majority of these respondents disapprove of such behavior. But a rather significant minority not only approve of them but also believe that tactics of this kind are efficacious. Specifically,

30% of the sample believe that a public protest is a way to get the government to do something; 45% approve of picketing, marching, and protesting; 39% believe that picketing, marching, and protesting are more likely to help solve the problems to which they are addressed than not; and 21% believe that some good comes from rioting.

At this point we logically suspected a great deal of overlap among this minority. For example, it seemed likely that persons who thought that one could get the government to do something by means of a public protest were also the same persons who approved of picketing, etc. In order to investigate this possibility, the answers to the questions above, together with fifteen other variables describing more conventional kinds of political activity such as voting, were factor analyzed.[1] The results are given in Table 1, which indicates that our initial supposition was correct. What emerges is a factor defined by the four high factor loadings on these four protest questions. Separate factors which were defined by the more conventional activities were extracted. The conventional activities are quite unrelated to the protest factor, as can be seen from their low factor loadings, many of which approximate zero.

Two comments are in order about this protest factor. In the first place, a question arises concerning why the variable entitled "participation in picketing, marches, and protests" did not load if, in fact, this is a protest factor. The answer to that question lies in the fact that only 32 of the respondents have ever actually engaged in such activities and, of those, 26 have picketed because they were members of striking unions. Apparently picketing as a means of influencing an employer and picketing as part of a civil rights protest are markedly different things in the eyes of our respondents. The variable here described as participation in picketing should not load on a protest factor because it overwhelmingly signifies walking a picket line in support of union demands, not picketing to protest social or political conditions.

The second remark is more significant. The fact that the variable "good comes from riots" emerges on the protest factor indicates that these respondents tend to view public protests, picketing, marching, demonstrating, and rioting in the same way. Whether they approve of, or believe in, such behavior, or they do not approve and do not believe in it, they nevertheless tend to view it along one response dimension. Usually a distinction drawn between marching or picketing on the one hand and rioting on the other tends to rest on the differences between nonviolent tactics as opposed to violent ones. Our respondents do not perceive much difference between the two. For them rioting, demonstrating, or protesting are much the same thing. If one favors one of these activities, he tends to favor them all. Similarly, to dislike one is to dislike them all.

TABLE 1
The Protest Factor

Questions	Factor Loadings
Believes personal connections influence government	−0.0894
Believes writing to officials influences government	0.1865
Believes forming a group influences government	−0.0824
Believes working through a political party influences government	0.0536
Believes organizing a protest demonstration influences government	−0.5937
Would act if a crossing guard was needed	0.0045
Would act if a friend was in trouble with the police	−0.0786
Approves of marches, picketing, protests	−0.8517
Believes marches, picketing, protests solve problems	−0.8544
Participation in marches, picketing, protests	−0.0398
Believes some good came of the riots	−0.5029
Believes the riots hurt neighborhoods	0.0730
Cares which party wins presidential elections	−0.0604
Interested in political campaigns	−0.1026
Registered to vote	−0.0231
Has ever voted	−0.0174
Votes regularly in primary elections	−0.0233
Votes regularly in general elections	0.0448
Voted in poverty board election	−0.0204

The next step in the analysis was to construct what we here term the protest index. Briefly, we calculated a factor score for each respondent, based on the four variables which defined the protest factor, weighting each item in proportion to its factor loading. The "factor score" is defined as

$$\sum_{i=1}^{4} FL_i \left(\frac{x_i - \overline{X}_i}{S_i} \right)$$

Where FL_i is the Factor Loading of the ith variable
x_i is the individual's score on the ith variable
$\overline{X}_i$ is the mean of the ith variable
S_i is the standard deviation of the ith variable.

Only the four variables which loaded on the factor were used in the calculations. This score is what we mean when we speak of a respondent's protest orientation; and it becomes, in the logic of our analysis, the dependent variable. What we seek to do is discover whether felt deprivation and personal mobility are related in any systematic way to that cluster of attitudes we have subsumed under the heading "an orientation toward protest."

DEPRIVATION AND PERSONAL MOBILITY

Our measure of deprivation and perceived mobility are derived from the use of the Cantril-Free Self-Anchoring Striving Scale.[2] Respondents were shown a diagram of a ladder with ten rungs. After answering preliminary questions which encouraged them to define for themselves what they considered to be the best possible life they might achieve and the worst circumstances they might encounter, they were asked to indicate the position on the ladder which they occupied at the present time. Then they were asked to place themselves on the ladder five years in the past and five years in the future. A majority of the poverty respondents tended to be optimistic. They reported moderate deprivation at the present time and expected to be upwardly mobile — 52% located themselves on rungs four, five and six at the present time, 40% expected to occupy positions nine and ten five years in the future, and an additional 29% expected to occupy positions seven and eight.

Our data, then, on deprivation and personal mobility consist of the respondents' positions on the ladder at three points in time, and of the amount and direction of their change in position through two time periods. The data is based on the respondents' own evaluations of their life situations and expected life situations, and they have themselves determined their position and mobility between the minimum and maximum values. In defining these minimum and maximum values, they were free to consider economic and noneconomic factors, thus yielding a measure of dissatisfactions and satisfactions in all spheres of experience which they consider important.

In operationalizing our measures of what we mean by *felt deprivation* and perceived mobility, we take the distance between the respondent's reported position on the ladder and the tenth or highest rung to be a measure of the gap between achievement and aspiration. This we have called felt deprivation. In the following pages we examine the relation between felt deprivation and the protest orientation at two points in time, the present and five years in the future. Since we have confined our discussion to the present and the future, we shall, therefore, operationalize *perceived mobility* by measuring the change, if any, reported by the respondent between his current position on the ladder and the position he expects to occupy five years hence.

THE FINDINGS

We turn first to our hypothesis that those individuals who perceive the greatest gap between the position they currently occupy and the top rung on the ladder, representing their best possible life, are those who are most likely to view unconventional forms of political activity with the greatest favor. They are, in short, likely to rank higher on the protest index than those for whom the gap is not as great. Table 2 indicates that the data does not disconfirm this hypothesis. There is a

statistically significant relation between current felt deprivation and positive rank on the protest factor.

TABLE 2

Protest Orientation by Present and Future Ladder Positions

	Ladder Position: Present		
	Low (0-3)	Medium (4-6)	High (7-10)
High Protest Orientation (above the mean)	43 (55.8)	124 (46.8)	60 (43.5)
Low Protest Orientation (below the mean)	34 (44.2)	141 (53.2)	78 (56.5)
	77	265	138 N=480

$$x^2 = 3.676$$
$$p = .10$$

	Ladder Position: Future		
	Low (0-3)	Medium (4-6)	High (7-10)
High Protest Orientation (above the mean)	31 (53.5)	40 (44.4)	156 (47.0)
Low Protest Orientation (below the mean)	27 (46.5)	50 (55.6)	176 (53.0)
	58	90	332 N=480

However, no such relation emerges when we examine the
respondents' evaluations of their positions five years in the
future. Those who believe they will rank relatively low five
years from now are very little more inclined than those who
think they will rank relatively high to favor protest activities.
That actual felt deprivation in the present leads to support of
protest activities seems altogether reasonable as a rational
reaction to an unfavorable environment. These individuals
have a stake in rapid social change, and the activities in
question — picketing, marches, public protests, and rioting —

are perceived as achieving change more rapidly than more conventional forms of political activity, such as voting. But why then does not the same relation hold in the future?

There are two possible answers to this question. The first arises from characteristics of the respondents. We deliberately drew a sample from a relatively impoverished and little educated population. And we know from numerous other studies that respondents with such characteristics are likely to be unwilling or unable to view the future with any degree of clarity whatsoever. The future is simply not salient for them.[3] For example, over 50% of these respondents replied in answer to a standard survey question that they did not think it possible to "plan ahead."

The second possible answer lies in the findings of other investigators who have worked with the Self-Anchoring Striving Scale. Briefly, analyses of responses to the scale in Latin America and the United States have yielded the conclusion that respondents' perceptions of their future positions is unrelated to present and past positioning. D. P. Bwy, for example, factor analyzed responses to the scale together with such activities as strikes, riots, internal war and coups for twenty-one sampling sites in Latin America, and while he discovered factors defined by present and past positions on the ladder and outbreaks of violence, he found that future position emerged on a separate factor from present and past positions.[4]

Before concluding that respondents have disregarded the future in arriving at judgments of protest activities, however, it is necessary to consider their projections into the future in relation to their present experiences. Three possibilities warrant exploration: respondents may expect to be upwardly mobile, they may expect to be downwardly mobile, or they may expect no change in level of satisfaction.

This brings us to our second hypothesis: that those individuals who perceive themselves as upwardly mobile are individuals with rising expectations and, in line with the "revolution of rising expectations" theory, they should be the individuals who most clearly and strongly support protests. The results (Table 3) do not require rejection of the null hypothesis. Those respondents who perceive themselves as upwardly mobile distribute themselves roughly in the same fashion as the whole sample along the protest index. It should be noted that these findings do not necessarily invalidate the general propositions concerning the revolution of rising expectations, since that proposition assumes that expectations outrun the capacity of the political system to meet them. Since we have here no measure of system capacity, we do not know if the rising expectations our respondents report are or are not outrunning the capacity of the system to satisfy them.[5]

TABLE 3

Perceived Upward Mobility and Protest Orientation

	Upward Mobile	Remainder of Sample
High Protest Orientation (above the mean)	169 (51.5)	67 (40.4)
Low Protest Orientation (below the mean)	159 (48.5)	99 (59.6)
	328	166 N=494

What we do know is that perceived upward mobility is somewhat related to a higher score on the protest index, but that the relation is not strong and is not statistically significant. We therefore turn to the third hypothesis, namely, is it the case that those who perceive themselves as downwardly mobile are more likely to favor protest activities? Or, put more generally, is there a revolution of falling expectations rather than rising ones? There is, of course, a wealth of research findings which suggest that persons who perceive their life situations as worsening will be prone to view demands for radical social change favorably and will support various kinds of activities directed toward such change.[6] This hypothesis is derived from that literature.

The results of this part of the investigation are similar to those just discussed. That is, there is no significant relation between expectations of downward mobility and support or non-support of protest activities. There is a slight tendency for those who are downwardly mobile to favor protest activities somewhat more than the whole sample, but, as was the case with those who perceived themselves as upwardly mobile, the relation is not strong and it is not statistically significant. We are, therefore, unable to reject the null hypothesis. It is not the case that persons who expect to be worse off in the future than they are at present will support such activities as riots or demonstrations much more than will the whole sample (Table 4).

TABLE 4
Perceived Downward Mobility and Protest Orientation

	Downwardly Mobile	Remainder of Sample
High Protest Orientation (above the mean)	30 (48.4)	206 (47.7)
Low Protest Orienattion (below the mean)	32 (51.6)	226 (52.3)
	62	432 N=494

We now come to the fourth and final hypothesis which is, in effect, a combination of the two just discussed: any degree of change will lead individuals to show greater support for the protest activities under examination here than no change whatsoever. In other words, we seek to learn whether, by combining both those who perceive themselves as upwardly mobile and those who perceive themselves as downwardly mobile, a strong differentiation will emerge when compared to those who expect no change whatsoever. In general terms we hypothesize that there may be both a revolution of rising expectations and a revolution of falling expectations which lead to support of protests. A brief examination of the results of this line of inquiry (Table 5) confirms its utility. Those who perceive change in their life situations between now and five years in the future support protest activities to a significantly higher degree than those who expect no change. This relation pertains whether the expected mobility is upward or downward.

TABLE 5

Perceived Upward and Downward Mobility and Protest Orientation

	Upward and Downward Mobility	No Mobility
High Protest Orientation (above the mean)	199 (51.0)	37 (35.6)
Low Protest Orientation (below the mean)	191 (49.0)	67 (64.4)
	390	104 N = 494

$$X^2 = 8.2466$$
$$p = .01$$

Why do respondents who expect no change in their personal situation in the next five years tend to disapprove of picketing, demonstrating, rioting, and like activities? The most immediate explanation which comes to mind in terms of the measures employed in this analysis is that the lack of expectation of mobility is an expression of general satisfaction. Persons who are satisfied with their present life situations would not be expected to support activities such as rioting, which are included in our protest factor. The difficulty with this explanation is that it should apply to all of those who expect no change. They should all be low on the protest factor if they are satisfied with life and feel no deprivation. But approximately 35% of the persons who report no expectation of mobility rank high on the protest factor. It cannot be true on the face of it that the expectation of a constant level of satisfaction indicates a lack of felt deprivation. Apparently some of those individuals who expect the same level of satisfaction five years in the future as they presently enjoy feel themselves to be deprived while others do not.

However, when respondents who do not expect change are categorized according to whether they report relatively high or relatively low positions on the self-anchoring scale, their scores on the protest factor are largely explained. Table 6 shows the pattern clearly. Those who claim to be near the tenth rung of the ladder, and are therefore presumably more satisfied, tend to evaluate protest activities unfavorably, and those who claim to be nearer the bottom rung of the ladder evaluate protest activities favorably.

TABLE 6

Lack of Perceived Mobility and Protest Orientation

	Reporting No Change:	Present-Future
	High Ladder Position (6-10)	Low Ladder Position (0-5)
High Protest Orientation (above the mean)	19 (25.3)	18 (62.1)
Low Protest Orientation (below the mean)	56 (74.7)	11 (37.9)
	75	29 N=104

$$X^2 = 11.26$$
$$p = .001$$

CONCLUSION

This paper began with the presentation of several hypotheses linking dissatisfaction to social and political unrest. We noted that disagreement existed about the nature of the relative deprivation which was assumed to be the root of such dissatisfaction, and proceeded to operationalize the term *relative deprivation* in four ways, viewing it as (1) a discrepancy between an individual's life experience and his definition of the good life, (2) a discrepancy between expected life situation and the good life, (3) discrepancy between present experience and the outcome of anticipated upward mobility, and (4) discrepancy between the outcome of anticipated downward mobility and present experience. We hypothesized that present felt deprivation, expectations of future deprivation, the rising aspirations associated with upward mobility, the despair derived from downward mobility, and any change in level of satisfaction through time would be associated with approval of protest activities. These hypotheses were tested using data from a survey of the residents of the poverty neighborhoods in Cleveland, Ohio. We found that, at least for a sample of the urban poor, some forms of relative deprivation are associated with protest approval while others were not.

Present feelings of deprivation, as measured by position on the Cantril-Free Self-Anchoring Striving Scale, predicted to protest orientation most strongly for those individuals who do not expect any change in position on the scale five years in the future. In addition, individuals who reported expectations of personal mobility, as measured by a change in position on the Self-Anchoring Scale from present to future, whether their mobility was in a positive or a negative direction, were more likely than others to approve of protest activities. These meanings of the term *relative deprivation,* then, yielded the expected relationship between dissatisfaction and unrest among the residents of Cleveland's poverty areas. Support exists both for the hypothesis linking felt deprivation with protest orientation and the hypotheses linking rising expectations and downward mobility with the approval of protest. Feelings of deprivation are associated most strongly with approval of protest activities for those persons who have no expectation of upward, or downward, mobility.

NOTES

1. The factor analysis was done by FACOM, a program of the Center for Documentation and Communications Research Computer Laboratory, Case Western Reserve University. All factors with eigenvalues exceeding 1.00 were orthogonally rotated to a varimax solution.
2. For a full discussion of the construction and application of the scale, see Hadley Cantril, *The Patterns of Human Concerns,* (Princeton Univ. Press, 1966).
3. See Don R. Bowen and Louis H. Masotti, "Spokesmen for the Poor: An Analysis of Cleveland's Poverty Board Candidates," Dept. of Political Science, Case Western Reserve Univ. (mimeo), 1966; 25pp.
4. D. P. Bwy, "Dimensions of Social Conflict in Latin America: Testing a Multivariate Model of Political Instability with Over-Time Data Among City and Provincial Units," *Am. Behav.*

Scientist, March 1968. See also Sheldon R. Gawiser, *Towards a Simulation Model of Urban Civil Disorder,* M. A. thesis, Dept. of Political Science, Case Western Reserve Univ., 1968, for a similar conclusion with data from the United States.

5. For a more exact treatment of the notion of the revolutionary impact of rapidly rising expectations, see Lawrence Stone, "Theories of Revolution," *World Politics,* XVIII (Jan., 1966), 159–176; or James Davies, "Towards a Theory of Revolution," *Am. Sociol. Rev.,* XXVII (Feb., 1962), 6.

6. While seldom described as falling expectations, it is nevertheless clear that perceiving oneself as worse off is an important ingredient in revolutionary movements, particularly, apparently, movements of the political right. See S. M. Lipset, *Political Man* (N. Y.: Doubleday, 1963); and H. Lasswell and D. Lerner (eds.), *World Revolutionary Elites* (N. Y.: Doubleday, 1964).

DIMENSION OF SOCIAL CONFLICT IN LATIN AMERICA

Douglas P. Bwy

■ This research proposes to take a close empirical look at political aggression, to extract from the extensive bodies of literature a theoretical model which might account for much of the variance about political instability, and to systematically apply the model within a common socio-cultural environment.

Politically relevant aggression, here, is defined as *behavior designed to injure* (either physically or psychologically) *those toward whom it is directed.*[1] Politically relevant violence, on the other hand, is defined as *any action* (attack or assault) *with intent to do physical harm* (injury or destruction) *to persons or property.* Violence, then, is on one end of the aggressive continuum (events such as riots, clashes, assassinations, and so on, appear to fall here); while more subtle forms of aggression, the nonviolent forms, fall at the opposite end of the continuum (and generally seem to find expression in such activities as threats, protests, boycotts, and so on).[2]

Because of a personal research interest in the politics of the Latin American republics, and because of the eclectic nature of the political aggression there, this region was selected as the natural laboratory in which to test the model of political instability.

AUTHOR'S NOTE: *In addition to the National Science Foundation, which partially supported this research under Grant NSF–GS789, the author is also indebted to the Vogelback Computing Center, Northwestern University, and the Andrew Jennings Computing Center, Case Western Reserve University.*

THE DIMENSIONS OF SOCIAL CONFLICT
IN LATIN AMERICA

Social conflict is so much a part of the Latin American
political process that "to treat violence . . . as [an] aberration,
places one in the awkward position of insisting that practically
all significant political events in the past half century are
deviations."[3] While such "deviations" are: (a) chronic, it is
equally true that the majority of them are (b) frequently
accompanied by limited violence, and that they (c) generally
produce no basic shifts in economic, social, or political policies.[4] The so-called "revolutions" of Latin America range from
the Chilean "revolution" of 1924 (when, in the throes of a
continuous cycle of cabinet instability, Arturo Alessandri re-
signed), to the rather violent removal of Porfirio Díaz, or
Jorge Ubico, or Enrique Peñaranda, and the kind of socioeconomic uprooting which took place in Mexico during 1911
and after, in Guatemala during 1945 and after, and in Bolivia
in 1943 and 1952. "Revolutions" have taken place after the
central decision-maker has spent as little as twenty-eight hours
in office (which was Arturo Rawson's tenure after being
"installed" by the Perón revolution), or as many as forty-four
years, in the case of Mexico's Díaz, or twenty-eight years, in
the case of Venezuela's Juan Vicente Gómez. They have been
as brutal and bloody as the guerrilla insurrection taking place
in Cuba, and as peaceful as the kind of game of "musical
chairs" played out year after year in Paraguay. They are
more often the result of precision and planning among elites
(as for example, the recent coup in Argentina, which saw
Arturo Illia's government fall prey to the militarism of General Juan Carlos Onganía) than events of mass participation
(as exemplified in Colombia's bogotazo, which cost the lives
of over 5,000 residents of Bogotá, or Guatemala's huelga
de los brazos caidos [the strike of the fallen arms], or the 1958
full-scale uprising in Caracas which brought down Pérez
Jiménez).

When looked at from the point of view of "revolution,"
then, the domain of political instability in Latin America
appears as eclectic as it is unmanageable. Underpinning each

of these "revolutions," however, is a character of interpersonal or intergroup conflict;[5] and whether each reflects an aggregate of conflict (e.g., civil uprising, guerrilla warfare) or a specific act (e.g., resignation, assassination), the conflict behavior itself is measurable. Thinking of the domain of political instability in terms of specific and aggregate instances of aggression, then, the conflict landscape can be reduced to: demonstrations and boycotts, protests and threats, riots, nonpolitical and political clashes, instances of *machetismo* [peasant rebellion], *cuartelazo* [barracks revolt], *imposición* [imposing oneself in office], and *candidato único* [single candidate], strikes and general strikes, acts of terrorism and sabotage, guerrilla warfare, plots, revolutionary invasions, military *coups,* civil wars, private warfare, banditry, and others.

In the belief that each of these instances of conflict could be empirically defined so as to yield mutually exclusive events for analysis, a set of over forty conflict events (generally involving either (a) aggressive activity from a populace to a government, or (b) that directed from a government to a populace) was developed,[6] and incorporated into a Domestic Conflict Code Sheet (see Figs. 1A and 1B). In applying the Code Sheet to journalistic data, it was decided that the conflict behavior sought should: (a) focus on aggressive activity taking place within nations on which detailed survey research data were also available;[7] (b) be applied to both comprehensive and comparable data sources;[8] and (c) have the quality of being able to be aggregated across specific city or provincial units of analysis.

Thirty-four of the events collected on the Code Sheet dealt with conflict directed by "individuals or groups within the political system against other groups or against the complex of officeholders and individuals and groups associated with them." It was these which were selected as measures of political instability. In an effort to reduce them to a smaller set of conceptual variables, the data were: (a) aggregated by provincial (state) units[9] for the total nine-year period over which they had been collected, and (b) factor analyzed.[10] The results of this analysis appear in Table I, below.

Each entry or "factor loading" of the matrix represents the correlation between the conflict measures and a given

CODE SHEET: DOMESTIC CONFLICT

Cols. 77-80

1 2 3

Actor
(sub-sets)

Distance from unit-
of-analysis actor

4 = 0   Actor loca-
ted in
sub-set

4 = 1   Actor loca-
ted in out-
er fringes
of sub-set

4 = 2   Actor loca-
ted in
same prov.
or state as
sub-set

Specific locality
where action took
place_____.

(Description)

Object

5 = 0   5 = 1
Natnl.   State/
Govt.   Prov.
      Govt.

5 = 2   5 = 3
Local   Other
Govt.   Authority

Specify # 3_____

6 7	8 9	10 11	12	13
Day	Month	Year	Data Source	Data Collector

14 = 0   14 = 1
Complete   Incomplete
Data   Data

15 = 0   15 = 1
Non-   Composite
Composite   (continuous)
(discrete)   event
event

Non-Violent Behavior

16 17 = 01

Non-violent
behavior
(Actor)

18

1 = Legal
2 = Extra-legal

19 20

01 = Anti-government demonstration   05 = Political boycott
02 = Anti-government demonstrating   06 = Anti-foreign demonstration
03 = Printed protests   07 = Anti-foreign threat
04 = Threat   99 = Other (specify)_____

16 17 = 02

Non-violent
behavior
(Object)

21

1 = Legal
2 = Extra-legal

22 23

01 = Imposición   03 = Continuismo
02 = Candidato único   99 = Other (specify)_____

Unplanned Violence

16 17 = 03

Unplanned
Violence

24 25

01 = Riot or Manifestaciones   05 = Anti-Foreign Riot
02 = Rioting   06 = Machetismo or Peasant Rebellion
03 = Political Clashes   99 = Other (specify)_____
04 = Non-Political Clashes

Figure 1A

Planned Violence

16 17 □□ = 04

26 27 □□

Planned Violence

01 = Strike
02 = General Strike
03 = Terroristic Act/Sabotage
04 = Terrorism/Continuing Sabotage
05 = Guerrilla Action
06 = Guerrilla Warfare
07 = Golpe de Estado/Coup d'Etat

08 = Cuartelazo
09 = Plots
10 = Assassination
11 = Civilian Political Revolt
12 = Private Warfare
13 = Banditry
14 = Revolutionary Invasion
99 = Other (specify)_____

Governmental Response

16 17 □□ = 05

28 29 □□

Governmental Response

01 = Limited State of Emergency
02 = Martial Law
03 = Arrest/Imprisonment of Politically Insignificant Persons

04 = Execution of Politically Insignificant Persons
05 = Governmental Action Against Specific Groups
99 = Other (specify)_____

16 17 □□ = 06

30 31 □□

Quality of Governmental Response

01 = Resignations of Political Elite
02 = Dismissals of Political Elite
03 = Dissolution of Legislature
04 = Cabinet Instability
05 = Mutiny
06 = Arrest/Imprisonment of Politically-Significant Persons

07 = Exile
08 = Execution of Politically Significant Persons
09 = Political Boycott
99 = Other (specify)_____

Individual Statistics

Duration

32 □ = 0 <Day   32 □ = 1 ≥Day   33 34 35 36 □□□□ Days   Or duration based on intuitive rating: 37 □ = 0 1-7 Days   37 □ = 1 8-30 Days   37 □ = 2 31-365 Days

Numbers Involved

38 39 40 41 42 43 □□□□□□ For Actor   44 45 46 47 48 49 □□□□□□ For Object   50 □ = 0 No's Involved for both Actor & Object

Numbers Injured

51 52 53 54 55 □□□□□   56 □ = 0 Injured for Actor   56 □ = 1 Injured for Actor & Object

Numbers Killed

57 58 59 60 61 □□□□□   Or if no data on killed or injured: 62 63 64 65 66 □□□□□ Casualties   67 □ = 0 Killed for Actor   67 □ = 1 Killed for Actor & Object

Number Arrested

68 69 70 71 □□□□   Amount of Property Damage (in $100's)   72 73 74 75 76 □□□□□

Figure 1B

factor. By squaring these factor loadings and summing them (h^2) we have an approximation of the amount of variance in social conflict (taken as the dependent variable) explained by the underlying factors. In addition to indicating the weight of each factor in explaining the observed measures, the matrix of factor loadings also provides the basis for grouping the measures into common factors. The various operational measures have clustered into two basic (and a third, primarily negative) configurations, with high intercorrelations within the clusters and relatively low correlations between them. By examining the nature of the operational measures, we are in a position to identify the basic dimension, or "latent variable," which "causes" the array of variables along the factor.

By adopting the value of $+.50$ as that necessary for assigning "significant" variables to any one dimension, the operational measures in Table 1 were shuffled into two interpretable factors (contained within the boxes).

Factor 1, Turmoil. Factors are computed in the order of their ability to explain the variation in the domestic conflict measures used. The first dimension to be extracted accounts for 58% of the common variance. Among the highest loading conflict measures on this dimension are: Anti-Government Riots [*Manifestaciones*], Political Clashes, Anti-Foreign Riots and Demonstrations, Arrests, and Deaths from Domestic Violence — most suggesting a kind of spontaneous, sporadic, and essentially non-organized conflict behavior dimension. The next highest loading measures — Strikes, Coups, *Cuartelazos*, Injuries, Assassinations, and Numbers Involved (all of which are not always spontaneous in nature) — also come out on this dimension.

Two inquiries are suggested by these data: first the methodological question (a) What about the stability of these findings? and secondly, the substantive question (b) Are they interpretable with such behavior as we know it within the Latin American context?

This finding is not unique. R. J. Rummel, and after him Raymond Tanter, reported[11] the existence of a basic dimension indexed by such things as Demonstrations, Riots, Strikes, Governmental Crises,[12] and Assassinations. With the possible exception of Governmental Crises (which may find partial expression in the measure "Plots" (9) from the present analysis), all of the variables in the Rummel and Tanter matrices also come out with strong factor loadings in the first dimension of Table 1. To use the name they applied to such a cluster, Factor 1 reflects the degree of Turmoil among the provincial units.

TABLE 1

ROTATED FACTOR MATRIX FOR TWENTY-FOUR DOMESTIC CONFLICT MEASURES, SIXTY-FIVE PROVINCIAL UNITS, NINE-YEAR TIME PERIOD

	Domestic Conflict Measures	Rotated Factor Loadings			h^2
		F_1	F_2	F_3	
1	Anti-Government Demonstration	.86	.35	-.16	.896
2	Riot or Manifestacion	.90	.21	-.18	.881
3	Anti-Foreign Demonstration	.96	-.01	-.02	.936
4	Political Clash	.93	-.03	-.15	.890
5	Anti-Foreign Riot	.94	-.02	-.05	.899
6	Strike	.82	.39	-.16	.849
7	Golpe de Estado / Coup d'Etat	.87	.12	-.01	.764
8	Cuartelazo	.69	-.05	.06	.493
9	Plots	.62	-.04	-.29	.476
10	Assassination	.72	.45	.09	.734
11	Deaths from Domestic Group Violence	.91	.27	.10	.908
12	Numbers Involved in Civil Violence	.84	-.03	-.37	.843
13	Numbers Injured in Civil Violence	.78	.11	-.12	.631
14	Numbers Arrested	.92	.10	-.19	.890
15	Terroristic Act/Sabotage	.52	.64	-.09	.705
16	Anti-Government Demonstrating	.22	.51	.39	.470
17	Terrorizing / Sabotaging	-.01	.92	.05	.855
18	Guerrilla Action	.48	.79	.08	.868
19	Guerrilla Warfare	-.07	.93	.03	.879
20	Revolutionary Invasion	-.04	.91	.04	.822
21	Threat	.49	.04	-.65	.678
22	Rioting	.22	-.03	-.79	.684
23	Non-Political Clash	.01	-.03	-.92	.842
24	General Strike	.07	-.14	-.85	.741
	% Total Variance	45.41	18.84	13.38	77.63
	% Common Variance	58.49	24.27	17.24	100.00

As to the substantive question of the fit of such findings to the Latin American scene, there seems to be little or no evidence challenging the fact that such a cluster of conflict events does not co-vary. The occurrence of one set of variables on the first continuum, however, may appear questionable: Coups (7), *Cuartelazos* (8), and Plots (9). That these events occur together (even though defined in a mutually exclusive manner), is not a point of contention; since we would indeed expect this to be the case. That they come out on the first dimension is, however, of legitimate concern. How can such loadings be explained in the light of the substantive literature which places such activities within the framework of highly organized and clandestine events? A previous analysis[13] has suggested the strongest correlate (negative) of highly organized violence (such as guerrilla warfare) to be the populace's perception of the legitimacy of the political system. Organized violence increased linearly as system legitimacy decreased. There was, however, little or no association with what was termed Anomic (or spontaneous) Violence and Legitimacy. That is, something other than challenges to the legitimacy of the system seemed to be at work in "causing" Anomic Violence. Since the vast majority of Latin American coups are relatively bloodless and of short duration (generally because of the lack of interest and participation of the masses), they rarely appear to be serious challenges to the legitimacy of the systems involved. Instead, they appear more in the form of a frequent game of rotation between the set of upper class "ins" and "outs." The coup d'etat, or palace revolution, then, as it is practiced in Latin America, appears to be as institutionalized a form of challenging governments as the ballot, or as viable a mechanism of protest (although more often practiced by an elite clientele) as the *manifestacion*, or demonstration, or strike. As such, its emergence on the first factor, in association with these events, is more than acceptable.

Factor 2, Internal War. The second basic factor computed for Table I accounts for 24% of the common variance. In the order of the strength of their association with the factor, the highest loading variables are: continuing Guerrilla Warfare, Terrorizing/Sabotaging, Revolutionary Invasion, Guerrilla Action, and discrete Terroristic Act/Sabotage. Only two other conflict measures produce factor loadings at or near the +.50 level established earlier: Anti-Government Demonstrating, and Assassinations. Together, the extreme loading conflict measures on Factor 2, then, generally refer to aggressive actions defined by high degrees of planning and organization. Or, to use Tanter's concept for such a cluster, Factor 2 reflects the degree of Internal War among the political units.

The two questions posed earlier, with respect to the first factor, can now be put to the results here, namely: (a) What about the stability of this second factor? and (b) Are these factor loadings interpretable within the context of Latin American behavior?

When Rummel factor analyzed data for 77 nations on nine domestic conflict variables gathered for the three-year period 1955-57, three basic dimensions emerged. The first was a dimension we have already described as having a nice fit to what we have also called a Turmoil dimension. The second and third dimensions were labeled by Rummel as Revolutionary and Subversive. In describing them, he notes that they "appear to represent organized conflict behavior, i.e., behavior that is planned with definite objectives and methods in mind."[14] When Tanter replicated Rummel's study, using 1958-60 data, the Revolutionary and Subversive dimensions were pulled together into a single basic factor which he labeled Internal War. Indexed by such variables as Revolutions,[15] Domestic Killed, Guerrilla War, and Purges, Tanter noted that "these activities are generally associated with organized conflict behavior of a highly violent nature."[16]

With respect to the second question asked about Factor 2, what we have labeled as Internal War has variously been referred to in the literature as "unconventional warfare," "protracted conflict," "irregular warfare," "paramilitary operations," or "guerrilla warfare."[17] Internal War is certainly not a new phenomenon to the Latin American scene. The most dramatic of the recent occurrences is, of course, Fidel Castro's 26th of July movement against the Batista regime from 1956 to 1959. Examples of guerrilla activities, while infrequent, go back to the 1800's and before, one of the earliest being

Antonio Conselheiro's open rebellion against the Brazilian government in the northeastern sectors of the country at the end of the nineteenth century. In analyzing the Cuban Revolution, however, Merle Kling notes that it was a case of the insurgents employing ". . . violence in a manner which deviated from the traditional Latin American practice."[18] The dimensions of .nis "deviation" appear to be embodied in the high loading variables on the Internal War factor in Table 1. The traditional pattern, to Kling, conforms more ". . . to the restraints inherent in a coup d 'etat or *golpe de estado* or palace revolution. Such revolts, while abruptly terminating the tenure of government personnel, do not disturb the prevailing pattern of social and economic relations."[19] The dimensions of this pattern are embodied in the high loadings on the Turmoil factor. With respect to the briefest review of the literature on the subject, then, the factor loadings in Table 1 appear interpretable within the context of Latin American conflict behavior.

THE PRECONDITIONS OF TURMOIL AND INTERNAL WAR: SYSTEMIC DISSATISFACTION, LEGITIMACY, AND RETRIBUTION

Psycho-Social Dissatisfaction and Political Instability. While the conflict literature has strongly suggested a (causal) linkage between discontent and political instability, it has often been at odds with respect to the *direction* of such a relationship. From Marx, for example, we can extract the proposition: "As a group experiences a *worsening* of its conditions of life, it will become increasingly dissatisfied until it eventually rebels."[20] A number of recent empirical studies have appeared which corroborate this proposition. Through a correlational analysis, Bruce Russett demonstrated that as the inequitable distribution of land (among 47 nation-units) increased, the number of violent political deaths also increased.[21] Through a regression analysis, he established the fact that an even stronger association existed when other indices of discontent (i.e., low GNP per capita, and high percentage of the labor force in the agricultural sector) were taken into consideration. In "A Theory of Revolution," Raymond Tanter and Manus Midlarsky also tested the relationship between land inequality and the occurrence of successful or unsuccessful revolution, and concluded that successful revolutions occurred in those polities with a higher degree of land inequality.[22] And finally, much of the work done by Lipset,[23] Cutright,[24] and Lerner[25] suggests, at least implicitly, that "satisfied" (i.e., wealthy) polities are stable polities. In propositional terms: Political instability increases as economic development decreases.

As reasonable as this argument seems, in both theory and empirical findings, it runs oblique, if not counter, to the propositional relationship between satisfaction and political instability often credited to Edwards, de Tocqueville, Brinton, Hoffer, and Davies; namely: "As a group experiences an *improvement* in its conditions of life, it will also experience a rise in its level of desires. The latter will rise more rapidly than the former, leading to dissatisfaction and rebellion."[26] De Tocqueville concluded, for example, ". . . so it would appear that the French found their condition the more unsupportable in proportion to its improvement." Eric Wolf has perhaps couched this relationship in its most dramatic form when he said: "Revolt occurs not when men's faces are ground into the dust; rather, it explodes during a period of rising hope, at the point of sudden realization that only the traditional controls of the social order stand between men and the achievement of still greater hopes."[27] James Davies finds that both Marx and de Tocqueville's notions have explanatory and possibly predictive value, if they are but juxtaposed and put into the proper time sequence: "Revolutions are most likely to occur when a prolonged period of objective economic and social development is followed by a short period of sharp reversal."[28] "Revolutions," according to these views, are not born in societies that are economically retrograde (downswing), but, on the contrary, in those which are economically progressive (upswing). And, as for the Marxian proposition before it, the evidence in support of the "upswing thesis" is considerable. Brinton finds it applies to the French, Russian, English, and American revolutions,[29] while Davies notes its unique fit to Dorr's Rebellion, the Egyptian Revolution of 1952, and the Bolshevik Revolution of 1917,[30] and Blasier sees it as a reasonably accurate description of conditions leading to the Mexican, Bolivian, and Cuban revolutions.[31]

While discontent, then, appears to be an important correlate of political instability, the direction of the association is very much in doubt. Even admitting a consistent finding as to direction, however, the highest correlations obtained in much of the quantified literature suggest that only a little over half of the variance about instability can be "explained" by measures of dissatisfaction. It appears, therefore, that the causes of political instability are numerous, and that the relationship is indeed complex. Let us look at another predictor variable — legitimacy — which may account for some of this unexplained variance.

Legitimacy and Political Instability. While it has been suggested that the effectiveness of a political system in satisfying demands is primarily an instrumental dimension, legitimacy is more an affective, or evaluative, dimension. Perhaps the most often quoted definition of legitimacy (or what has also been termed "political allegiance") has been given by Lipset, who noted that it involved the capacity of a political system to engender and maintain the belief that existing political institutions were the most appropriate ones for the society.[32] The strength of this variable in predicting instability is emphasized by Lipset, who claims that the political stability of any given nation depends more on this factor than on its effectiveness in satisfying wants.

Despite their separate treatment here, certainly the model's first two variables — discontent and political legitimacy — cannot be considered independent. Actually, psycho-social satisfaction and notions of legitimacy are closely related subsystems of phenomena, which can only be separated for analytic purposes. For example, when explaining how political systems manage to maintain a steady flow of support (legitimacy), Easton concluded that it is (a) through a process of "politicization" (by which attachments to the political system are built into the maturing member), and (b) through *outputs* that meet the demands of the members of society, as well.[33]

It should be emphasized, however, that legitimacy (or allegiance) is not the exclusive province of Western democracies, or what have been referred to as "participant" political cultures. Many closed or hierarchically organized systems, or "subject" political cultures as Almond and Verba[34] would call them, enjoy positive affect, or high feelings of legitimacy. For example, many American Indian political communities or African tribal communities are traditionally oriented, and more often than not hierarchically organized and authoritarian; nevertheless, their inhabitants feel that the systemic arrangements are morally right and proper. The simple fact of the matter is that the members of any type of political system may or may not take pride in it or like it; in short, may or may not ascribe legitimacy to it.

It appears clear, however, that the members of a poliitcal system will ascribe legitimacy to the system if the political structure is congruent with the political culture. According to Almond and Verba, when the political structure (regardless of whether traditional, centralized-authoritarian, or democratic) is cognized, and when the frequency of affective (or positive feeling) and evaluative orientations are high, a congruence between culture and structure occurs and is accompanied by high amounts of allegiance or legitimacy. The congruence between culture and structure is weak when the political structure is cognized, but the frequency of positive feeling and evaluation approaches indifference or zero. Here, in place of allegiance, one finds apathy or anomie. Incongruence between political culture and structure begins when the indifference point is passed and negative affect and evaluation grow in frequency. The end product of this mechanism is alienation. Almond and Verba suggest further that such a continuum can also be thought of as one of stability/instability. As political systems move toward allegiant or legitimate orientations, they also tend to become more stable; while movement away from legitimacy, toward apathy and alienation, is often associated with instability. And furthermore, if forced to choose, as correlates of political instability, either low system output (what could roughly be equated with dissatisfaction) or low system legitimacy, Almond and Verba suggest, as did Lipset earlier, that "long-run political stability may be more dependent on a more diffuse sense of attachment or loyalty to the political system — a loyalty not based specifically on system performance."[35]

Retribution: The Correlates of Force. Some notions freely translated from psychology, and particularly those of Arnold Buss, indicate the relationship between force (punishment) and aggression to be curvilinear. From this premise, therefore, it is hypothesized that very little political instability is found at the two extremes of a permissive-coercive continuum, but great quantities of instability should be observed at the center. Buss notes, for example, that low levels of punishment do not serve as inhibitors; it is only high levels which are likely to result in anxiety or flight. Punishment in the mid-levels of intensity acts as a frustrator and elicits further aggression, maintaining an aggression-punishment-aggression sequence.[36] Robert LeVine, in his study of African violence against colonial regimes, came to similar conclusions. He found that if colonial policy is consistently repressive toward African self-rule (as was supported by the cases of the Union of South Africa, Portugal's Mozambique, and Angola), or if it is consistently permissive toward self-rule (as seemed to be the case in Nigeria, Ghana, Sudan, and Uganda), then violence against Europeans was relatively low. Only if colonial policy toward self-rule was ambivalent, therefore arousing conflicting expectations of political autonomy (as was the case in Nyasaland and Kenya), did LeVine find violence to be greater.[37] And finally, some recent research which distinguishes between basic types of domestic conflict is beginning to suggest that such a curvilinear model may only apply to non-organized (or disorganized) and spontaneous violence (such as riots or demonstrations); and that the linear model is a more accurate reflection of more organized types of violent behavior (such as guerrilla warfare and armed rebellions).[38]

OPERATIONALIZING THE MODEL AND TESTING IT WITH OVER-TIME DATA

SATISFACTION: AN ASSESSMENT THROUGH SELF-ANCHORING SCALING

The Self-Anchoring Striving Scale is a survey research technique developed by Hadley Cantril and F. P. Kilpatrick[39] which attempts to locate an individual on a scale in terms of a spectrum of values he is preoccupied or concerned with, and by means of which he evaluates his own life. The respondent describes, as the top anchoring point, his wishes and hopes as he personally conceives them, the realization of which would constitute for him the best possible life. At the other extreme, he describes the worries and fears, the preoccupations and frustrations, embodied in his conception of the worst possible life he could imagine. Then, utilizing a nonverbal ladder device, he is asked where he thinks he stands on the ladder today, with the top (or tenth rung) being the best life as he has defined it, and the bottom being the worst life as he has defined it. He is also asked where he thinks he stood in the past (five years ago), and where he thinks he will stand in the future (five years from now). Similar questions are then asked about the best and worst possible situations he can imagine for his country, so his aspirations and fears on the national level can be learned. And again, the ladder is used to find out where he thinks his country stands today, where it stood in the past, and where it will stand in the future.[40] By avoiding the pitfalls of imposing a predetermined set of structures on the respondent, and talking in terms of each individual's perceptions of reality (in terms of his own "reality world," as Cantril and Kilpatrick would say), the scale provides a remarkably comparable cross-cultural tool for measuring similar phenomena (namely, frustrations or dissatisfactions) across often divergent populations.

Among the fourteen national surveys in which the scale was administered under the direction of the Institute for International Social Research, four were in Latin American polities: Brazil, Cuba, the Dominican Republic, and Panama. These interviews (numbering over 8,000 units) were obtained,[41] "cleaned,"[42] and aggregated by city[43] and provincial[44] units of analysis.

The Striving Scale produces at least six numerical ladder ratings, which in turn can be used to generate additional measures. One can move from static to dynamic measures, for example, by calculating the difference (or change) in ladder ratings given by respondents from one period to the next. While many different operationalizations were employed, the descriptions of the five below should suffice as a general guide for the procedures followed.

P-SAT$_2$. Personal satisfaction at t_2 (where t_1 = five years ago; t_2 = present, the time of the interview; t_3 = five years into the future). The datum, then, represents the mean ladder rating (or personal standing), ranging from 0–10, on the Self-Anchoring Striving Scale, aggregated for provincial units for the present time (the time of the interview, which for Brazil and Cuba was 1960, while for the Dominican Republic and Panama it was 1962).

P-SAT$_1$. Same as above, with the exception that the ladder rating given by the respondent refers to his perceptions of where he stood on the (self-anchored) ladder approximately "five years ago."

PSAT$_{21}$. $\triangle$ P-SAT$_1$−P-SAT$_2$; the amount of *change* from the respondent's personal ladder rating "five years ago" to the personal ladder rating "at the present time" (the time of the interview).

NSAT$_{23}$. The first unrotated factor score of the mean ladder ratings for the nation for three time periods (N-SAT$_1$ = five years ago, N-SAT$_2$ = present, N-SAT$_3$ = five years in the future); where the highest loading variables on the factor were: N-SAT$_2$ (+.83), N-SAT$_3$ (+.92), N-SAT$_1$ (−.56). The factor scores are, therefore, measuring a combination of the *present*, plus *future*, satisfaction at the national level.

PN-SAT. The first unrotated factor score of an analysis composed of the following four input variables: $\triangle$ P-SAT$_1$−P-SAT$_2$ (personal satisfaction), $\triangle$ P-SAT$_2$−P-SAT$_3$ (personal aspiration), $\triangle$ N-SAT$_1$−N-SAT$_2$ (national satisfaction), $\triangle$ N-SAT$_2$−N-SAT$_3$ (national aspiration). The factor scores are measuring personal (+.89) and national (+.84) *satisfaction,* since personal (−.78) and national *aspiration* (−.59) came out with negative factor loadings on this dimension.

LEGITIMACY: MEASURING POSITIVE AFFECT
TOWARD POLITICAL STRUCTURES

The basic measure of legitimacy consists of, first (a) separating from the total number of individuals in any one sampling site (province) those responding in terms of political considerations[15] when asked to describe their worries and fears (or wishes and hopes) for the future of the nation; and of this group, (b) a calculation (mean) of their perceptions of the *nation's* ladder standing at the present time. The higher the perception of the nation's standing, presumably the less concerned the respondent (with respect to the political hope or fear he may have mentioned), and the higher the feeling of positive affect.

LGLAD1 represents the mean ladder rating of individuals responding only in terms of "national worries and fears," and LGLAD2 the ladder rating of individuals responding only in terms of "national hopes and wishes." LGLAD3 and LGLAD4 correspond identically to the first two measures, with the exception that the political response "Political Stability (Instability), Internal Peace (Chaos), and Order (Civil War)" has been eliminated. The variable LGTMCY consists of the first unrotated factor scores resulting from a factor analysis of the four individual measures of legitimacy; and since the loadings were: $+.96$, $+.93$, $+.84$, and $+.95$, respectively, all four measures participate equally in this composite variable.

OPERATIONALIZING MECHANISMS OF SOCIAL CONTROL: FORCE

An inspection of the fourteen different types of "governmental response to domestic conflict" gathered across the sixty-five provincial units in Brazil, Cuba, the Dominican Republic, and Panama (see Figure 1B), suggested that the application of social control may be a more complex phenomenon than was envisioned in the earlier theoretical discussion of the model. In order to test whether these measures would empirically break down into a smaller set of independent clusters, the force data were inter-correlated and factor analyzed,[46] and the results appear in Table 2, below.

When the original operational indices are "assigned" to factors on the basis of high loadings (i.e., the correlation between an index and a given factor), and made more visible by their location within boxes, a definite picture of three independent dimensions emerges. Once again, taking the factor loadings as a clue to the identity of the basic factor, or latent

TABLE 2

ROTATED FACTOR MATRIX FOR THIRTEEN VARIABLES MEASURING
GOVERNMENTAL RESPONSE, SIXTY-FIVE PROVINCIAL UNITS,
NINE-YEAR TIME PERIOD

Governmental Response Measures	Rotated Factor Loadings			h^2
	F_1	F_2	F_3	
1 Resignations of Political Elite	.90	.01	.28	.890
2 Dismissals of Political Elite	.67	.18	.62	.864
3 Dissolution of the Legislature	.56	.13	.37	.499
4 Cabinet Instability	.92	.02	.28	.932
5 Mutiny	.49	-.47	-.29	.556
6 Arrest/Imprson Politically Signfcnt Prsns	.71	.23	.59	.914
7 Exiles	.65	.63	.33	.932
8 Limited State of Emergency	.13	.54	.48	.534
9 Martial Law	.07	.91	-.09	.847
10 Arrest/Imprson Pol. Insignfcnt Persons	.31	.23	.82	.857
11 Execution of Politically-Insignificant Prsns	.12	-.07	.89	.807
12 Governmental Acts Against Specific Groups	.37	.15	.76	.739
13 Execution of Politically-Significant Persons	.41	.08	.71	.675
% Total Variance	31.29	15.21	30.75	77.25
% Common Variance	40.50	19.70	39.80	100.00

variable, the three dimensions have been named: (a) Elite
Instability, (b) Non-Violent, and (c) Violent Governmental
Response.

Factor 1, Internal Response: Elite Instability. The operational measures loading highest on the first factor, itself accounting for a large portion (31.3%) of the total variance,
are: Cabinet Instability, Resignations of Political Elite, Arrest/
Imprisonment of Politically Significant Persons, Dismissals of
Political Elite, Exiles, Dissolution of the Legislature, and
Mutiny. It appears, therefore, that Factor 1 may be interpreted
as representing a kind of elite insecurity, fractionation, and
reshuffling that often accompanies political instability. It reflects both the small-scale individual behavior in both resignations and dismissals, as well as the more macro-behavior of

dissolutions of the legislature and cabinet instabilities. It seems worth noting that Mutiny is positively associated with Elite Instability ($r = +.49$), but negatively associated with both Factors 2 and 3, which appear to index instances of solidarity among elites.

Factor 2, External Response: Non-Violent. The second rotated factor is related to three of the operational indices — Martial Law, Limited States of Emergency, and Exiles — and unrelated to the rest. Although Mutiny also comes out on this dimension, it is negatively ($r = -.47$) related to the latent factor itself. As one observes the declaration of more and more Limited States of Emergency and instances of Martial Law across units, one also observes the less frequent occurrences of Mutiny. It is important to recall that the factor matrix in Table 2 has been rotated to fit the three clusters with perpendicular, or orthogonal, factors; and, therefore, that Non-Violent External Response on the part of the government (Factor 2) occurs independently of Elite Instability (Factor 1).

Factor 3, External Response: Violent. The third factor also accounts for a large proportion of the total variance (30.7%), and therefore continues to support strong inferences. This factor is related to four of the operational measures, and unrelated to the rest: Executions of Politically Insignificant Persons, Arrest/Imprisonment of Politically Insignificant Persons, Governmental Action Against Specific Groups, and Execution of Politically Significant Persons. What differentiates this dimension from the rest is the violent nature of the response itself. Once again, it should be noted that the operational measures were defined so as to make them mutually exclusive, and thus the associations discovered are functions of the observed occurrences of the phenomena themselves.

In order to take advantage of over-time data, and to provide a closer approximation to a "causal" test of the instability model, the nine years of conflict data were aggregated: (a) by the number of specific events (see Figures 1A and 1B), (b) by over 21 provincial units, (c) by two four-year time periods (yielding a "before" and "after" measure about the available survey research data). Thus, conflict data for Brazil and Cuba (from which survey data were available for 1960) were aggregated from 1956-59 for the first time period, and from 1960-64 for the second. For the Dominican Republic and Panama (which were sampled in 1962), data corresponding to the first time period were aggregated across the years 1958-61, and, for the second time period, across the years 1962-66.

A series of composite variables representing the three basic force dimensions were created out of the first unrotated factor scores from separate factor analyses, which were themselves composed of variables selected on the basis of the results presented in Table 2. *Elite Instability* for both the first (ELITE$_1$) as well as the second time period (ELITE$_2$), therefore, consists of the first unrotated factor scores across the variables: Resignations, Dismissals, Dissolutions, Cabinet Instability, and Mutiny. *Non-Violent Governmental Response* (NVLNT) is a composite of: States of Emergency, Martial Law, and Exiles.[47] And the third dimension, *Violent Governmental Response* (VIOLNT), consists of factor scores from the analysis of: Arrest/Imprisonment of Politically Insignificant Persons, Execution of Politically Insignificant and Significant Persons, and Governmental Actions Against Specific Groups.

Finally, the dependent variables themselves, the Turmoil and Internal War dimensions discussed earlier, were "created" out of the first unrotated factor scores from separate analyses for the two time periods among the 21 provincial units. The most compatible factor solutions across the two time periods for the *Turmoil* dimension were yielded from an analysis of the following variables: Anti-Government Demonstration, Anti-Government Demonstrating,[48] Anti-Foreign Demonstration, Anti-Government Riot or *Manifestacion*, Anti-Government Rioting, Political Clash, Strike, and the Number of Deaths from Domestic Violence. The composite-variables TURMOIL$_1$ and TURMOIL$_2$, then, consist of the sum of: (a) the occurrences of these separate events (within each of the 21 provincial units) in standard score form, (b) weighted by the respective factor loading for that event.

The same procedures were followed in the "creation" of the measures of *Internal War* for the provincial units, with the following variables participating in the index: discrete Guerrilla Action, continuing Guerrilla Warfare, Revolutionary Invasion, Terroristic Act/Sabotage, continuing Terrorizing/Sabotaging, and the Number of Violent Political Deaths.

With each of the model's variables operationalized over time, we are in a position to input these dimensions directly into multiple regression (or predictive) equations. Such mathematical equations provide linear "explanations" of a dependent variable, such as Turmoil, as the sum of separate contributions from several "independent" variables, such as Elite Instability, Violent Governmental Response, Discontent, and so on. In the two-variable cases, the regression coefficient represents the degree of list, or "slope," the dependent variable (Y) has on the independent variable (X). In raw score form, this slope is known as a b-weight; while in standard score form it is known as *beta*. A b-constant larger than unity indicates a steeper slope. The steeper the slope, the larger the change in Y for a given change in X. And, likewise, if the b-constant is less than one but greater than zero, it will take a larger change in X to produce a given change in Y. For example, in terms of the first regression coefficient in Equation 1A, every unit change in X_1 (or Elite Instability) is accompanied by a $-.435$ unit loss in Y (or Turmoil).

When faced with a number of independent variables, as we are with the provincial data, it makes sense to look at the effects of these independent variables on each other, as they cause changes in the dependent variable. In other words, we are interested in observing the effect of Elite Instability on Turmoil, while controlling for Legitimacy, Satisfaction, Violent Governmental Response, and so on.[49] In addition, since we have admitted that theoretically it is impossible to explain all the variance about one variable (Internal War, for example), in terms of only one other (Discontent, for example), the technique of *multiple* regression analysis seems especially appropriate. Now instead of explaining the variance in the dependent variable by just one independent variable, the multiple *correlation* coefficient allows one to indicate how much of the total variation in the dependent variable can be explained by all the independent variables acting together. By squaring the multiple correlation (R^2), as is done in the two-variable case (r^2), therefore, we can determine the explanatory strength of the linear combination of the independent variables. The multiple R's for the following regression equations are unusually high by the normal standards of social research, and indicate that in most cases over 90% of the variance (or practically all the variation) in both Turmoil and Internal War occurring in the 21 provinces can be "explained" in terms of the various independent variables within the equations.

The b-coefficients in the following equations are given with the independent variables, while their corresponding *beta* weights appear in parentheses below. Since all but the last two variables of "satisfaction" and "legitimacy" represent factor scores (all with zero means, and standard deviations ranging from .5 to .7),[50] the b and *beta* weights for these variables are comparable. Since the two dependent variables (Turmoil and Internal War) were in standard score form, *beta* coefficients for the remaining variables (P-SAT$_1$, P-SAT$_2$, PSAT$_{21}$, LGLAD1, LGLAD2, and LGLAD3) were obtained by multiplying the b-coefficients times the standard deviation of their respective independent variables.[51] B-weights identify changes in the dependent variable "produced" by changes in the independent variables in terms of the measurement units involved, and therefore are not comparable. If we wish to compare the independent variables as to their *relative* abilities to bring about changes in the dependent variable, we must correct for the scale differences involved. In standardizing the variables, we obtain adjusted slopes, or what we have called *beta* weights, which are comparable from one variable to another. The *beta* weights, then, indicate how much of a change in the dependent variable is produced by a standardized change in one of the independent variables, when the others are controlled.

$$1A \ (R = .97)$$

$$TURMOIL_2 = \underset{(-0.43)}{-.435 \, ELITE_1} \quad \underset{(+0.27)}{+.275 \, ELITE_2} \quad \underset{(+0.06)}{+.062 \, NVLNT_1} \quad \underset{(-0.41)}{-.413 \, VIOLNT_1}$$

$$\underset{(+1.16)}{+1.16 \, VIOLNT_2} \quad \underset{(+0.02)}{+.025 \, NSAT_{23}} \quad \underset{(-0.01)}{-.001 \, LGLAD1}$$

$$1B \ (R = .95)$$

$$INTERNAL \ WAR_2 = \underset{(+0.18)}{+.176 \, ELITE_1} \quad \underset{(-0.14)}{-.140 \, ELITE_2} \quad \underset{(+0.48)}{+.477 \, NVLNT_1} \quad \underset{(-0.87)}{-.869 \, VIOLNT_1}$$

$$\underset{(+1.30)}{+1.30 \, VIOLNT_2} \quad \underset{(+0.18)}{+.181 \, NSAT_{23}} \quad \underset{(-0.01)}{-.001 \, LGLAD1}$$

Elite Instability (ELITE) during both time periods is strongly related to the occurrence of Internal War$_2$ as well as to Turmoil$_2$, but the *signs* of the coefficients are reversed for the two conflict dimensions. Elite Instability at t$_2$ is positively related (+.27) to Turmoil during the same time period; that is, Resignations, Dismissals, and Dissolution of Governmental Bodies come during periods of Domestic Turmoil (i.e., Demonstrations, Riots, Strikes, and Political Clashes). Just the opposite occurs with respect to Organized Violence, however, and here elites cohere (Elite Instability is low, −.14) when Internal War events (such as Guerrilla Warfare, Revolutionary Invasion, and Terrorism) take place.

The signs of the coefficients for Elite Instability at t$_1$ are again reversed for the two dimensions of social conflict. Organized Violence at t$_2$ seems to be a partial product of Elite Instability at t$_1$, for Internal War increases (+.18) in relationship to the fragmentation of elites. And, likewise, it decreases if elites coalesce. The greater elites are characterized as coalescing at t$_1$ (low Elite Instability), however, the greater provincial units seem to experience domestic turmoil at t$_2$ (−.43). These conclusions appear to be extremely stable, with the direction (and often the strength) of the regression coefficients of Elite Instability for the two time periods on the two dimensions remaining essentially the same for a number of regression equations (see Equations 2 and 3, below).

One of the consistent features Crane Brinton finds occurring *prior* to basic revolutions (a type of conflict we have suggested to be more adequately reflected in the Internal War dimension), is what he calls "the disorganization of the government," or "a loss of self-confidence among many members of the ruling class."[52] As we have seen from the first equation, Elite Instability does seem to occur prior (at t$_1$) to Internal War. Harry Eckstein has also concluded that ". . . internal wars[53] are unlikely wherever the cohesion of an elite is intact, for the simple reason that insurgent formations require leadership and other skills and are unlikely to obtain them on a large scale without some significant break in the ranks of an elite."[54] Again, the regression coefficients seem to support this assertion. The greater the Elite Instability at t$_1$ (the greater the recruitment potential for the insurgent forces, according to Eckstein's notion), the greater the occurrence of Internal

War at t_2. The same does not hold true for the occurrence of Turmoil. In mathematizing an Internal War Potential model, Manuel Avila hypothesized that as Elite Cohesiveness increased linearly, Internal War potential decreased.[55] The relationship, in other words, was negative. Again, these data seem to support such an inference; elites cohere (Elite Instability is low, $-.14$) as Internal War events take place.

Non-Violent Governmental Response (NVLNT) at t_1 (indexed by the first unrotated factor scores of variables: Limited States of Emergency, Martial Law, and Exiles) is much more important in explaining Internal War than Turmoil events. In Equations 2 and 3 (below), in addition to Equation 1, as Non-Violent Governmental Responses at t_1 increase, both organized (Internal War) and spontaneous (Turmoil) events occur. However, Acts of Terrorism, Guerrilla Insurrections, and Revolutionary Invasions (Internal War events) break out at higher levels in provinces experiencing States of Emergency, Martial Laws, and a high Exile rate in *earlier* time periods. Again, these findings are extremely stable across all regression equations.

$$2A \quad (R = .96)$$

$$TURMOIL_2 = +.628 \, ELITE_2 + .132 \, NVLNT_2 - .056 \, VIOLNT_2 - .000 \, P\text{-}SAT_1$$
$$(+0.63) \quad\quad (+0.13) \quad\quad (-0.06) \quad\quad (-0.00)$$
$$- .013 \, LGTMCY$$
$$(-0.01)$$

$$2B \quad (R = .92)$$

$$INTERNAL \, WAR_2 = +.384 \, ELITE_1 + .543 \, NVLNT_1 + .003 \, VIOLNT_1 + .052 \, PSAT_{21}$$
$$(+0.38) \quad\quad (+0.54) \quad\quad (+0.00) \quad\quad (+0.07)$$
$$- .009 \, LGLAD3$$
$$(-0.04)$$

$$3A \quad (R = .89)$$

$$TURMOIL_2 = -.417 \, ELITE_1 + .285 \, NVLNT_1 + .722 \, VIOLNT_1 - .036 \, PSAT_{21}$$
$$(-0.42) \quad\quad (+0.28) \quad\quad (+0.72) \quad\quad (-0.05)$$
$$- .028 \, NSAT_{21} + .026 \, LGLAD1 - .012 \, LGLAD2$$
$$(-0.06) \quad\quad (+0.14) \quad\quad (-0.07)$$

$$3B \quad (R = .92)$$

$$INTERNAL \, WAR_2 = +.435 \, ELITE_1 + .614 \, NVLNT_1 - .126 \, VIOLNT_1 + .059 \, P\text{-}SAT_2$$
$$(+0.43) \quad\quad (+0.61) \quad\quad (-0.13) \quad\quad (+0.29)$$
$$- .055 \, LGLAD1$$
$$(-0.28)$$

Violent Governmental Response (VIOLNT) (indexed by the unrotated factor scores of the extremely high loading variables of: Arrests of Politically Insignificant Persons, Executions of Politically Insignificant Persons, and of Significant Persons, and Governmental Actions against Specific Groups) taking place at t_1 inhibits both Organized ($-.87$, Equation 1B) and Spontaneous ($-.41$, Equation 1A) Violence at t_2. This finding is not stable, however, since the direction of the regression coefficient did change with a change in the other variables composing separate regression analyses. Any conclusions on the relationship between Violence and the two dimensions of Political Instability, then, will have to remain extremely tentative. One possible reason for the instability of the findings with respect to Violence is the fact that a *linear* model (regression analysis) is being applied to what has earlier been theoretically specified as a curvilinear function. If the data are indeed curvilinear (and there is some preliminary evidence to indicate that Violence is curvilinearly related to Turmoil, but linearly related to Internal War), then the slight est change in any one province's position in a scatter plot, vis-a-vis the others, might easily change the direction of the relationship (than if the association were linear).[56]

In an earlier study,[57] the strongest correlate (curvilinear) of what was referred to as Anomic Violence (indexed by such events as Riots, Strikes, and Demonstrations) was found to be a measure of Retribution (i.e., Expenditure on Defense as a Percentage of GNP). When force was both very permissive as well as very restrictive among nation-units, Anomic Violence was found to be negligible. Force in the mid-levels of intensity, however, elicited high levels of Anomic Violence. No such relationship was found when associating the same force measures to the occurrence of what was called Organized Violence (i.e., Armed Rebellions, Guerrilla Warfare). Not respecting a series of caveats pointed out at the time (the primary one being the problem of "ecological" inferences), one could conclude that: "Guerrilla Warfare, Terrorism, and Sabotage . . . occur and continue irrespective of the extent and quality of governmental force. That is, they may break out just as readily among militarily strong as militarily weak regimes; and they may continue in the face of what would appear to be overwhelmingly adverse inhibiting power." There is some suggestion that these conclusions might tentatively be applied to the present data. In Equation 2B, for example, Violent Governmental Response$_1$ yields little or no association with Internal War; while in Equations 2A and 3A we find a tighter relationship between Violent Governmental Response and Turmoil. As a matter of fact, the higher *beta* weight of Violent Governmental Response$_2$ on Internal War (1B) (than Violent Governmental Response$_2$ on Turmoil, Equation 1A), suggests that when violence is higher at t_2 there is a greater chance of Internal War than Turmoil occurring.

While the results with respect to Violent Governmental Response are not at all conclusive, this cannot be said to be the case for the remaining variables: (a) Satisfaction (P-SAT, N-SAT), and (b) Legitimacy (LGTMCY, LGLAD). Equations 2 and 3 offer the clearest examples of a consistent pattern: Legitimacy is negatively related to both Internal War and Turmoil, but the stronger effect appears to occur among Organized Violent Activity. The higher the illegitimacy (or negative affect) ascribed to the government (low Legitimacy), the higher the prospect of Internal War ($-.28$, Equation 3B) than Turmoil ($-.07$, Equation 3A). Satisfaction, on the other hand, is positively related to Internal War ($+.29$, Equation 3B; $+.07$, Equation 2B), but negatively related to Turmoil ($-.05$, Equation 3A; $-.00$, Equation 2A).

The finding that Legitimacy is more tightly related to Organized than Spontaneous Violent Behavior also finds support in an earlier study. On the basis of findings reported in *The Civic Culture* (namely, that the ability to participate in a system leads directly to the building of positive affect toward it),[58] Legitimacy was operationalized across nation-units, as the degree of change a polity experienced in ratings of "system openness," prior to experiencing either Anomic or Organized Violence. In this and other operationalizations of the concept, Legitimacy proved to be the strong negative correlate of Organized Violence. In all cases, as Political Legitimacy decreased, Organized Violence increased. When the effect of Political Legitimacy on Anomic Violence was tested, however, the negative association proved to be considerably weaker. Once again, dismissing the pitfalls of ecological correlations, it was concluded that ". . . the participants in Organized Violent Activity [i.e., Guerrilla Warfare, Terrorism, Armed Rebellion] seem to be challenging the Legitimacy of the political systems involved . . ."[59]

The Personal Satisfaction of respondents (and their perception of the national standing) among 21 provinces was measured in a variety of ways. P-SAT$_2$, it will be recalled, represents the mean ladder rating (personal standing) on the Self-Anchoring Striving Scale aggregated across province-units for the time of the interview (which for Brazil and Cuba was 1960; for the Dominican Republic and Panama, 1962). P-SAT$_1$ represents the same measure, with the exception that the data consist of the respondent's evaluation of his ladder rating "five years ago." PSAT$_{21}$ is the difference between P-SAT$_2$ and P-SAT$_1$, or the amount of *change* in personal standing among respondents from "five years ago" to the present.

Regardless of the operationalization, however, Satisfaction is negatively related to Turmoil, but positively related to Internal War. That is, in provinces where respondents rate themselves *low* on the Self-Anchoring Striving Scale, or where they rate themselves as "worse off today than in the past," Demonstrations, Riots, Strikes, and Political Clashes are *high*. On the other hand, instances of Guerrilla Warfare, Revolutionary Invasion, Terroristic Acts, and Sabotage appear at times (a) when people rate themselves "better off today than in the past," and (b) in areas where people tend to rate themselves high on a scale of best and worst possible existences.

While the *beta* weights indicate these relationships are not the strongest predictors of Organized (i.e., Internal War) and Spontaneous (i.e., Turmoil) Violence, the consistent *directions* of the associations are highly suggestive of the two directions Satisfaction appears to take when associated with the occurrence of Revolution in the conflict literature. One of the dependent variables here, Internal War, is admittedly not the same as the variable Revolution discussed by Edwards, de-Tocqueville, Brinton, and Davies. An affinity is, nevertheless, proposed, if by "revolution" these authors meant "behaviors attempting (or succeeding) to bring about basic social restructuring." If such is the case, and we have suggested that it is, then Guerrilla Warfare and Revolutionary Invasion (i.e., Internal War) do come closer to an approximation of attempts at basic social change than Demonstrations, Riots, and Strikes (i.e., Turmoil). The positive regression coefficients then, ap-

pear to support the notion that "revolutions are born in societies on the upswing."

The work done by Russett, Lipset, Cutright, and Lerner,
on the other hand, has supported the "downswing thesis," e.g.,
that "satisfied" (or wealthy) polities are stable (low on Political Instability). An affinity of the findings to this proposition
is likewise suggested, if by Political Instability these authors
were considering behavior on the order of Riots, Demonstrations, Strikes, and Political Clashes. The negative regression
coefficients indicate that as Satisfaction (regardless of operationalization) goes up, Political Instability, indexed by Turmoil
events, goes down.

SUMMARY AND CONCLUSIONS

Domestic conflict in Latin America (a) empirically dis-
tributes itself into two basic clusters of activities: (b) Turmoil
and Internal War, (c) which are generally independent of
each other, and (d) which can generally be differentiated on
the basis of structure, direction, and spontaneity. When 24
operational indices of domestic conflict behavior occurring
within 65 Latin American provinces over a nine-year period
were intercorrelated and factor analyzed (see Table 1), these
two "families of conflict" emerged as the first two factors. This
finding provided the basis for the construction of composite
representations of these two dimensions: (a) the spontaneous,
disorganized, or *Turmoil* dimension (which was indexed by:
Demonstrations, Riots, Clashes, Strikes, and Deaths from
Domestic Group Violence), and (b) the *Internal War* dimension, involving conflict behavior of a more planned and organized nature (and indexed by instances of Guerrilla Warfare, Revolutionary Invasion, Terrorism/Sabotage, and Domestic Killed).

When a general model isolated (a) Discontent, (b) Force,
and (c) Legitimacy, as the possible *causes* of Turmoil and
Internal War, and the effects of these independent variables
were tested, using (a) over-time data, (b) among the 21 Latin
American provinces (for which detailed survey research data
were also available), the following conclusions emerged from
a series of multiple regression analyses:

Force was viewed as a special type of conflict — that directed by a government to a populace. Governmental response to domestic conflict is itself highly structured in terms of three independent clusters of activities (see Table II): (a) Elite Instability, (b) Non-Violent and (c) Violent Governmental Response.

1. Elite fragmentation and disorganization consistently takes place *prior* to the occurrence of Internal War; but elite cohesion is clearly the mode of behavior as Internal War events are occurring. Elite Instability (t_1), therefore, appears to be a contributing condition to Internal War (t_2).

2. Elite Instability at t_2, on the other hand, is positively related to the occurrence of Turmoil during the same time period (t_2); that is, Resignations, Dismissals, and Dissolutions come during periods of Domestic Turmoil (i.e., Demonstrations, Riots, Strikes, and Clashes).

3. Non-Violent Governmental Response (indexed by: States of Emergency, Martial Law, and Exiles) at t_1 is positively related to both the amount of Turmoil and Internal War which occurs in the following time period; but appears to be far more important in explaining the occurrence of Internal War.

4. Although the measure Violent Governmental Response (a composite of the variables: Arrests, Executions, and Actions Against Specific Groups) appeared in a number of regression equations, both the weight and the signs of the coefficients were too unstable to support firm conclusions.

5. When *Legitimacy* is operationalized through survey responses measuring the degree of positive and negative affect among provincial inhabitants, Legitimacy (or positive affect) proves to be the negative correlate of both Internal War and Turmoil. The stronger affect appears to occur with Organized Violent Activity, however; for the higher the Illegitimacy, the greater the prospect for the occurrence of Internal War, than Turmoil, events.

6. Whether *satisfaction* is operationalized (through survey research data also aggregated by provinces) as either: (a) the mean ladder standing at which respondents place themselves on a scale of best and worst possible existences, or

(b) a change in these ladder ratings from an estimate of where they stood in the past compared to where they now stand, the results are the same. Satisfaction is (a) positively related (upswing) to the occurrence of Internal War events, but negatively related (downswing) to the occurrence of Turmoil events. That is, in provinces where respondents rate themselves *low* on the Self-Anchoring Striving Scale, or when they rate themselves as *"worse off* today than in the past," Demonstrations, Riots, Strikes, and Clashes are high. When measures of discontent are used as predictors of Internal War, however, instances of Guerrilla Warfare, Revolutionary Invasion, Terroristic Acts, and Sabotage appear at times (a) when people rate themselves *"better off* today than in the past," and (b) in areas where people tend to rate themselves *high* on a scale of best and worst possible existences.

NOTES

1. Ivo Feierabend provides the linkage to "political instability," when he notes that it is "... the degree or the amount of *aggression* directed by individuals or groups within the political system against other groups or against the complex of officeholders and individuals and groups associated with them. Or, conversely, as the amount of aggression directed by these officeholders against other individuals, groups, or officeholders within the polity." See Ivo K. and Rosalind L. Feierabend, "Aggressive Behaviors within Polities, 1948-62: A Cross-National Study," *J. Conflict Resolution,* X (Sept., 1966), 250.

2. The phrase "civil violence," which has also been used in this connection, might have better illustrated one of the distinctions being made; namely, that between (a) aggressive activity directed from a populace (individuals or groups) to a political system (either at the community, regime, or government level), and (b) criminal aggressive activity (generally, though not always, of an individual nature), such as murder and robbery. The latter is not the concern of this study.

3. James Payne, "Peru: The Politics of Structured Violence," *J. Politics,* XXVII (May, 1965), 363.

4. Merle Kling, "Toward a Theory of Power and Political Instability in Latin America," *Western Political Q.,* IX (March, 1956), 21.

5. Mack and Snyder note that "conflictful behaviors are those designed to destroy, injure, thwart, or otherwise control another party or other parties. . . ." See Raymond W. Mack and Richard

C. Snyder, "The Analysis of Social Conflict: Toward an Overview and Synthesis," *J. Conflict Resolution*, I (June, 1957), 218.
6. D. P. Bwy, "Systematizing the Collection of Data on Aggressive Activity: A Domestic Conflict Code Sheet" (Dept. of Political Science, Case Western Reserve Univ., July, 1967), 26pp. An example of the detail that was attempted in differentiating over forty instability events can be seen in one of the definitions, that of "Riots or Manifestaciones": "A *violent* Anti-Government Demonstration (1). The presence of accompanying violence differentiates "riots" from "demonstrations." A mob clashing with the police or military, or attacking private property, is considered a riot as long as such violence appears to be spontaneous; otherwise it may qualify for consideration as a Civilian Political Revolt (27). Riots are considered discrete events, limited to a given group of people. Violence between rival political and nonpolitical groups is classified as either a Political (12) or Non-Political Clash (13)."
7. Sampling sites within four Latin American nations qualified under this criterion: Brazil, Cuba, the Dominican Republic, and Panama. Detailed and comparable national probability samples were drawn for these nations by the Institute for International Social Research, in 1960 and 1962. A research commitment was made to these nations primarily because the survey data provided much more sensitive measures of (a) a population's "satisfaction" (or "dissatisfaction"), and (b) feelings of legitimacy — two core variables of the political instability model.
8. The primary source which most adequately satisfied this criterion was the *Hispanic American Report,* which is itself a synthesis of major news media. *Hispanic American Report: A Monthly Report on Developments in Spain, Portugal, and Latin America* (Ronald Hilton, ed.); Institute of Hispanic American and Luso-Brazilian Studies, Stanford University), Vol. I (1948)–Vol. XVII (1964).
9. *Brazil*: Acre, Alagoas, Amapá, Amazonas, Bahía, Ceará, Espírito Santo, Fernando de Noronha, Goiás, Guanabara, Maranhão, Mato Grosso, Minas Gerais, Pará, Paraíba, Paraná, Pernambuco, Piauí, Rio Branco, Rio de Janeiro, Rio Grande do Norte, Rio Grande do Sul, Rondônia, Santa Catarina, São Paulo, Sergipe, Serra do Aimores, Distrito Federal.
   Cuba: Pinar del Río, Havana, Matanzas, Las Villas, Camagüey, Oriente.
   Dominican Republic: Distrito Nacional, Azua, Bahoruco, Barahone, Dajabón, Duarte, El Seibo, Espaillat, Independencia, La Romana, La Vega, Maria Trinidad, Monte Cristi, Pedernales, Peravia, Puerto Plata, Salcedo, Samaná, Sánchez, Santiago, Santiago Rodriguez, Valverde.
   Panama: Chiriquí, Colón, Darién, Panamá, Veraguas.

   In order to reduce the role of aberrations on what were meant to be general findings (thereby preventing any of the correlation input into the factor analysis from being dependent on too few cases), the criterion was established that each variable must have at least 5% participation among the units, to be included in the analysis. On the basis of this criterion, 24 variables qualified for inclusion in the analysis. The measures failing to satisfy the 5% criterion were: Printed or Broadcast Protests, Political Boycott, Anti-Foreign Threat, *Imposición, Candidato Único, Continuismo, Machetismo* or Peasant Rebellion, Civilian Political Revolt, Private Warfare, and Banditry.

10. The computer program which monitored these calculations was MESA₁, a 95 x 95 Factor Analytic Program with Varimax Rotation. The lower limit for eigenvalues (i.e., a proportion of variance which may vary from near zero to n, where n is the number of variables entering a factor matrix) to be included in rotation was 1.00. Rotation is carried out in order to obtain a solution which is not entirely dependent upon each particular variable in the analysis. Orthogonal rotation is the fitting of factors to clusters of variables, with the restriction that the correlation between factors is zero. The *varimax* criterion is used to rotate orthogonally to "simple structure," that is, the maximization of high and low loadings. Thus, this form of rotation continues to maintain independence among the factors.

11. R. J. Rummel, "Dimensions of Conflict Behavior Within and Between Nations," *General Systems Yearbook,* VIII (1963), 1–50.

Raymond Tanter, "Dimensions of Conflict Behavior Within and Between Nations, 1958-60," *J. Conflict Resolution,* X (March, 1966), 41–64.

R. J. Rummel, "Dimensions of Conflict Behavior Within Nations, 1946-59," *J. Conflict Resolution,* X (March, 1966), 65–73.

12. Although the other measures in the Rummel and Tanter analyses are generally comparable (by definition) to the domestic conflict measures used here, the variable "Major Governmental Crises" appears to be somewhat different. Rummel and Tanter defined this event as: "Any rapidly developing situation that threatens to bring the downfall of the present regime — excluding situations of revolt aimed at such an overthrow."

13. D. P. Bwy, "Political Instability in Latin America: The Cross-Cultural Test of a Causal Model," *Latin Am. Research Rev.,* III (Spring, 1968), forthcoming.

14. Rummel "Dimensions of Conflict Behavior Within and Between Nations," *op. cit.,* p. 35.

15. "Revolutions" were defined by both Rummel and Tanter as: "Any illegal or forced change in the top governmental elite, any attempt at such change, or any successful or unsuccessful armed rebellion whose aim is independence from the central government."

It should be noted that their data on "civil wars" and "social revolutions" were too infrequent to be used alone, and so apparently both Rummel and Tanter merged these data with those on "palace revolutions" into the general category, above, "Number of Revolutions." In view of this fact, a comparison across studies on this measure becomes somewhat difficult.

16. Tanter, *op. cit.*, p. 43.
17. To Andrew Janos, "Unconventional warfare may be said to exist if and when the adversaries confronting each other have grossly disproportionate capabilities, whether in manpower, resources, or organizational base. In unconventional warfare, one of the participants possesses either an army of inferior size, equipment, and organization, or no army at all. Whereas one side can rely on a regular army, the other will have to fight the war with scratch military units, 'part-time soldiers,' non-military organizations, unorganized masses, or with tightly organized but numerically inferior groups of political activists." See Andrew C. Janos, "Unconventional Warfare: Framework and Analysis," *World Politics*, XV (July, 1963), 637–638.
18. Merle Kling, "Cuba: A Case Study of a Successful Attempt to Seize Political Power by the Application of Unconventional Warfare," in J. K. Zawodny (ed.), "Unconventional Warfare," special issue of *The Annals, Am. Acad. Polit. Soc. Sci.*, CCCXLI (May, 1962), p. 44.
19. *Ibid.*, p. 43.
20. From: James Geschwender, "Social Structure and the Negro Revolt: An Examination of Some Hypotheses," *Social Forces*, XLIII (Dec., 1964), 249. (Italics added.)
21. Bruce M. Russett, "Inequality and Instability: The Relation of Land Tenure to Politics," *World Politics*, XVI (April, 1964), 442–454.
22. Raymond Tanter and Manus Midlarsky, "A Theory of Revolution," *J. Conflict Resolution*, XI (Sept., 1967), 264–280.
23. Seymour Martin Lipset, "Some Social Requisites of Democracy: Economic Development and Political Legitimacy," *Am. Polit. Sci. Rev.*, LIII (March, 1959), 69–105.
24. Phillips Cutright, "National Political Development: Measurement and Analysis," *Am. Sociol. Rev.*, XXVIII (April, 1963), 253–264.
25. Daniel Lerner, "Modernizing Styles of Life: A Theory," in Lerner, *The Passing of Traditional Society: Modernizing the Middle East* (N. Y.: Free Press, 1958), 43–75.
26. Lyford P. Edwards, *The Natural History of Revolution* (Univ. of Chicago Press, 1927). (Italics added.)
27. Eric R. Wolf, *Sons of the Shaking Earth* (Univ. of Chicago Press, 1959), 108–109.
28. James C. Davies, "Toward a Theory of Revolution," *Am. Sociol. Rev.*, XXVII (Feb., 1962), 6.
29. Crane Brinton, *The Anatomy of Revolution* (N. Y.: Vintage Books, 1952).
30. Davies, *op. cit.*
31. Cole Blasier, "Studies of Social Revolution: Origins in Mexico, Bolivia, and Cuba," *Latin Am. Research Rev.*, II (Summer, 1967), 28–64.
32. Lipset, *op. cit.*, p. 86.
33. David Easton, "An Approach to the Analysis of Political Systems," *World Politics*, IX (April, 1957), 383–400.
34. "Subject" political cultures are those in which individual cognitions are primarily oriented toward *output* structures and the "system as a general political object." See Gabriel Almond

and Sidney Verba, *The Civic Culture: Political Attitudes and Democracy in Five Nations* (Princeton Univ. Press, 1963).

35. *Ibid.*, p. 246.
36. Arnold H. Buss, *The Psychology of Aggression* (N. Y.: John Wiley, 1961), p. 58.
37. Robert A. LeVine, "Anti-European Violence in Africa: A Comparative Analysis," *J. Conflict Resolution,* III (Dec., 1959), 420–429.
38. Bwy, *op. cit.*
39. F. P. Kilpatrick and Hadley Cantril, "Self-Anchoring Scaling: A Measure of Individuals' Unique Reality Worlds," *J. Individ. Psychol.,* XVI (Nov., 1960), 158–173.
40. Hadley Cantril, *The Pattern of Human Concerns* (New Brunswick: Rutgers Univ. Press, 1965), pp. 22–26.
41. The raw data, interview schedules, and code books were obtained from the Roper Public Opinion Research Center, Williams College, Williamstown, Mass. The on-site interviewing for the Institute for International Social Science Research was conducted by: Instituto de Estudios Sociais e Economicos, Ltda. (Brazil); International Research Associates, S.A. de C. V., of Mexico City (the Dominican Republic and Panama); and anonymous (Cuba).
42. See: D. P. Bwy, "RECODIGO: An All-Purpose Computer Program for 'Cleaning' Multiple-Punched Data," *Behavioral Science,* forthcoming.
43. City units: *Brazil:* Fortaleza, Niteroi, Porto Calvo, Recife, Rio de Janeiro, and São Paulo; *Cuba:* Cardenas, Havana, Remedios; *the Dominican Republic:* Santiago de los Caballeros and Santo Domingo; *Panama:* Colón, David, and Ciudad de Panamá.
44. Province units: *Brazil:* Alagoas, Amazonas, Bahía, Ceará, Guanabara, Maranhão, Minas Gerais, Paraíba, Paraná, Pernambuco, Rio de Janeiro, Rio Grande do Sul, and São Paulo; *Cuba:* Havana, Matanzas, and Las Villas; *the Dominican Republic:* Distrito Nacional and Santiago; *Panama:* Chiriquí, Colón, and Panamá.
45. Coding scheme of "political" considerations, mentioned by at least 5% of all respondents:

NATIONAL HOPES AND ASPIRATIONS

National Unity — absence of unrest, tensions, and antagonisms based on regional, class, caste, religious, etc., differences.

Honest Government — fair and just; no corruption or nepotism.

Efficient Government — competent leadership and administration; effective party system; no excessive bureaucracy.

Socialistic Government — aspiration to become a socialistic or welfare state.

Balanced Government — adequate system of checks and balances; no excessive power in the hands of government; less central government; more power to states and provinces.

Freedom — with specific reference to freedom of speech, of religion, of occupation, of movement, etc.

Political Stability, Internal Peace, and Order.

Law and Order — maintenance of the public peace; decrease or no increase in crime, juvenile delinquency, etc., fair courts, good or improved juridical practices, penal system, etc.

Representative Government — maintain present democracy, or become a democracy.

NATIONAL WORRIES AND FEARS

Disunity among People of the Nation — unrest, tensions, antagonisms, based on regional, class, caste, religious, etc., differences.

Dishonest Government — unfair and unjust; corruption and nepotism.

Inefficient Government — weak, indecisive leadership and administration; no effective party system; excessive bureaucracy.

Fear Country Will Become Socialistic

Central Government Too Big and Too Powerful — no adequate system of checks and balances; not enough power for states and provinces. Also coded: Fear of Communist danger or the consequences of Communist control from within.

Lack or Loss of Freedom — in general, or with specific reference to freedom of speech, of religion, of occupation, of movement, etc.

Political Instability, Chaos, and Civil War.

Lack of Law and Order — failure to maintain public peace; prevalence of or increase in crime, juvenile delinquency, etc., unfair courts, poor or unfair juridical practices, penal system, etc.

No Representative Government — loss of democracy; totalitarianism.

46. The measure "Governmental Boycott" was the only variable which failed to qualify under the previously established criterion of at least 5% participation among the units, and was dropped from the analysis presented in Table II.

47. The two sets of factorial solutions (across the two time periods) were generally compatible. One of the more comparable solutions, presented below, shows the remarkably similar factor loadings for the only extracted, unrotated factor. The unrotated first factor was selected to index these composite variables, since it explains the maximum amount of variance about the original operational indices.

t_1: Brazil/Cuba: 1956-59
Dominican Rep/Panama: 1958-61

Variable	Factor Loading
Limited States of Emergency	.874
Martial Law	.887
Exiles	.890

t_2: Brazil/Cuba: 1960-64
Dominican Rep/Panama: 1962-66

Variable	Factor Loading
Limited States of Emergency	.861
Martial Law	.851
Exiles	.716

48. The same as "Demonstration," with the exception that "Demonstrat*ing*" refers to an indeterminate number of continuous demonstrations taking place (either within or across coding units) simultaneously.

49. Furthermore, in *multiple* regression, as in the two-variable case, b's and *beta's* which appear in the regression equations represent slopes, or the amount of change in Y that can be associated with a given change in one of the X's, *with the remaining independent variables held constant.* This technique, then, allows us to determine the distinct explanatory power of each of the independent variables in bringing about a change in the dependent variable, with the others held fixed.

50. With an N of 21, one might expect the standard deviation range about the mean of a normal distribution to be half (3.0) that for the large-sample case. Therefore, by extension, the standard deviation itself will probably be half its normal size of (1.0), which accounts for the .5 to .7 standard deviations among the variables in the present analysis. In addition, because of (a) the N of 21, and (b) a problem of unequal variances, two of the *"beta"* coefficients are greater than unity (Equation 1A: $VIOLNT_2$, and Equation 1B: $VIOLNT_2$).

51. Thus, a movement of 2.0 raw score b-units should equal a 1.2 standard deviation movement, if the standard deviation for the independent variable is .6.

52. Crane Brinton, "The Anatomy of Revolution," in Harry Eckstein and David Apter (eds.), *Comparative Politics: A Reader* (N. Y.: Free Press, 1963), pp. 560–569.

53. "Internal War," for Eckstein, was defined as: ". . . attempts to change by violence, or threat of violence, a government's policies, rulers, or organization." See Harry Eckstein (ed.), *Internal War: Problems and Approaches* (N. Y.: Free Press, 1964), p. 1.

54. Harry Eckstein, "Internal War: The Problem of Anticipation," in: Ithiel de Sola Pool et al., *Social Science Research and National Security,* a report prepared by the Research Group in Psychology and the Social Sciences. (Smithsonian Institution, Washington, D. C., 1963), pp. 102–147.

55. Manuel Avila, "Preliminary Model of Internal War Potential," in *Project Camelot: Report on Research Design* (Special Operations Research Office, American Univ., Washington, D. C., April, 1965), p. D-4.

56. This could account for the considerable change about $VIOLNT_1$ in Equation 3A; while no such changes occurred in the Internal War equations.

57. Bwy, "Political Instability in Latin America," *op. cit.*

58. ". . . the opportunity to participate in political decisions is associated with greater satisfaction with that system and with greater general loyalty to the system. . . . Everything being equal, the sense of ability to participate in politics appears to increase the legitimacy of a system and to lead to political stability." See Almond and Verba, *op. cit.,* p. 253.

59. Bwy, *op. cit.*

BLACK POWER, COMMUNITY POWER, AND JOBS

Harry W. Reynolds, Jr.

■ Like their counterparts in numerous other cities during recent summers, the extensive civil disturbances which convulsed Omaha for three days and nights in July, 1966, and for two days and nights in August, were precipitated by seemingly trivial incidents and prolonged humid weather. Like most of their counterparts, too, the Omaha disturbances were not by precise definition racial conflicts, i.e., predominantly Caucasian and Negro groups contending against each other. And as with other disturbances of this kind, the sources of the Omaha commotion were deep-seated and filled with bitterness. The most obvious consequences of these disturbances—numerous arrests, considerable property damage, and the need for militia reinforcements to sustain law and order—

AUTHOR'S NOTE: *The present study, as well as inquiry into other facets of the employment problem in Omaha, was financed by a grant from the U.S. Labor Department to the Municipal University of Omaha and the writer. The portion of the research dealing with the riots is the first to be published. The Omaha Police Department, the Nebraska Department of Employment, and the Omaha* WORLD-HERALD *and* SUN, *in making their records and personnel available to aid in this study, is gratefully acknowledged. The assistance of several faculty members in the Omaha University Department of Sociology is gratefully acknowledged in carrying out portions of this field research. Errors of fact or interpretation are the author's alone.*

tell little of the underlying reasons for such trouble. An examination in depth of the sentiments and explanations offered by participants in those disturbances, and of the relevant behavior of other persons and organizations that were in various ways associated with the participants, affords a better understanding.

BACKGROUND AND METHOD

The underlying cause of the Omaha riots of 1966 has been identified as stemming from inadequacies in the field of employment. This conclusion is derived from extensive interviewing of riot participants, and of a random sample of residents in the census tract (the city's poorest) where the riot occurred and practically all of the rioters lived. In addition, a sample of residents in a different census tract having very nearly identical socioeconomic characteristics has been scrutinized in some depth for analogous disparate disclosures. By utilizing the latter category as a control group vis-a-vis the second, and considering the second as a probably fuller mirror of the thinking of the first, it was deemed rather certain that a more complete and representative picture accounting for the behavior of the participants in the summer disturbances would emerge. Certainly this approach affords a meaningful way for relating the participants in the disturbances to the Negro and Caucasian communities of which they are a part, and to the social and economic leaders and values controlling those communities.

In support of the observation that inadequacies in the field of employment underlay the disturbances, evidence was amassed from interviews with 147 of the 163 individuals arrested as participants in the looting and arson accompanying the rioting. (See Table 1.) All of the arrestees were Negroes; six were women. Responding to the question of what prompted their active participation in the civil disturbances, 114 without qualification identified difficulty and repeated disappointment in seeking satisfactory and continuing employment. All of these respondents, it should be noted, were over 18 years of age. The only significant variance from this very preponderant first preference by the rioters came from the arrested who were 16-17 years of age. Thirty-one of those apprehended were in this age group. Their first choice as to causation was lack of recreational facilities near their places of

TABLE 1

First Choice Reasons Given by Riot Participants for
Taking Part in the Omaha Riot of July, 1966

Difficulty in acquiring or holding jobs	114
Lack of sufficient recreational facilities*	31
Trouble with police	1
Unsuitable housing	1

*All teen-agers	N=147

residence. Given their youth and greater lack of economic responsibilities, this attitude is not difficult to understand.

Not without relevance here, however, is the low importance accorded housing and the police. Housing placed consistently third and fourth as a grievance in the rioters' choices, well behind job needs and recreational inadequacies. To 84 of the arrestees, housing was a third choice; to the remainder, a fourth choice. With respect to the matter of the police (more specifically, grievances against the police), the rioters' choices were similarly ordered. To only 31 were police actions even a second-place grievance; to 82 they ranked third; and, to the remainder, fourth. Economic exploitation by neighborhood grocers and businessmen figured almost uniformly last in importance. In this context, the pre-eminence of the factor of jobs as a motivation to riot is not much in dispute. And there is further revealing corroboration.

THE SALIENT FINDINGS

The length of time on their most recent job, among the rioters, was a mean of 21 months. (See Table 2.) Fully one-third (46) had worked on their present job no longer than about one calendar year. Thirty-eight had worked in their current position five years or longer. Approximately one-fifth of the rioters were unemployed at the time of their being arrested, most of them teen-agers. Earnings averaged $1.30 an hour, with 80 of those arrested receiving more than that amount, and 34 averaging a $2.00 hourly

TABLE 2

Length of Time on Present Job for Riot Participants*

Five years or more	38
Three years to five years	14
One year to three years	46
Less than one year	20
Unemployed	29

*As of date of arrest   N=147

rate or more. (See Table 3.) Information acquired for many of the
participants in the riots, again through interviews, suggests that

TABLE 3

Earnings of Rioters in Most Recent Job

$3.00 per hour	24
$1.50 - 2.99 per hour	55
$0.50 - 1.49 per hour	39
Unemployed	29

N=147

these earning averages and mean periods of employment have
remained relatively constant for the persons concerned during
their preceding periods of compensated labor. For 106 participants electing to answer the question of how long they served and
how much they earned in their immediately preceding employment, but with only their recollections to go by, length of service
averaged 19½ months and hourly compensation $1.24. Only 24
interviewees recalled earning more than $1.50 an hour in their
immediately previous jobs. As to reasons for leaving their previous
employment, 26 gave illness, 70 said they were fired, and 24
admitted that they did not like where they worked or what they
did.

Undergirding this scheme of things, and to a considerable extent responsible for it, was a highly elaborate but very informal job-finding network. So, too, was a minimal and deficient level of educational attainment. This latter consideration is an almost universal concomitant of those participating in the urban disturbances of recent summers. A majority of the Omaha rioters had no more than a junior high school education (77). Only 39 were high school graduates, although about four dozen had had at least one year of high school. (See Table 4.) The details of the informal job-finding network require more elaboration.

TABLE 4

Level of Educational Attainment of Rioters

No more than 12 years of school	39
No more than 9 years of school	46
No more than 7 years of school	31
Less than 7 years of school	31

N=147

What is particularly significant about the informal job-finding network is the light it throws upon the ways in which employment is attained among the Negro poor in this urban center. Stated simply, it is characterized by the preeminence of family, personal friends and contacts, neighbors, and (occasionally) newspaper want ads as sources of information about employment opportunities. And by definition, the usefulness of public or private employment services, or of other institutionalized sources of information such as labor unions, welfare agencies, and churches, is very much discounted. Less than one-third of the arrested participants in the Omaha disturbances had ever applied to a place of employment directly (i.e., store, factory) for a job. Almost 90% of these individuals acknowledged that the jobs they currently held, or last held, had been acquired by leads supplied through one or another of the souces noted previously. Employment so obtained did not consist only of menial tasks such as car washing, delivering groceries, or loading and unloading trucks, although many of the jobs acquired were of this type. Numerous semi-skilled positions also were secured by this means—sign painting, automobile repairing at service stations, and some forms of radio and television repairing. Approximately one-third of the positions held by the employed rioters were of this semi-skilled type. When placed against the backdrop of a random sample of residents in the same census tract in which the bulk of the participants in the disturbances resided, these findings reflect a considerable measure of continuity and prevalence.

Interviews with 447 residents of the poorest census tract in Omaha (where, as noted, practically all of the riot participants lived) revealed dimensions of job tenure, acquisition, compensation, and termination which closely paralleled the material emanating from interviews with the rioters themselves. This particular census tract, it may be noted in passing, also contained the scenes of the disturbances. Of the total in this second group of respondents, all but 11 were Negroes, the remainder being Caucasians and one American Indian. For this larger sample, the mean period of employment in the jobs currently held by respondents (378 replying) was almost 23 months, two months greater than the mean of the riot participants. (See Table 5.) Mean hourly

TABLE 5

Length of Employment on Present Job of Random Sample of Residents in Same Census Tract Where Riot Participants Lived; Compared with Rioters

Census Tract Residents	Rioters	
Five years or more	136	38
Three years to five years	37	14
One year to three years	71	46
Less than one year	134	20
Unemployed or no answer	69	29
	N=378	N=147

earnings were $1.37, slightly higher than that of the riot participants (291 responding). (See Table 6.) Among 182 respondents in the census tract sample, 66 recalled having earned no more than $1.00 an hour in their immediately preceding position, 64,

$1.01-$1.50; 20, $1.50-$2.00; and 25, $2.00-$3.00. Only five recalled earning over $3.00 an hour. Forty-nine respondents in the

TABLE 6

Earnings on Present Job of Random Sample of Residents of Census Tract Where Rioters Lived, Compared with Rioters

Census Tract Residents		Rioters
$3.00 per hour	26	24
$1.50 - 2.99 per hour	122	55
$0.50 - 1.49 per hour	143	39
Unemployed or no response	158	29
	N=291	N=118

larger sample admitted being discharged from their preceding employment, 43 left because of low pay, 21 gave illness, and 57 said they wished to move to other locations. The informal job-finding network was also much in evidence. Among respondents in the broader sample, friends, relatives, and neighbors, rather than employment agencies or replies to want ads, accounted for over four-fifths of the jobs currently held or immediately previously held. In all, the riot participants and the rank-and-file of ghetto residents closely resemble each other in education, earnings, job histories, and length of residence in Omaha.

Comparing this data with that for 75 families in a census tract in Omaha having like socioeconomic characteristics, but a more evely divided racial composition, brings forth more specific contrasts. These accent more clearly the Negro disadvantages vis-a-vis Caucasians, even where the job skills are comparable. Among 31 families which were Negro, 8 were not employed at the time they were interviewed; the remainder (principal wage earners) were working in selected lines of their current jobs nearly 19 months and earned an average of $1.41 an hour. For 44 Caucasian families whose heads were comparably employed, however, job tenure averaged 30 months, hourly earnings approached a mean of $1.94, and only two families were lacking employment. Other disparities (in education, etc.) could be noted, but the point nevertheless stands out, in terms of the racial variable—the Negro consistently has fared less well even in identical occupations than the Caucasian. Interestingly, Caucasians acquired their jobs through the state employment service, the Negroes overwhelmingly by the informal network. Clearly, when these data are compared with that in Tables 5 and 6, the disadvantages of slum residency augmented by racial miniority status are starkly dramatized.

INTERPRETING THE FINDINGS

From this evidence it is certain that the persons apprehended as riot participants have much company, socially and economically speaking. Their condition and behavior, including the potential to beget civil disturbances, were not peripheral considerations involving only several dozen individuals of riffraff caliber. Nor were they the peculiar attributes of newcomers to Omaha. Among the rioters, 114 had lived in Omaha more than a decade; among the larger sample, 336 resided there 9 years or more.

Subscription to the riffraff theory has probably become the most fashionable means of emplaining the extensive urban riots of recent summers. Its starting premise is to regard participants as a fringe group actuated by conditions of poverty more than race. Often this fringe group is alleged to consist of a large number of recent arrivals in a city. Poverty, by traditional American thinking (upper and middle-class) can be altered, while race quite obviously cannot. Thus, by ascribing the propensity to riot to the personal disabilities of the participants, and viewing the looting, violence, and arson as simply senseless acts (of newcomers, conceivably) rather than expressions of legitimate grievances, the true meaning of these distrubances tends to be played down and their real causes overlooked. Future disturbances of this kind can be prevented, according to such thinking, by merely improving the lot of the riffraff, and disregarding or not perceiving the need for radical changes in Negro ghettoes, thereby supporting the continuation of the status quo of the Caucasian majority.

From the foregoing commentary, one should not conclude that identifying the matter of employment as the key consideration in explaining Omaha's riots lends credence to the riffraff theory. In two respects it falters as an explanatory mechanism: (1) it fails to account for the lack of motivation to riot on the part of the vast majority of Negro ghetto residents who have the same educational, social, and employment deficiencies as the actual rioters; and (2) it disregards the length of time that deprivations were endured by the riot participants in Omaha, all of whom are long-time residents, not very recent arrivals, in this city. Its discounting of the factor of race has already been commented upon, unrealistic as that perception is in the case of Omaha's experiences and in those of other convulsed municipalities. A more realistic explanation of Omaha's disturbances has to be found in the matter of the job needs and expectations of Negro ghetto residents as affected by the handicaps of race, the breakdown in the customary restraints against rioting, an uncertain leadership, and the informal job-finding network.

SHARPENING THE FINDINGS

The handicap of non-Caucasian racial status in the United States scarcely requires much elaboration relative to its effect on securing good-paying, long-lasting jobs. As one writer has phrased the matter, "It's hell to be black, poor, and ghettoed in rich, white America. . . ."[1] Special emphasis should be focused, however, upon the particular features of the ghetto which bear upon riot propensities. The factor of numbers is one consideration, particularly as it bears upon the question of the breadth of support among slum residents for rioting. While fewer than eight-score persons were actually apprehended by the police in Omaha as demonstrable participants in the rioting (i.e., looters, assaulters, arsonists), both police and newspaper reporters' estimates of the sizes of crowds roaming the streets during the height of the disturbances in successive eveneings ran much higher. By and large, they agree that these crowds contained from three to four thousand persons each evening. How much support and identity did these larger throngs lend to the riot participants? A precise answer is not possible in any detail. The random sampling of the residents in the census tract in which the rioters and the scenes of the riots per se are located has brought unmistakable evidence of a kindred sympathy on the part of these individuals with some of the things the rioters did, and the justification for such behavior. The usual disclaimers against supporting *all* that the rioters did or stood for were, of course, much in evidence.

But on the question of whether either the riot participants themselves or their sympathizing neighbors and friends perceived the riots as an instrument of mass political action, through which large numbers of Negroes would seek fundamental changes in social and economic relationships between classes, one finds virtually no basis for an affirmative answer. The objective of such action would have been to create a new majority coalition capable of exercising political power in the interest of new social policies. Clearly, this kind of a coalition would have to be interracial. Neither in Omaha nor in any other American city, in all probability, do Negroes by themselves have the power to bring about such a social revolution. Their numbers, however, would allow them to participate (in this city and elsewhere) in such a coalition as a powerful and stimulating force. The political direction such a coalition would take is highly uncertain. Of central importance to this discussion is the nature of the rioters and their census tract supporters. As preceding data have indicated, they constitute a slum proletariat which is distinct from the working class. This observation is important. The working class is employed, has a relation to the production of goods and services, and is unionized, cohesive, and possesses stability. The proletariat, by contrast, lacks cohesiveness, stability, and organization, not to mention a sense of having no stake in society. Within this context, the proneness to riot needs to be seen, as so many commentators about this grievous phenomena have seen it, as an act of ultimate futility and desperation, not an ideologically actuated cataclysm.

WHY THE RIOTS?

Nevertheless, this ascription of motives for rioting needs to be sharpened. For one thing, it does not adequately explain why traditional restraints against rioting have broken down in a number of cities where Negroes are concerned, but not where other components of the slum proletariat are concerned. e.g., Puerto Ricans, Mexicans. Admittedly, Omaha has no appreciable numbers of non-Negro slum dwellers, except a minuscule but inert American Indian population. Still, the question of why Negroes in particular riot has merit. For another thing, this ascription of motives for rioting does not explain why Negroes, unlike other ethnic minorities preceding them, chose to confront the public authorities directly, disregarding cardinal restraints upon civil rioting in the United States. The restraints are numerous, however imperfectly understood: fear of arrest and possible conviction and imprisonment; concern for personal safety; and belief in the superiority of legitimate channels for effecting desired changes. In this sense, the Negroes' rioting in Omaha—despite the grave dangers inhering in a situation in which all available police and national guardsmen acted under instructions to fire when fired upon and to take all steps necessary, short of indiscriminate slaughter, to subdue the rioters—reflects something of the crushing and virtually unendurable weight of their greivances. It reflects as well the weakening of the restraints inhibiting the tendency to riot. Indeed, these distinct considerations of actuation to violence and eroded constraint complement each other. Both are indispensable conditions for spawning civil violence.

The use of the Srole test measuring alienation, as applied to rioters and their census tract neighbors, produced a scoring of maximal alienation for nearly one-half of those interviewed in both groupings, with another quarter of each grouping in the next immediate bracket.

Fear of arrest did not restrain the riot participants because virtually all of them had previous arrest records. At least 100 individuals apprehended by Omaha police because of their involvement in the 1966 rioting had previously been arrested. Approximately one-third of these individuals were known to the Omaha police as repeated law violators with lengthy records of arrest. (In interviews, a number of the rioters theretofore lacking arrest records indicated that they wholly expected to attain such dubious distinctions sooner or later.) It was altogether consistent with their families' or friends' prior experiences. Apprehension about personal safety similarily exerted no restraint upon riot participants, because they had long been hardened, probably to a point of indifference, by the unrelenting assaults upon life and limb in the Negro ghetto. And as for placing faith in the processes of orderly social change, the rioters readily revealed, when interviewed, that such trust was unwarranted in their eyes.

As in Watts and a number of other riot locations, Negro arrests in the wake of preventive police patrolling in the Omaha ghetto traditionally have run high. Resentment of this practice easily boiled over into destructive civil disobedience when the suspicion and harassment which Negroes have endured so long in the name of such surveillance seemingly found no other outlet. Yet in interviews, the rioters, while acknowledging harassment from the police, remained consistent by a high margin in subordinating this grievance to the one surrounding lack of jobs.

Whether this is really a correct assessment, however, may be subject to some doubt. On March 4, 1968, George Wallace appeared in Omaha to boost his presidential candidacy. He made a speech and organized a drive for signatures in order to qualify for a place on the Nebraska ballot. His speech in the city auditorium, widely publicized in advance, touched off a considerable commotion which was instigated by an impromptu group of civil rights activists and indignant citizens. The demonstrations of this group occurred mainly in the auditorium. They were prone to much boisterousness and placard waving. After a few firm warnings to cease their disturbances, the police acted vigorously to eject a large number of the demonstrators. Although order was restored inside the auditorium, the zealousness of the police evoked considerable indignation and reaction throughout the Negro ghetto. Roving gangs of Negro youth appeared in that area of the city. Store windows were broken. There was some looting. Several fires were started and automobiles were overturned. These acts heightened the agitation from property owners and businessmen, Negro as well as Caucasian, that the police intensify their preventive patrolling. Off-duty policement were hired to guard various properties.

One such policeman shot to death a Negro juvenile allegedly breaking into a store in the ghetto. Tensions heightened further as a consequence, absenteeism in adjacent public schools skyrocketed as parents (Negro and Caucasian, but primarily Caucasian) kept children at home to avert incidents, and police patrols were enlarged to control roving gangs of Negro and Caucasian youths. A self-constituted committee of Negroes complained to the mayor and the city council about these heavyhanded actions of the police, as they saw things. The mayor took no action on the complaint; the city council complimented the police on their effective actions both in the auditorium and afterward ("for preventing things from getting out of hand"). The policeman who shot the Negro was exonerated after a departmental investigation.

And there matters have come precariously to rest. By the start of the following week, school enrollments had returned to normal and the police patrolling of the ghetto was somewhat scaled down. The resentment of Negro residents of the city (mostly living in the ghetto) remains largely unassauged, however.

This episode suggests the obvious absence of rapport between ghetto residents and the Omaha police department, and between them and the official centers of authority in the city government. It reflects, too, the likely exacerbation of a sensitive nerve among ghetto Negroes which, while not badly inflamed as a result of earlier police contacts, has been rendered increasingly flammable as a result of such contacts during and since the 1966 riot.

RIOTS, RACE, AND JOBS

Of the approximately one-fifth of the rioters, mostly teen-agers, who were unemployed at the time of their apprehension, half (24 of 49) indicated that they had not bothered to seek jobs during the preceding two months. Fifteen had not bothered to seek jobs during the preceding six months.

These individuals in effect had dropped out of the labor market. Further seeking was useless, evidently. Interestingly, 112 of the rioters felt that it was largely futile to seek better-paying jobs, and indicated no inclination to do so, primarily because they did not feel they could secure them even though acknowledging that such jobs existed. Of the larger census tract sample, 175 indicated that they were seeking better-paying jobs, or had made that effort in the preceding few months, while 215 were not inclined toward that end. (See Table 7.) In the larger sample, 235 were employed full-time at the time of the rioting, whereas 206 were not. Of this latter figure, nearly one-third neither had work nor sought any. Two hundred and thirteen felt that better-paying jobs were available in Omaha, although 141 did not. Approxi-mately one-quarter of those not seeking better-paying jobs also believed such jobs were not available. Overwhelmingly, both rioters and their census-tract neighbors indicated their willingness to move to other locations outside Omaha if *jobs or better-paying jobs* were available. For both groups of interviewees, it scarcely needs mentioning, the mean earnings cited previously placed most of them in the poverty category, viz., $60 a week or less. As one

TABLE 7

**Attitudes of Rioters and of Random Sample of Residents
in Same Census Tract Where Rioters Lived
Regarding Better Job Prospects in Omaha**

	Rioters		Census Tract Residents	
	Yes	No	Yes	No
Believed better jobs available	112	35*	213	141*
Seeking better-paying jobs	35	112	175	215

*negative opinion or no opinion

writer has noted, "It is only in the slums, though not only in. . .Negro slums, that preventative patrolling is practiced so intensively, illegitimate enterprises operate so openly, and minority groups are frustrated so frequently in their quest for a better life."[2] Whereas the overall rate of unemployment in Omaha in 1966 was calculated at 3%, that for the ghetto was set at 25% and possibly higher.

In Omaha, as in other cities, the gap with respect to joblessness between Negro and Caucasian males over twenty years of age has narrowed somewhat since 1961. For male teen-agers of each race, however, the gap has widened. Caucasian unemployment among male teen-agers (ages 16-19) in Omaha stands at about 9%, according to the best-informed state labor department sources (as of mid-1967), whereas Negro male unemployment in the same teen-age brackets has been estimated to be 35%. This latter figure was found not to have been affected by the generally downward trend in the metropolitan area's (and the nation's) unemployment rate last year. Regarding the very obvious question of whether Negro teen-agers will work, one of their more plausible apologists, Bayard Rustin, has written:

Nor is there any evidence that Negro teen-agers do not want to work. Whenever job programs have been announced, they have turned out in large numbers, only to find that the jobs weren't there. In Oakland, a "Job Fair" attracted 15,000 people; only 250 were placed. In Philadelphia, 6,000 were on a waiting list for a training program. What Negro teen-agers are not inclined to accept are dead-end jobs that pay little and promise no advancement or training. Many would prefer to live by their wits as hustlers or petty racketeers, their version of the self-employed businessman or salesman. That their pursuit of this distorted entrepreneurial ideal only mires them deeper in the slum proletariat is not the point. They want to be part of the white collar organization man's world,. . .not trapped behind brooks and pushcarts.[3]

For the most part, these observations hold true with respect to Omaha's experiences vis-a-vis the Negro teen-age jobless. When, for example, an emergency employment office was opened in the Omaha ghetto immediately following the rioting in July, 100 teen-age Negroes registered for jobs the first day, and 55 were employed. Within a week both figures climbed, the latter to 105, of whom 28 were under 16 years of age. Lacking adequate training, however, 165 registrants could not be given continuing, remunerative employment.

RACE, JOBS, AND COMMUNITY POWER

One must bear in mind that the bifurcated character of the way of life of Negroes in Omaha's ghetto, involving persons in the working class and in the slum proletariat, has created tensions and myriad difficulties in relationships between the members of each group. By and large, those in the former category have reached their present, more affluent condition by movement upward from the latter category—by "passage through the Golden Door," as one writer has expressed this development.[4] The principal sources of tension and difficulty in relations between the two categories, in this as in other cities, emanate from the facts that (1) movement through the Golden Door is uncertain and precarious, and (2) those who pass through ordinarily show little interest in, and in turn are little importuned by, those who are left behind, to assist in facilitating like passage for them. "It can be particularly painful," one commentator has observed, "if, just when a man thinks he's moving toward the Golden Door, the door begins to close."[5]

In Omaha this happens repeatedly. Yet the door is open far enough, at least to the prepared Negro, and sometimes to some half-prepared ones, so that those who do not get through can find no solace in saying, "I can't make it because I'm black." They have to face the pain of being black.

Accepting for this discussing the fact that the grievances and expectations of young Negroes are immediate and urgent, and that they will not wait for their grandchildren to enjoy the solutions of their problems, a question arises—how have opportunities for passage through the Golden Door been contrived and effected in Omaha? (The standard for measuring successful passage in the context of this narrative, it might be noted, has been the securing of employment by Negro aspirants which carried with it prospects both for steady or rising earnings and a lengthy incumbency.)

A number of ways may be identified toward this end in this city. One has been selection and placement of individuals (for present discussion purposes, Negroes) under prominent civil rights groups' auspices in industrial training programs run by business organizations. Another has been the hiring of Negroes directly by Caucasians for training and eventual employment in the latter's firms. A third method has been the funding of training programs by civil rights groups or charity organizations, with attempted placement later of trainees in private enterprise. Still another approach involves government sponsorship or funding of training opportunities, with trainees placed in private employment in due course. And, finally, as a variation on the foregoing, government-sponsored or -funded training has been undertaken in conjunction with subsequent public employment for trainees. Common to all of these approaches are three important debilities, viz., actual opportunities for training and subsequent job placement are far more meager than prevailing needs; the foci of training are not always synchronized well with the skills of likely trainees of the requirements of the marketplace; and the sustenance and the supplying of trainees for such programs have always depended upon, respectively, the generosity of the better-financed business and the willingness of old-line, respected charity or civil rights groups to sponsor matriculants.

This last consideration is of particular importance. Its functioning in fact signifies a meaningful, long-standing, and indispensable relationship which obtains in Omaha between elements of the

white power structure and the leadership of the Negro community. This latter leadership, it should be noted, traditionally has
embraced a loose coalition of the better educated merchants,
professional persons, clergy, and sports/theatrical celebrities,
whose only claims to such preeminence among members of their
race were simply that they were vocal, visible, better endowed
educationally and financially, and—most important—recognized by
the white power structure as the plausible spokesmen to and from
their race in its dealings with the Caucasians. This Negro leadership
in Omaha already had passed through the Golden Door. So far as
most of the slum proletariat is concerned, these individuals are
part of the white power structure. For them to presume to speak
for the proletariat is simply folly.

COMMUNITY POWER AND THE SLUM NEGRO'S
JOB TRAINING

An analysis of the ways in which the aforementioned training
activities were set up and run affords some useful insights into the
relationships which have obtained between the Caucasian and
Negro power structures. Prior to the riots of 1966, there had been
three manpower training programs in Omaha which were con-
ceived by white-owned or -managed firms. More specifically, a
committee of the local chamber of commerce prevailed upon a
number of industrial establishments to undertake certain kinds
of training for Negroes, who experienced mounting joblessness in
the early and mid-1960's. In each program, the types of skill-
training to be offered were led to the firms doing the training.
Negro civil rights groups in the city endeavored with some
persistence to keep the support of white business leaders, some on
the chamber committee, from flagging. But prominent Negro businessmen and a few professionals partook of these plans and
programs in only two ways, viz., in being asked to serve on a
committee designed to publicize each training venture prior to its
launching; and, in one instance, in being asked to serve on a
committee seeking (successfully) federal funds to defray some of
the training costs. Nominations of trainees were dependent upon
initiation and clearance within a committee of Negro influentials.
This procedure was acquiesced in as a selection method by the

white-run firms doing the training largely in deference to the Negro emphasis of the individual undertakings. The number of individuals trained altogether did not exceed 50; the number actually receiving employment on a steady basis did not exceed one-third that number.

One fact seems abundantly clear. From private conversations with some of the officers in the civil rights group which helped make this kind of training program a reality, and with individuals known to have been on the sponsoring chamber of commerce committee, it is clear that many business leaders, including those in the firms handling the training, desired no greater participatory voice for the Negroes. These business leaders were doing the Negroes a favor, and discharging a commendable civic duty. It was their time, equipment, and resources which were being applied to the task. What more was there for the Negro to do but be grateful and aspiring?

Much the same story unfolds with respect to most of the other types of training programs. Where Negroes were hired for training and eventual placement by white-owned or -managed firms directly, without the intercessory participation of civil rights groups or Negro selection committees, this was particularly true. Where training programs have been run by government agencies, notably as part of the renewed interest in vocational education in the md-1960's and of the War on Poverty, the access of interested civil rights groups to the individuals shaping curriculum and instruction has been greater than in the other contexts noted. Nevertheless, the selection of trainees, the acquisition and use of certain facilities for training, and the placement of trainees still have depended primarily upon the continued indulgence of the community's white power structure.

COMMUNITY POWER AND RIOT RESPONSES

The foregoing commentary is simply a detailed exposition of the fact that the influence wielded by the visible and recognized Negro power structure in its relations with the white power structure—or, more correctly, that portion of it comprised by economic dominants—in the area of job training and job attainment for Negroes has been negligible. And what influence that Negro power structure does have is, in these matters, almost exclusively what the white power structure wishes to accord to it. The point is important because, not surprisingly, it was during the period when the riots were occurring in Omaha that the many prominent Negroes in that power structure who strove to restrain the participants quickly perceived that they could not make effective contributions toward that end. All that the Negro power structure could do in relation to the riots was to denounce their causes and caution against a weakening of the civil rights movement generally because of the civil excesses.

For the white power structure, however, the alternatives in coping with the riots were neither so limited nor innocuous. Very shortly after the early July disturbances a committee of eight prominent Negroes, all possessing non-controversial credentials in the Negro power structure of Omaha, was appointed at the behest of the mayor and the chamber of commerce and other white influentials to assay the causes and implications of the riots. To this committee the riots were attributable to insufficient job and training opportunities, and to frustrated, hot-headed youths who became overzealous on a hot night. Then it observed:

> The committee does not accept as fact the claims that Omaha's city officials have failed to face the problems of the Negro community fairly. Rather the committee members believe that the leadership of the Negro community is at equal fault for not clearly and concisely outlining a program for improving bad conditions wherever they are found.[6]

This commitment had the effect of exonerating the white power structure for any of its omissions, while placing most of the blame on the Negroes themselves—a useful, if not deliberate, obeisance to the whites' conception of their own responsibility in this matter.

LIMITATIONS ON THE EFFICACY OF COMMUNITY POWER

Attempting to explain the riots in these terms points up a number of interesting considerations. For one thing, it reveals the intractable inclination of the white power structure to continue thinking of the Negro power structure as an important force with which to assuage rioters' grievances. For another, it reveals the tactical inclinatin of the white power structure to keep some freedom of movement for itself, apart from reliance just upon the Negro power structure, to effect a satisfactory solution to the causes of the riots, as understood by the whites. It was from this perspective that, within a few days after the cessation of the July disturbances, influential members of the business community (and of the community power structure) prevailed upon some of their members who owned, or who were known to have access to those who did own, certain pieces of real estate in the riot area to lease them to the city for recreational purposes. When these pieces of land were in due course turned over to the city, the mayor succeeded in having the appropriate municipal department quickly outfit them for recreational uses.

Simultaneously, the business influentials responded favorably to a suggestion arising from the mayor that some of their members comprise a committee to hear suggestions (and complaints) from Omaha citizens generally about what steps could be taken under both public and private auspices to overcome some of the causes of the riots. Officially constituted with 50 members, this committee held only a few hearings. These brought forth scarcely a corporal's guard of interested commenators, whose ideas for action were woefully limited in scope and utility relative to getting at root issues (e.g., keeping playgrounds open longer hours was the favorite recommendation).

A more constructive step, responding to the impact of the riots, was the leadership taken by the white power structure to facilitate the establishment of an application office for aspiring ghetto workers in the riot area. Such an office was opened by the Omaha branch of the state labor department a few days after the disturbances ended in July. Its accomplishments, initially satisfactory, have already been detailed. What jobs were acquired, however, were as a rule only for the remainder of the summer of 1966. Significantly, all but 9 of the job opportunities lay in the public sector. In these facilitative roles, the white power structure has acted as a loosely-knit instrument reflecting a somewhat narrow consensus among business influentials, the mayor (himself long a businessman), and prominent members of the city council, and a few labor leaders. The state labor commissioner's office has been privy to the sentiments and ideas for action favored by this unstructured amalgam, but, through its state employment offices, has followed rather than led in effecting the resultant group decisions.

Steadfastly, one may note, the white power structure has declined to deal with the indigenous leadership of the ghetto Negroes, in part because some uncertainty lingers as to such a group's claims to authenticity and in part because direct dealings mentally offend that power structure's conceptions of who should speak for, and be rewarded in, the ghetto. Thus when, close by the July riots, these kinds of individuals requested opportunities to meet with the mayor and business leaders to discuss job needs and keeping peace in the ghetto, they were—except for an initial courtesy meeting—spurned.

In all, the various constructive (as contrasted with punitive) steps which have been taken in Omaha since the 1966 riots to forestall recurrences of trouble have moved through the hands of, and largely been subject to the pleasure of, the city hall-chamber of commerce leadership group, i.e., the white power structure of the municipality. In the summer of 1967, another and larger job-finding effort was undertaken for ghetto youth aged 16-19 by this source. Designated Project YES, it found employment for 3200 youths for six-to eight-week periods.

A THEORY OF RIOTING-CONSCIOUS DISPARITIES IN POWER RELATIONSHIPS

How can one assess the Omaha riots, then? At bottom, they were an exercise in violence induced by a conscious disparity in power relationships. Race has been the conspicuously identifiable variable in this disparity, joblessness the principal symptom. Reflecting deep-seated disappointments, futility, and impatience, these disturbances were fed by long-standing causes that required little provocation to ignite them. Steady and better jobs were their aim. They were neither the results of a conscious design by participants to realign classes so as to build a new urban power base for long-run remedial programs, nor of agitation by hot-headed rabble-rousers.

Re-interviews with 74 rioters during the summer of 1967 indicate that the aim of better, more sustained employment is still a long way off. One-fourth of those interviewed had no summer job this year or last, or had worked no more than thirty days either summer. One-half of those interviewed said they felt things were a bit better than last year. They presently had jobs which had run or were expected to run for three months or more. Eleven interviewees were working on the same job six months or longer. It is clear from the experiences in Omaha that, to effect the kind of sustained improvement in the root cause of the riot that its scope and depth portray, a new generation of Negro leaders must be created in whom the young Negroes will place their trust and with whom the white establishment can live and work. Clearly, the power of the white community is so superior that the security of the Negroes lies in the determination of the Caucasians not to let the conflict go to extreme limits. The question then becomes how and, indeed, whether the white power structure can be induced to pay the costs, financial and otherwise, of the reform and reconstruction which can assuage Negro grievances.

From Omaha's experiences, it would appear that as long as the advance of the Negro is presented as a form of white philanthropy, nothing on the scale needed will be practical politically. Walter Lippmann has observed that the uplifting of the Negro cannot be accomplished as a pro-Negro enterprise, for large communities of men are not that generous and unselfish. Rather, the advance of the Negro will seemingly have to be part of a much greater, more general effort to uplift the whole community, carrying the Negro portion thereof with it. This would mean creating, in effect, something like the Great Society.[7] It means, too, absorbing Negro grievances into the greater needs of the whole community. Such an undertaking requires an overwhelming desire and intention on the part of the white power structure to reform and rebuild its own social order.[8]

The hope of the Negro is to participate in such a movement. Considering the scale needed for such an undertaking, one may note, opportunities for efficacy on so broad a front occur only now and then in the life of a city—any city. Omaha's experiences indicate one or two promising leads which can be taken into this deeper, broader venture, notably the mastery and reversal of the informal job-finding network. Its transformation can redeem some of the alienation traditional in job matters with ghetto Negroes. It is less ephemeral and unwieldy, for example, than waiting out educational voids or the recasting of the Negro male's image of himself by other means. Preliminary evidence amassed in this study has indicated, for example, that job tenure and earnings run higher for positions acquired by Negro slum-dwellers through the state employment service than for positions acquired by them informally. Such evidence merits the fullest exploitation.

"So the mine is dark," Morgan Evans observes to Miss Moffatt in *The Corn Is Green.* "But when I walk through the shaft, in the dark, I can touch with my hands the leaves on the trees...." American cities will be coping with future riots, in part by walking in the dark, but in part by reaching out for various answers that seem to promise solutions to their difficulties, too. One city's forte may be another's foil; still, all that can be done is to put promising approaches to the test. The stakes are too high not to try; the prospects for eventual success in thwarting or containing riots cannot be maximized by any other attitude.

NOTES

1. *New York Times Magazine,* September 3, 1967, p. 42.
2. Robert M. Fogelson, "White on Black: A Critique of the McCone Commission Report on the Los Angeles Riots," *Polit. Sci., Q.,* LXXXII (Sept., 1967), p. 361. Apart from inexperience and a paucity of numbers, it may be noted, the police performance in Omaha has not, during the actual riots, been the subject of unseemly comment in the press or among white influentials or non-rioting ghetto residents. Bitter disagreement between the police chief and the director of public safety over police tactics during the riots, however, led to the latter's resignation. The preventive policing in the ghetto was widely denounced, however, before and since the riots. That it contributed to the riots is not in dispute. The issue did not involve police effectiveness.
3. Bayard Rustin in *New York Times Magazine,* Aug. 13, 1967, p. 60.
4. The phrase is J. Anthony Lucas' *New York Times Magazine,* Aug. 13, 1967, p. 60.
5. *Ibid.,* p. 43.
6. *Omaha World-Herald,* July 7, 1966, p. 2.
7. As quoted in *Omaha Sun,* Sept. 14, 1967, p. 24.
8. Implicit in this statement, of course, is the consideration that the Negro, in Omaha or elsewhere in the United States, has no future as a member of a separate minority living apart from whites. Separatism in its various forms, as advocated by Black Muslims, Nationalists, etc., was not found to be widely espoused or accepted among Omaha rioters. Rather, advocacy of such extremism in this city, one commentator has observed, is really an after-the-fact justification utilized by persons looking for a constituency.

GHETTO-AREA RESIDENCE, POLITICAL ALIENATION, AND RIOT ORIENTATION

Jay Schulman

■ Urban sociologists, following Louis Wirth,[1] have had an unfortunate tendency to speak and write of the black ghetto as if it were one indivisible entity, an ideal type into which all blacks located in a central city area could be neatly fitted. It is hardly a surprise to blacks who live in ghettos that blacks, like other people, constitute different social groupings and that these different social groups tend to congregate in different areas within the ghetto.[2]

The purpose of this paper is to consider whether residence in "better-off" and "worse-off" ghetto areas makes any appreciable difference in the extent of political alienation among different categories of black people *and,* in their attitudes toward urban riots. A related concern of mine is to show that neither the riffraff theory of riots[3] nor the relative deprivation theory[4] are sufficient explanations of urban disorders.

I will try to show—in a tentative way to be sure, for the data are crude and the numbers are small[5] —that people who have little to lose and people who feel relatively deprived hold attitudes that are on the one hand conducive to riot participation and on the other hand antithetical to riot participation. These findings lead me to speculate that the eruption of riots in American cities is due to a multiplicity of factors that are not adequately expressed or comprehended in any of the riot theories now circulating.

DATA COLLECTION

The data for this paper were gathered as one part of a larger investigation of how different groups in Rochester, New York, have reacted to the riot of 1964.[6] Two hundred seventy-one blacks were interviewed by three black students who themselves had grown up in the ghetto and were self-reported militants. The interviews were conducted in bars, dives, pool parlors, and on the streets—wherever blacks between 14 and 30 tended to congregate. The population is therefore not a sample, but rather a population of 271 black ghetto-dwellers.[7] Twelve schedules were unusable, either because the respondent terminated the interview with a rhetorical comment like "You working for Whitey, ain't you?" or "It's for you to ask and me to know," or because the interviewers were unable to or neglected to obtain answers to important background questions.[8] Forty-five respondents turned out to live in transitional areas, outside of the two main Rochester ghettos. Thus the total population that is worked with in the body of this paper consists of 214 persons, of whom 100 lived in the better-off ghetto and 114 lived in the worse-off ghetto.

DEPENDENT VARIABLES AND THEIR MEASURES

The interviews were conducted in early June, 1967. At that time, FIGHT,[9] an Alinsky-initiated black community organization, was engaged in a protracted and bitter struggle with the Eastman Kodak Company over recognition of FIGHT as the bargaining agent for the Negro poor in Rochester. One question asked of the respondents was whether they felt rioting was preferable to the type of pressure FIGHT was trying to exert upon Kodak. Answers that stated a preference for throwing bricks or "shooting Whitey" were taken as an indication of a *prima facie* alienation from rational politics for the following reasons: I considered FIGHT's conflict with Kodak to be a limiting case of a black power operation conducted within the bounds of rational politics.

Therefore, I viewed rejection of a black power type of instrumental politics in favor of street rioting as probably indicative of a rejection of the framework of instrumental politics as such.

It would be nice to have data from an unbiased population, i.e., a population that is not in prison for having participated in a riot, no reported riot participation and/or willingness to participate in future riots. There are serious problems in collecting such data, and equally serious problems in using such data if they have been collected. One of my Negro ex-interviewers who is often mistaken for Roosevelt Brown, the professional football player, and is now a leading black power spokesman, was forced to leave a home at gunpoint for asking far more innocuous questions than were asked in this survey. Several years of observation of ghetto dynamics have persuaded me that most black ghetto-dwellers are prone to lie in one of two directions when they are asked about riot participation: either they say they were participants when they were not; or they say they were not participants when, in fact, they were.

Questions were used to establish measures of the extent to which respondents felt riots to be legitimate instruments of protest and/or of accelerating opportunity. One question asked if the 1964 riot helped the effort of Rochester Negroes to get ahead. A second question asked about the chances of another riot in Rochester in the summer of 1967. A third question asked, if there is a riot in Rochester in the summer of 1967, will it help or hurt? Answers that the 1964 riot had helped and that the hypothesized 1967 riot would help were taken as indirect measures of riot predisposition or orientation.

ROCHESTER SUB-GHETTOS

In Rochester both blacks and whites alike speak of two different black ghettos. One ghetto is described as the "better-off" or "not-so-bad" ghetto, while the other ghetto is described as the "worse-off" or "bad" ghetto. The 1964 Rochester riot, which compares to the Watts riot of 1965 and the 1967 Newark and Detroit riots in seriousness when number of arrests and property damage are the enumerators and number of "riot-eligible"[10] Negroes is the denominator, began in the worse-off ghetto and then rapidly spread to the better-off ghetto. Exactly the same pattern was repeated during the so-called Rochester riot of 1967.[11]

In the everyday speech and thought of Rochesterians, the better-off ghetto is more or less associated with the third ward, an area that was one of the fine residential districts of the city[12] and is but a mile or two from the heart of downtown. The worse-off ghetto is similarly associated with the seventh ward, an area that is considered blighted, the hub of prostitution and the rackets, the haven of corrupt Democratic Party machine politics, and the landing place for Negroes migrating to Rochester directly from the South or coming off the migrant labor stream. The "better-class" Negroes thus are believed to be concentrated in the better-off ward, and the "undesirable" Negroes, or "dregs," are thought to be concentrated in the seventh ward.

As Table 1 shows, these verbal descriptions correspond generally with certain aspects of Negro ghetto realities as these realities are depicted in the 1960 Census statistics for those tracts in which the great majority of Rochester Negroes live.[13] The third ward (better-off) black tracts show a substantially lower median rate of male unemployment for nonwhites than do the seventh ward (worse-off) Negro tracts. The third ward black tracts contain significantly more males and females of relatively high occupational status than do the seventh ward tracts. Family income and income of individuals is higher in the third ward tracts, as is median years of education for persons 25 and older; more third ward Negroes own their homes and own at least one automobile. Finally, more third ward single- and multiple-unit houses are sound structures, presumably with a greater aesthetic appeal. Obviously, the third ward black ghetto, for all of its deficiences, is the area in which relatively successful blacks prefer to live.

Table 1 also points up how common-sense perceptions, even when they are in touch with some aspects of reality, miss or gloss over other aspects of reality at least as important and possibly more important. Thus, however important the relative differences in socioeconomic status between Negroes living in the two ghettos, the central fact not to be forgotten is that both ghetto populations are substantially deprived when compared with whites living in the same areas of the city, not to mention comparisons to almost all-white areas outside of the central city and in the suburbs. Table 1 also shows that the overwhelming majority of Negroes in both

TABLE 1

Selected Economic and Social Characteristics for
Third Ward "Black"[a] Tracts and Seventh Ward "Black"
Census Tracts Aggregated,[b] 1960

1960 Characteristics	Third Ward Tracts	%	Seventh Ward Tracts	%
% unemployed, nonwhite males	219 / 1792	12.2	285 / 1399	20.4
% high occupational status,[c] nonwhite males	453 / 1573	28.8	192 / 1114	17.2
% high occupational status, nonwhite females	183 / 1149	15.9	69 / 730	9.5
Median income, nonwhite families	1,577	$4,659	1,286[d]	$3,616
Median school years completed, nonwhites	3,037	9.3	2,656	8.1
% home ownership, nonwhites	668 / 2691	24.8	192 / 1268	15.1
% sound housing units	2932[e] / 3968	73.9	930[e] / 2034	45.7

a. A "black" tract is an area in which blacks had become a majority of the population, i.e., had succeeded the whites; or had become a significant proportion of the total population, i.e., had invaded an area.
b. The third ward black tracts are 3, 4, 64, 65; the seventh ward black tracts are 11, 12, 13, 14, 15. **Source:** 1960 U.S. Bureau of Census Report on Monroe County and City of Rochester.
c. All occupations from semi-skilled through upper white collar and proprietors are considered high status occupations.
d. Does not include tract 11.
e. Only tracts which have a black majority are included in these calculations.

ghettos are of relatively low socioeconomic status, constituting a subjugated social class. Table 1 also confirms the everyday observation that there are low status Negroes living in the better-off ghetto and high status Negroes living in the worse-off ghetto.

Table 2 brings to light a subtle relative deprivation process that operates differently in the two sub-ghettos. There are significantly more white males of relatively high socioeconomic status who live in the third ward Negro tracts. Yet the white male unemployment rate is also higher in these tracts. Even if, as is likely, this anomaly is in part the result of the presence of a larger number of older white men in these tracts, more of the socially mobile third ward Negroes live in close proximity to a larger number of whites of equivalent or higher socioeconomic status who have chosen to or have been forced to remain in a ghetto area, and a large number of whites who can be considered occupational failures.

SOME HYPOTHESES

One proposition concerning the relationship between ghetto location and political alienation and riot orientation can be derived from the riffraff theory of riots. A converse prediction concerning these relationships can be derived from the relative deprivation theory of riots.

The riffraff theory suggests that there will be a greater degree of political alienation and riot orientation among seventh ward Negroes than among third ward Negroes.

The relative deprivation theory suggests that there will be a greater degree of political alienation and riot orientation among third ward Negroes than among seventh ward Negroes.

INTERPRETATION OF FINDINGS

The use of confidence-level measures to test the relative strength of an association is not warranted, since the respondents constitute a population and not a sample. The question that arises is whether findings from a population have any meaning. The obvious answer is that they have meaning for that population, but not necessarily for other populations, and certainly for the total universe of Rochester blacks. The findings thus have limited value until and unless they can be shown to correspond with findings from other populations and samples. A further question is how, in a population, findings are to be established. I shall assume, for the purposes of this analysis, that 15% or larger differences indicate a substantive association that is due to other than chance. In other words, I shall assume that differences of 15% or more point to real differences and are not due to artifacts; but, of course, this is an assumption and nothing more.

TABLE 2

Selected Social and Economic Characteristics
of Whites Living in Third Ward and Seventh
Ward Black Tracts,[a] 1960

Characteristics	Third Ward Tracts		Seventh Ward Tracts	
	(N)		(N)	
% unemployed, white males	$\frac{370}{2274}$	16.3	$\frac{112}{1996}$	5.6
% high occupational[b] status, white males	$\frac{1053}{2052}$	51.3	$\frac{822}{1884}$	43.6
% high occupational status, white females	$\frac{776}{1328}$	58.4	$\frac{395}{1150}$	34.3
Median years of education	5,248	8.5	3,164	8.4

a. Third ward tracts are 3, 4, 64, 65. Seventh ward tracts are 11, 12, 13, 14. **Source:** 1960 U.S. Bureau of Census Report on Monroe County and City of Rochester.
b. For purposes of this analysis, all occupations from semi-skilled through upper white collar and proprietors are considered high status occupations.

FINDINGS

Table 3 shows that ghetto area is *not* related to the measure of high alienation and is barely related to the measure of low alienation. In brief, conditions that give rise to high alienation are endemic to both ghettos, while there appear to be slightly stronger countervailing conditions or immunity factors at work in the better-off ghetto. However, there are substantially more residents of the poorer ghetto who would not or could not locate themselves in respect to the item purporting to measure political alienation.

Table 3 also shows the substantial extent of alienation from rational politics existing in both ghettos. These data make it abundantly clear that there is sufficient negative energy, even in the Rochester black ghettos, to trigger spontaneious acts of collective violence if precipitating incidents occur. Assuming that there are few black ghettos in cities of 100,000 population or greater that have less negative energy than the Rochester ghettos, it is not surprising that there have been very few American cities of this size that have not experienced collective violence in recent years.

TABLE 3

Percent of High and Low Alienated[a]
Blacks by Sub-ghetto Residence

	Third Ward		Seventh Ward	
	N	%	N	%
High	(43)	43	(43)	38
Low	(40)	40	(33)	29
Don't know	(17)	17	(38)	33
	N=100	100%	N=114	100%

a. The measure of alienation is a preference for riots in lieu of using political channels or working things out with white leaders. Low alienation is indicated by a preference for political channels or talking things out.

Although alienation from rational politics is almost equally distributed in the two Rochester ghettos, Table 4 shows that, with one exception, certain better-off ghetto population segments are substantially more alienated than their equivalents in the worse-off ghetto. Two distinct patterns are shown in Table 4. First, females, persons 26 years and older, and persons who take their religion seriouslv who live in the better-off ghetto, are more likely to be alienated than females, persons 26 or older, and religious persons living in the worse-off ghetto. Second, better-off ghetto residents of low occupational status and those who identify with the lower classes are more alientated, as are worse-off ghetto residents of high occupational status.

It is interesting to find that social integration, as indicated by attributing value to religion and by high occupational rank, is associated with minimal or low alienation in the better-off ghetto; while another indicator of social integration, marital status, is associated with low alienation in the worse-off ghetto. Another paradoxical finding is that young black adults living in the better-off ghetto are substantially more likely to show low alienation than are their age peers in the worse-off ghetto.

These findings fall into three logical groupings which are hardly implied by either the riffraff or relative deprivation theory of riots.

THE LABOR MARKET CONDITION

Young adults living in the poorer ghetto are more likely to be high on alienation, while young adults living in the better ghetto are more likely to be low on alienation. This difference in alienation may have to do with how young adults encounter the labor market as well as the type of labor market they encounter in different ghetto situations.

THE SOCIAL SOLIDARITY CONDITION

Some population segments that are socially integrated along specific dimensions and live in the better-off ghetto are simultaneously more alienated and less alienated than equivalent segments living in the worse-off ghetto. Yet at least one state of social integration is associated with less alienation and residence in the worse-off ghetto. These seeming paradoxes or contradictions underscore the Durkheimian contention that "social facts" have different meaning and weight in different social circumstances. They also point up the influence of interactional contexts in the formation and reinforcement of social attitudes.

TABLE 4

Positive Differences[a] in Alienation Between Population Segments
Living in Two Rochester Black Ghettos

Population Segments	High Alienation	
	Better-Off Ghetto	Worse-Off Ghetto
	%	%
Females	24	
26 and over	21	
Consider religion important	21	
High occupational status,[b]		19
Low occupational status	17	
Low social class identifiers[c]	16	

	Low Alienation	
	Better-Off Ghetto	Worse-Off Ghetto
	%	%
17-25	21	
Married		19
Consider religion important	21	
High occupational status	21	

a. Positive differences equal variations of 15% or greater between equivalent population segments residing in the two Rochester black ghettos or in transitional areas, on one measure or another. The denominator in all cases is the N for the particular population segment. The dividend is the frequency distribution within a particular population segment for a particular measure. The proportions expressed are a measure of the variation between identical population segments that are known to differ by ghetto residence on a particular variable.

b. High occupational rank includes those jobs from semi-skilled upwards in the occupational system; low occupational rank includes unskilled jobs, no jobs, and who hustle or live off the rackets.

c. Lower-class identifiers are those persons who say that they are members of the working or lower classes.

THE "RANK INCONSISTENCY" CONDITION[14]

People of low occupational rank living in the better ghetto are likely to be higher on alienation, and persons of high occupational rank who live in the poorer ghetto are also higher on alienation; but persons of high occupational rank who live in the better ghetto are lower on alienation. Living in the better-off ghetto but identifying with the "fellow at the bottom of the heap" is associated with a greater sense of alienation. In short, rank inconsistency, or living in a condition that is discrepant, is associated with a high state of alienation, while rank consistency, or living in a concurrent condition, is associated with a low state of alienation.

Tables 5 and 6 report three sets of consistent findings. One is that blacks living in the better-off ghetto are significantly more likely than blacks living in the worse-off ghetto to regard riots, past and hypothetical, as instrumental to the social changes necessary to improve the lot of black people in Rochester. These findings appear to support the hypothesis derived from the relative deprivation theory.

The second is that, though blacks living in the worse-off ghetto are a bit more opposed to riots than blacks living in the better-off ghetto, these are not substantial relationships. Opposition to riots as instruments of social change is more or less equally distributed across both ghettos. Thus it is crucial to ask what sorts of people in both ghettos favor or oppose the use of riots to force social change.

The third is that blacks living in the worse-off ghetto are substantially more likely than blacks living in the better-off ghetto to respond with "don't know" or "no comment" replies. Persons living in the poorer ghetto seem to have less information at thier disposal or are less motivated to reply to questions, even when posed in a non-threatening manner on their home grounds. This finding, which by the way remains when education is controlled for, suggests that the psychological capacity for instrumental or political leadership is found more often in the better-off ghetto.

Tables 5 and 6, if compared, reveal that in both ghettos there is an overall decrease in the tendency to view riots as instrumental. Fewer persons in both ghettos consider a second Rochester riot as desirable. This may be because more third ward residents have benefited from the programmatic reactions to the 1964 riot or

TABLE 5

Perception of 1964 Riot by Sub-ghetto Residence

1964 riot	Third Ward		Seventh Ward	
	N	%	N	%
Helped	(64)	64	(55)	48
Didn't help	(21)	21	(26)	23
Don't know	(15)	15	(33)	29
	N=100	100%	N=114	100%

TABLE 6

Perception of Hypothetical 1967 Riot
by Sub-ghetto Residence

"1967" riot	Third Ward		Seventh Ward	
	N	%	N	%
Helped	(52)	52	(42)	37
Hurt	(18)	18	(22)	19
Don't know	(30)	30	(50)	44
	N=100	100%	N=114	100%

because they are more concerned about the white "backlash," or it may be due to a combination of the two. This wariness may be a clue to why cities that have had one major riot are not likely to have another too soon after.

Tables 7 and 8 reveal that the observed association between area of residence and perception of riot utility holds firm when a number of important social attributes are considered. Third ward residents are more likely than seventh ward residents to see riots as useful in every instance in which there is a substantial difference between population segments of different ghetto residence.

When a significant proportion of the members of a third ward population category see *both* the 1964 riot and the hypothetical 1967 riot as useful, that population category can be considered to be riot-disposed. Such are youngsters 16 and under, single persons, persons who attach little or no importance to their religion, and persons whose class identification is with the working or lower social classes. Persons born in either the South or the North and persons of either high or low educational status are equally riot-prone.

What is most important about this social profile is that for the most part it is a profile of the wanton youngster without social ties, but who lives in the better-off ghetto. It may well be that the youngster with a "failure" self-image who also lacks social ties responds favorably to the idea of riot because he is able to recognize that he is at odds with or shut out from souces of opportunity in his own immediate ghetto environment.

Some may be surprised that third ward residents of high educational status are prone to favor riots and that this is true of persons born in both the South and the North. There are two likely explanations of why these people of high educational status favor riots: one is that they have come to a reasoned conclusion that riots are in fact helpful to the achievement of social change; or, as is more likely, their high educational status has not yet led to equivalent occupational status, resulting in an embitteredness that tends to lead them to favor riots as a way of retaliating against the "White Power Structure." That persons born in both the South and North living in the better-off ghetto favor riots may mean nothing more than that region of birth is not as important to how a black person views the world as, say, where he spent his formative years.

Riot disposition for any one category of people is subject to change if changes occur in their real-life situation. Although the changes that are reported in Tables 7 and 8 may be due to sociological vocabulary or methods of data collection, they are tantalizing enough that I will speak of them as if they correspond to real changes in the situation of the population segments concerned.

TABLE 7

Positive Differences in Perception of Value of 1964 Riot Between
Population Segments Living in Two Rochester Black Ghettos

Population Segments	1964 Riot Helped	
	Better-Off Ghetto	Worse-Off Ghetto
	%	%
16 or less	34	
Without spouses	20	
Born in South	17	
Born in North	18	
Consider religion to be of low value	26	
High educational status[a]	16	
Low educational status	21	
High school students	42	
Lower-class identifiers	43	

Population Segments	1964 Riot Didn't Help	
	Better-Off Ghetto	Worse-Off Ghetto
	%	%
16 or less		16
Married		18
Students	24	
Lower-class identifiers		26

a. High educational status includes all those persons with at least a high school diploma.

TABLE 8

Differences in Perception of Value of Hypothetical 1967 Riot Between
Population Segments Living in Two Rochester Black Ghettos

Population Segments	Hypothetical 1967 Riot Will Help	
	Better-Off Ghetto	Worse-Off Ghetto
	%	%
Females	28	
16 or less	30	
26 or more	16	
Married	23	
Without spouses	21	
Born in North	23	
Consider religion of low value	25	
High educational status	21	
Low educational status	17	
Low occupational status	21	
High school students	26	
High original social class	21	
Low original social class[a]	17	
Lower-class identifiers	26	

Population Segments	Hypothetical 1967 Riot Will Hurt	
	Better-Off Ghetto	Worse-Off Ghetto
	%	%
16 or less		18
26 or more	16	
Married		26
Low occupational status		24
High school students		16

a. High original class includes those persons whose fathers were employed as semi-skilled
workers or better in the occupational system. Low original class includes persons
whose fathers were employed as unskilled or farm laborers or were shifters.

The people who found the 1964 riot useful and were less likely
to regard the hypothetical 1967 riot as useful were either those
people whose original social class or occupational status were at
odds with residence in the better-off ghetto or people whose social
status was high enough to form a bridge to at least the illusion of
opportunity. The people who regarded the hypothetical 1967 riot
as useful and were less likely to see the 1964 riot as useful were
older adults, women, and married people.

It may well be that the people who have benefited from the
aftermath of one riot or have experienced its costs are less likely
to view riots as instrumental, while persons whose social ties at
first disposed them to oppose riots are more likely to see a second
riot as instrumental, as a result of their having observed other less
cautious people in their vicinity benefiting directly from their riot
participation or from the resulting programs.

Turning now to those categories of people who opposed the use
of riots as instruments of social change, the only two seventh ward
population categories that consistently and significantly denied
the value of riots were youngsters 16 and under and married
people. The two third ward population segments that were consistently opposed to riots were women and those persons to whom
religion was important.

One inference that emerges from these data and data already
reported is that while some social ties turn people away from
approval of collective actions like riots, this is by no means so for
all types of social ties. Indeed, there appear to be some types of
social ties that under certain conditions lead people to favor riots
as instruments of social change. Thus any theory of collective
behavior worthy of the name must specify the conditions under
which particular social constraints have conservatizing, neutral-
izing, or radicalizing effects.

The dubious attitude toward riots of youngsters 16 and under
living in the poorer ghetto comes as something of a surprise,
especially as this is the main source of flotsam and riffraff. One
possible reason for this skepticism is that these youngsters may
very early in their lives acquire from their surroundings some of
the dominant moods of a poor ghetto such as resignation, defeat,
inferiority, and cynicism. It seems that other dominant moods of
the poor ghetto such as truculence and burning hate are directed
toward whites at a slightly later age, after there has been some

direct experience of the meager opportunities open to young adults mired in a poor ghetto.

Karl Marx was not the first to observe that people with empty bellies seldom make revolutions. Yet the riffraff theory appears to assert otherwise. What little evidence there is on this point in Tables 7 and 8 supports Marx. Thus persons living in the poor ghetto who identified themselves with the lower classes questioned the value of the 1964 riot. Likewise, it was persons of low occupational rank living in the poor ghetto, "drifters," high school students, and those who attached little value to their religion who believed that another riot would hurt the Negro cause and presumably threaten their own survival adaptations. These data tend to support the contention that persons of low rank with failure self-images who live in poor circumstances are not likely to be the initial source of energy for ghetto riots.

Table 9 underscores findings already noted: not only are residents of the better-off ghetto more disposed to riot, but they are more likely to predict the occurrence of a riot than are

TABLE 9

Estimation of Chances of a 1967 Riot by Sub-ghetto Residence

Chances of a riot	Third Ward		Seventh Ward	
	N	%	N	%
Good	(69)	69	(65)	57
Small	(17)	17	(16)	14
Don't know	(14)	14	(33)	29
	N=100	100%	N=114	100%

residents of the poorer ghetto. However, there is no appreciable difference in the extent to which residents of the two ghettoes discount the possibility of a riot. Yet again, the residents of the poorer ghetto show themselves to be less able to or less willing to respond to questions asked of them.

Table 10 gives a clear picture of the third ward population segments that expected a riot last summer. It was the teen-agers, single people, persons who attached low value to their religion,

TABLE 10

Positive Differences in Estimations of Chances of a 1967 Riot
Between Population Segments in Two Rochester Black Ghettos

Population Segments	High Chances of a Riot[a]	
	Better-Off Ghetto %	Worse-Off Ghetto %
Females	18	
16 or less	38	
Without spouses	18	
Born in North	22	
Consider religion of low value	22	
High original social class	24	
Low educational status	15	
High school students	32	
Lower-class identifiers	16	

a. Estimates of a low chance of 1967 Rochester riot were distributed uniformly among all population segments, so that no positive differences emerged between population segments living in the two ghettos.

persons of low educational status, social drifters, persons born in the North, high school students, and persons whose identification was with the lower classes. Generally speaking, it was the rabble living in the better ghetto who fantasized about a riot and often gratuitously mentioned that they were "going to help things along."

The one and only seventh ward population segment more likely to anticipate a riot was the young adults from 17 to 25. This population segment has been the major exception to the rule throughout. That is, its members have reported alienative and riot attitudes that more resemble attitudes of people living in the third ward than they do attitudes of other population categories living in the poorer ghetto. Obviously there is something explosive in the situation of these young adults. It seems reasonable to assume that a significant proportion of riot participants and some riot leaders come from their ranks.

DISCUSSION

The data on the ecological distribution of alienation and attitudes concerning the legitimacy of riots unequivocally indicate that there is sufficient alienation and approval of riots in both ghettos to support the outbreak of a riot in either area. Where and when a riot breaks out is likely to be contingent on the behavior of political officials and their agents, or a matter of accident. The presence, however, of proportionately more people in the better-off ghetto with some commitment to the political system presumably acts as a deterrent to the triggering of riots in the better-off ghetto and to lower participation rates by third ward residents. The significantly larger proportion of poorer ghetto residents of an unknown state of alienation implies the existence of a large reserve army in the poorer ghetto whose members are not likely to be the leaders of a riot, but who are quite likely to join such an action after one has started.

It seems a reasonable inference from both the gross findings and the more refined findings dealing with population segments that ghetto milieu contributes in an important if unknown way to the crystallization of attitudes concerning the legitimacy and utility of collective action, and thus is an influence upon the undertaking of such actions. The central question to be answered is why should people living in the better-off ghetto and particular population segments living there be more riot-oriented than persons living in the poorer ghetto?

The riffraff theorist might respond that rabble become ready for collective action when they can readily observe the degree to which they are rabble. But such an explanation begs the question on two counts. First, it substitutes a relative deprivation explanation for one that directly links rabble status to collective action. Second, it offers no explanation for the variations in riot orientation between rabble living in the different ghettos.

The relative deprivation theorist seemingly stands on firmer ground. He might respond that the findings of differences in riot orientation between the two ghettos have to do with differences in discontent stemming from differences in relative deprivation. His argument might take the following form: Blacks living in the better-off ward, generally speaking, have higher expectations than blacks living in a poorer ghetto; opportunities are less commensurate by white middle-class standards for third ward residents than for seventh ward residents; thus third warders are likely to harbor more riot orientation than seventh warders.

This analysis also has logical and empirical limitations: it assumes that expectations are higher among persons living in a better ghetto; it assumes that opportunities are less commensurate for third ward residents; and it further assumes that the relative deprivations experienced by residents of the better ghetto are more depriving than the relative deprivations experienced by residents of the poorer ghetto. The crux of my objection to the relative deprivation theory, as it is usually elucidated, is that it does not suggest mechanisms by which feelings of discontent are directed against institutional targets rather than becoming self-directed.

I am not proposing that propositions dealing with relative deprivation and its correlates be thrown out with the bath water. I am arguing that these propositions by themselves do not constitute an adequate explanation of riots or revolutions. I am arguing for an approach that seeks to combine and integrate propositions that refer to ecological processes, cultural and structural differentiation, as well as those that deal with aspects of relative deprivation.

A combined approach is more likely to contribute to an understanding of sub-ghetto differences in riot orientation. Take, for example, the notion of dynamic density, an ecological notion which refers to the ratio of interaction between different population categories in a defined space. The noxious effects of dynamic density are likely to be greater for residents of the better-off ghetto for three reasons. First, residents of the better-off ghetto, almost regardless of population category, are more aware, have more information at their disposal, and are more conscious of its implications than are residents of the poorer ghetto. Second, although third ward residents appear to have fewer direct contacts with gouging landlords and shopkeepers, with officials such as policemen and welfare workers, and with small-time racketeers and other criminal elements, they seem to be more likely to complain of harrassment and exploitation and to be more resentful of these enforced contacts.[15] Third, residents of the better-off ghetto also appear to be more likely to have regular contacts with blacks of high rank and with whites of both low and high rank in and outside the ghetto. The net result of these interactions, especially when they occur within a value climate in which middle-class American values are more emphasized than in a value climate in which survival values are more emphasized, may well be to establish a mood of free-floating, externalized aggressivity.

It is worth noting that blacks living outside either ghetto, in transitional areas, were also significantly less riot-oriented than residents of the better-off ghetto, showing about the same proportion of riot disposition as residents of the poorer ghetto. Yet the out-of-ghetto blacks were about as likely to be high on alienation as the residents of the better-off ghetto.[16] These findings, besides showing the inhibiting effects of social-mobility-generated marginality, suggest that living in a better-off ghetto, but a ghetto nevertheless, with middle-class aspirations and values, but limited social as well as economic opportunitites, has catalyzing effects.

Any "adequate" explanation of the differences reported in riot orientation between ghetto areas must also deal with why residents of the poorer ghettos are less likely to see riots as instrumental events, i.e., as bringing them social or psychological utilities. I would like to mention four conditions, hypotheses if you wish, that may have a bearing. First, residents of a poorer ghetto may be so entrapped by the care-taking institutions upon which they have become dependent for survival and/or by their own "deviant" arrangements that they have too much to lose from a riot identification. Second, these people may devote so much of their energy to survival activities that they have too little energy left over to aggress towards external targets. Third, these people may bring to their ghetto condition a tradition or style of torpor, of ennui, of despair, of hopelessness learned from generations of forefathers, which is then reinforced by the experience of living in a poor ghetto. The miserable conditions of life in a poor ghetto may themselves create a mood of despair, of anomie, that is absorbed very early in life by the very young, leading to "drop-out" reactions rather than to the direction of aggression toward social targets.

The findings presented in this paper, when considered together with the findings reported for Watts residents[17] and findings reported for Newark and Detroit blacks,[18] appear to demonstrate conclusively that no one theory can explain urban riots, predict their occurrence, suggest practices for their control, or, most importantly, in my view, suggest either the exact scale or type of social changes required to prevent the future occurrence of urban disorders. What is wanted is a sociological perspective that is able to take into account the reasons for both the discontent of the black underclass and the discontent of the black middle classes, separate them out, show how each form of discontent is related to specific social circumstances, and fathom the range of different political responses that are likely to be correlated with these disparate life circumstances. Perhaps the initial step toward such a perspective is to recognize that urban riots, whatever else they are, are political acts, i.e., a primitive form of rebellion, that have a potential for societal transformation as well as social amelioration. The pity is that so many officials and social scientists are more concerned, in a mindless way, with problems of social control that are raised by the recent rounds of urban disorders than they are with them as portents of an emerging political force in a society in which contradictions are sharpening, not abating!

NOTES

1. Wirth himself dealt with the differences between European and American Jewish ghettos, but sociologists dealing with the black ghetto have been less sensitive to or interested in variations in ghetto organization and way of life. See Horace Clayton and St. Clair Drake, *Black Metropolis* (N.Y.: Harcourt, Brace, 1945), and Kenneth Clark, *Dark Ghetto* (N.Y.: Harper, 1965).
2. Claude Brown, *Manchild in the Promised Land* (N.Y.: Macmillan, 1965), shows an array of characters who appreciate the subtleties of internal differentiation in Harlem, as do the novels of James Baldwin and Ralph Ellison.

3. The riffraff theory asserts that riots are precipitated by people who are at the bottom of the heap, who have nothing to lose since they have no stake in the society, who do not value middle-class civic virtues, and who are largely sociopaths or psychopaths. This theory of riots is essentially a police chief's theory, seeking solutions through repressive action in the name of law and order; see the testimony of Chief Parker before the McCone Commission and the testimony of a number of police chiefs before the National Advisory Commission on Civil Disorders. Robert Fogelson remarks on the riffraff theory in similar terms; see his "White on Black: A Critique of the McCone Commission Report on the Los Angeles Riots," *Polit. Sci. Q.*, Sept., 1967. The riffraff theory is also held by some psychiatrists and social psychologists. For example, Franz Alexander, the noted psychoanalyst, has commented: "Underprivileged members of an organized group who profit least from social cooperation will be the main offenders of peace and order."

4. The relative deprivation theory, in one or more of its forms, is favored by behavioral scientists. The theory is based upon the acknowledged tendency of people to evaluate their own situations through comparisons with others with whom they identify or interact. For a theoretical treatment of reference groups, see Robert Merton, *Social Theory and Social Structure* (Glencoe, Ill.: Free Press, 1957). For a recent example of how relative deprivation theory is applied in the explanation of collective behavior, see Denton E. Morrison and Allan Steeves, "Deprivation, Discontent, and Social Movement Participation," *Rural Sociology*, Dec., 1967, pp. 414-435.

5. There are at least two very sound reasons why the findings presented in this paper must be considered as tentative and interpreted with caution. First, the cross-tabulation method of investigating associations between variables is not the best, when the total number of cases is relatively small, nor even a very appropriate technique for testing associations. What is needed are stronger tests of association, like partial correlations or multiple regressions. Second, 214 cases is not a large population, especially since it is not a sample of a larger universe in any technical sense.

6. The funds for this study were provided by the Foundation for Voluntary Service. Interviewers were made available through the Cornell University work-study program and O.E.O. This study, however, could not have been accomplished had not the New World Foundation supported the overall research program.

7. I wished to interview a riot-potential population—black people varying in age from 13 to 35. Previous studies carried out by myself and others in the Rochester ghettos had revealed that "deviance-prone" people were seldom to be found at home at any time, but especially so in the summer. While the strategy that I adopted has obvious deficiencies, so too have all those strategies that base a probability sample upon households or other units of social space.

8. The interviews were all conducted by three Afro-American male Cornell undergraduate students. Two of the students were natives of Rochester ghettos, and the third grew up in Harlem. The interview schedule was largely open-ended. The structured items had as few alternatives as possible, in order to minimize problems of interpretation for both the interviewers and the respondents. The interviewing situation was precarious and produced considerable anxiety for all three interviewers. The use of militant black students to represent a white professor in carrying out systematic

interviews with ghetto populations is not a panacea.

9. FIGHT is an acronym: F stands for Fight; I initially stood for Integration, but was recently changed to stand for Independence; G stands for God; H stands for Honor; and T stands for Today. The organization was begun, with the organizing services of Saul Alinsky, in March, 1965. The funds for the first three years were provided by the major Protestant denominations through the Board of Urban Ministry, a semi-autonomous division of the Rochester Area Council of Churches.

10. I include in the riot-eligible population persons from 11 to 50. My assumption is that persons under 11 and over 50 are sufficiently ineligible to participate in riots that they should be excluded from the denominator.

11. I was on the ghetto streets July 25-27,1967, as much as any white man. I also had a Negro participant-observer in the streets. We saw the "riot" more as a riot situation that grew into a near riot as a result of repressive police action, than as a riot that was controlled by aggressive police action.

12. See Black McKelvey's monumental four-volume history of Rochester, *Rochester: The Quest for Quality, 1890-1925* (Cambridge: Harvard Univ. Press, 1956).

13. The importance of disaggregation of Census data for social analyses of ghetto dynamics is pointed to by Daniel P. Moynihan, "Urban Conditions: General,," *Annals*, Vol. 371 ("Social Goals and Indicators for American Society," Vol. I; May, 1967), pp. 159-177.

14. "Rank inconsistency" describes a situation in which an individual holds a number of positions of unequal or incongruent rank. For example, a person of high income but low education would be rank-inconsistent, as would be a person of high occupational status, with high education, who is Negro. There is some evidence that rank inconsistency is associated with social-change orientation, while rank consistency is associated with a status-quo orientation. See Gerhard Lenski, *The Religious Factor* (N.Y.: Doubleday, 1961).

15. Gloria Joseph, *A Study of Group Conflicts and Leadership in a Riot-Victimized Northern City* (unpub. Ph.D. thesis, Cornell University, 1967), has documented the tendency of higher-status Negroes to grieve more than lower-status Negroes in respect to police harrassment and exploitation by merchants and landlords. These differences in the expression of grievances appear to be related to different expectations of social equity and capacities for articulation.

16. Forty-five persons living outside of the two major Rochester ghettos fell into this study. The fact that residents of the better-off ghetto are significantly more riot-oriented than out-of-ghetto residents points to the role of ghetto factors in the buildup of riot orientation. Out-of-ghetto residents are less likely to be riot-oriented, because they are caught in a double contradiction: they are renegades, from the point of view of some ghetto people, but they are seen as undesirable by many of the whites with whom they or their families identify.

17. See the study of Watts by Raymond J. Murphy and James M. Watson, *The Stucture of Discontent: The Relationship Between Social Structure, Grievance, and Support for the Los Angeles Riot* (mimeo'd report MR-92; Institute of Government and Public Affairs, University of California, Los Angeles, 1967).

18. See the *Report of the National Advisory Commission on Civil Disorders* (N.Y.: Bantam Books, 1968).

SOCIAL STRAIN AND URBAN VIOLENCE

Everett F. Cataldo
Richard M. Johnson
Lyman A. Kellstadt

Acts of urban violence take place within the context of serious strain within the community.[1] In part, strain is likely to exist whenever a sufficiently large number of people within the community view their life situation unfavorably, have very low evaluations of government's performance of important functions, expect more from the political system than they regularly receive, are sharply divided with significant others over basic issues, and

AUTHORS' NOTE: *The research reported herein was performed pursuant to a contract with the Office of Economic Opportunity. The opinions expressed, however, are those of the authors and should not be construed as representing the opinions or policy of any agency of the United States government.*

The research has also received support from the State University of New York. Computing time and facilities were contributed by the Computing Center of the State University of New York at Buffalo, which is partially supported by NIH Grant FR-00126 and NSF Grant GP-7318. The authors gratefully acknowledge the assistance of James Hottois in the preparation of this paper.

have a willingness to endorse extreme forms of political participation to dramatize their situation or get what they want. Where these attitudes persist for a long while, strain is acute and the probability of hostility is high. The main purpose of this article is to identify the strains that contributed to hostile outbursts in Buffalo.

In June, 1967, Buffalo experienced the first serious hostile outburst in its east-side Negro ghetto. In the winter of 1966-67, the authors, as part of a larger research team, conducted a survey that focused, in part, on the dimensions of strain outlined above. While the purpose of our research was not to forecase whether Buffalo would have disturbances, our "pre-riot" data are useful for understanding subsequent events.

PERCEPTIONS OF SELF AND ENVIRONMENT

The objective conditions of one's life may be importantly related to his behavior; however, the way in which the individual perceives the objective situation is perhaps even more important. An individual relates to his environment in terms of perceived social status, aspirations, significant reference groups, a sense of mastery or impotence in the face of his environment, personal hopes and fears, etc. These basic attitudes, often called self-conceptions, are of particular importance in an attempt to understand the hostile behavior which took place in Buffalo this past summer.

The concept of self is difficult to operationalize under the best of conditions, and a survey instrument at the present state of the art is not entirely adequate. The importance of discovering where our respondents stood in terms of hopes, fears, and aspirations led us to use the technique devised by Hadley Cantril called the "Self-Anchoring Striving Scale."[2] Respondents placed themselves on a nonverbal ladder consisting of ten "rungs" on the basis of the following question:

> Think for a moment of the best possible life you could imagine. The very best and worst life could be seen as the top and bottom of a ladder. Imagine that the top of the ladder represents the best possible life for you and the bottom represents the worst possible life for you.

Respondents placed themselves on the ladder in terms of the present, five years ago, and five years in the future. Table 1 compares the mean responses of Negroes and whites. Negroes place themselves lower on the ladder at the present than do whites. From a temporal perspective, both whites and Negroes tend to feel they are better off presently than they were five years ago, and seem to view the future in optimistic terms. Negroes, however, see themselves as having made more progress in the past five years and anticipate more progress in the coming five years. In terms of percentage distributions, 34% of the Negro respondents see themselves as having been on the lowest three rungs of the ladder five years ago, 14% place themselves there presently, and only 4% do so five years from now.

TABLE 1

PERSONAL LADDER RATINGS: GROUP MEANS

	Past	Present	Future
Whites	5.78	6.50	7.67
Negroes	4.82	5.99	7.72
	Change	Change	Total Change
Whites	+ .72	+1.17	+1.89
Negroes	+1.17	+1.73	+2.90

Only 15% of the Negroes see themselves as having been on the top three rungs of the ladder five years ago, while 26% place themselves there now, and 61% look forward to being there in five years. Whites do not exhibit as high a sense of progress, hope, and optimism.

A further control for education produces some interesting differences. While personal progress is sensed by both groups in each educational category (grade school, high school, college), Negroes sense a greater amount of progress up the ladder than do whites. Even among the most poorly educated Negroes, only 5.3% anticipate being on the bottom three rungs five years from now, in comparison with 30.1% who feel that they were there five years ago. Poorly educated whites, however, are far less hopeful. Less than 14% see themselves as having been on the lower three rungs five years ago, whereas over 18% see themselves there five years from now.

TABLE 2

UNITED STATES ON THE LADDER

	Past	Present	Future
Whites	7.39	6.82	7.38
Negroes	6.59	7.04	7.62

	Change	Change	Total Change
Whites	- .57	+ .56	- .01
Negroes	+ .45	+ .58	+1.03

BUFFALO ON THE LADDER

	Past	Present	Future
Whites	5.43	5.39	6.58
Negroes	4.57	5.69	7.28

	Change	Change	Total Change
Whites	- .04	+1.19	+1.15
Negroes	+1.12	+1.59	+2.71

Do Negroes generalize their sense of personal progress to a sense of progress for the entire United States and the Buffalo community? To ascertain perceptions of the nation and community in terms of respondents' "greatest hopes" and "worst fears," the ladder technique was again employed. Table 2 demonstrates that Negroes and whites differ remarkably in their assessments. Negroes sense more progress on the part of both the United States and the Buffalo community than do whites. In fact, whites feel that the United States will be no higher on the ladder five years from now than it was five years ago.

The picture that emerges here may run counter to normal expectations. Certainly one does not see in these data a Negro community which is pessimistic about their present personal standing and the prospects for the future. On the other hand, one must not take a prematurely sanguine view of these findings. It is understandable that Negroes should perceive an improvement in their life situation when they compare today with five years ago. It is generally conceded that strides have been made. Projections into the future, however, constitute hopes and aspirations which may or may not be fulfilled. This gap between what is and what is expected—considerably greater for Negroes than for whites—may be a source of persistent strain in the community, if indeed this gap is not narrowed in the coming years.

GOVERNMENTAL FUNCTIONS: EVALUATIONS AND EXPECTATIONS

As noted above, it is one thing to express hope about the future; it is quite another thing to believe that one's hopes will be fulfilled. To discover the extent of satisfaction with governmental performance among our respondents, we asked them first to evaluate government in terms of how effectively it performed a series of basic functions and then to indicate their expectations of governmental responsibility to perform these functions. A factor analysis of twenty-four items dealing with governmental functions produced an identical and theoretically significant factor on both the "evaluation" and "expectation" dimensions. This factor has to do with governmental activities such as providing opportunities for a good living, social mobility, employment, and the like. Factor scores, ranging from 1.00 to 4.00, were given each respondent for both of the dimensions.

Considerable differences were found between whites and
Negroes (see Table 3).

TABLE 3

HOW WELL DOES GOVERNMENT PROVIDE OPPORTUNITIES FOR
THE INDIVIDUAL? EVALUATIONS AND EXPECTATIONS

	Evaluations		Expectations	
	White	Negro	White	Negro
Education: Grade	3.12	2.67	3.24	3.19
High	3.00	2.37	3.01	3.29
College	2.88	2.36	2.82	3.10

Gap Between Evaluations and Expectations

	White	Negro
Grade	.12	.52
High	.01	.92
College	-.06	.74

At each educational level, the gap between evaluations and expectations is consistently and significantly higher for Negroes than for
whites. Whites appear quite satisfied; indeed, college-educated
whites appear to feel that government is actually doing more than
it ought to. Somewhat surprisingly, Negro dissatisfaction is not
greatest among the lowest-educated but among the high school-
educated, followed by the college-educated.

An inspection of some individual items dealing with governmental functions (see Table 4) reveals consistently lower evaluations and higher expectations for Negroes than for whites. It
also reveals that Negroes are least satisfied with governmental
performance with respect to housing and job opportunities. An
indication of the gulf between whites and Negroes is that three-
quarters of the whites think government is doing an effective job
on housing opportunities, while only 39% of the Negroes think so.
Among whites, in fact, more think that government is effective

TABLE 4*

EXPECTATIONS ABOUT WHAT GOVERNMENT SHOULD DO AND EVALUATION OF GOVERNMENTAL PERFORMANCE

ITEM	RESPONSE	EVALUATION		RESPONSE	EXPECTATIONS	
		White	Negro		White	Negro
1. Provide a chance to make a good living	Effective	87%	71%	Important	77%	94%
	Ineffective	13	29	Unimportant	23	6
2. See to it that every man who wants a job can	Effective	64	44	Important	83	91
	Ineffective	36	56	Unimportant	16	9
3. Secure civil rights and civil liberties	Effective	80	55	Important	93	97
	Ineffective	20	45	Unimportant	7	3
4. Provide justice for all	Effective	81	59	Important	98	99
	Ineffective	19	41	Unimportant	2	1
5. Make it possible for people with the means to live where they want to	Effective	75	39	Important	72	94
	Ineffective	25	61	Unimportant	27	6

* Sample sizes for this table range from 526 to 536 for whites, and from 245 to 252 for Negroes

with respect to housing opportunities than think that government has an important responsibility to help provide them.

The source of Negro dissatisfaction appears to be a failure of government to perform up to Negro expectations. The hostile outbreaks of last summer may only be a prelude of things to come if Negroes continue to feel that government is performing at a level far lower than it ought to. Keeping in mind that dissatisfaction is greatest among the higher-educated Negroes, it is possible to suggest that any rise in the educational level of Negroes may, in the short run, produce even more dissatisfaction. This will be the case especially if gains in that area are not accompanied by more than token gains in other areas such as housing and job opportunities. Improving education without improving other things, therefore, may exaggerate rather than reduce community conflict.

BATTLEGROUND ISSUES

On what issues are whites and Negroes most deeply divided? What will be the battle lines of future disputes between racial groups? Our respondents were asked to evaluate a series of symbols concerned with race and poverty and to indicate their extent of endorsement of these concepts by telling us whether they felt they were "helpful," "harmful," or "neither helpful or harmful" to the Buffalo community. Responses to such poverty symbols as "welfare payments," "aid to dependent children," and the "War on Poverty" produced no significant differences between races; whites endorse these notions about as highly as Negroes. This is the case even though these symbols are often linked with race in the mass media and in other public sources.

Other symbols, more directly related to race, however, produced wide differences between whites and Negroes, as can be seen in Table 5. The differences between the races on each of these symbols suggest important sources of strain and cleavages in the community. Open housing is supported by practically all Negroes but by only a bare majority of whites. Marches in Milwaukee, Chicago, and other large cities of the North to encourage the passage of open-housing ordinances and/or open-housing policies have met with a good deal of white resistance. While open-housing marches have not taken place in Buffalo, our data suggest that

TABLE 5

ATTITUDES TOWARD RACIAL SYMBOLS

Symbol	White				Negro			
	Helpful	Neutral	Harmful	N	Helpful	Neutral	Harmful	N
Open housing	53%	26%	22%	482	95%	3%	1%	236
Integration	46%	31%	23%	468	87%	7%	6%	238
School bussing to facilitate integration	23%	18%	59%	484	67%	16%	17%	222

they might be given a similar reception. While we found that our white respondents were highly favorable to "equal opportunity" as an abstract symbol, they found it harder to support when it was brought closer to home.

Even fewer whites approve of the concept of integration than of open housing. This is somewhat surprising, for one might expect the more general conception to be less threatening. The problem is clarified to some extent when a control for education is instituted (see Table 6). Clearly, white endorsement for integration increases steadily with education. By far the greatest hostility for the concept is among the lowest-educated.

TABLE 6

WHITE ATTITUDES TOWARD INTEGRATION BY
LEVEL OF EDUCATIONAL ACHIEVEMENT

Level of Education	Helpful	Neutral	Harmful	N
Grade school	26%	34%	40%	65
High school	44%	33%	23%	250
College	57%	29%	14%	153

While Negro support for school bussing falls below their support for integration and open housing, white support for school bussing is barely at the level of one out of every four. Interestingly enough, lower-educated Negroes are less supportive of this concept than more highly educated ones. Strongest opposition to school bussing is found among highly educated whites. These people may be supportive of integration as an abstract symbol, but they are quite unwilling to support school bussing to promote that goal.

These data point to willingness among whites to pay the price for segregating Negroes through large public expenditures for welfare and related programs, but far less willingness to support goals and measures that might ease Negroes out of their ghetto existence. Unless a change of attitudes by whites is forthcoming or Negroes lower their expectations, strains within the community are likely to increase when issues of open housing, integration, and school bussing are raised. At this point it is unrealistic to expect Negroes to lower their expectations. Therefore, reduction in strain will occur only when white attitudes shift to a more favorable position on these issues.

POLITICAL PARTICIPATION

For Negroes, what effect does dissatisfaction have on their rates and kinds of political participation? Does dissatisfaction among Negroes lead to a disposition to substitute extreme forms of

TABLE 7

POLITICAL PARTICIPATION MEAN SCORES BY EDUCATION

Education	Whites	Negroes
Grade school	4.59	4.56
High school	5.14	4.93
College	6.24	6.86

participation for more conventionally acceptable forms–rioting for voting, for example? An index of political participation was devised on the basis of responses to five items: frequency of discussing politics, political advice-giving, attempts to influence local decisions, registering to vote, and voting. (Scores range from 1 to 10; the higher the score, the more intense the participation.)

The mean score for whites on this index is 5.38; it is 5.12 for Negroes. As can be seen in Table 7, although the mean score for whites is higher than for Negroes, the college-educated Negro reports the highest rate of participation. Differences between Negroes with college education and those with high school education are striking, with the former participating at much higher rates with respect to conventional means of participation. This is the case even though data presented earlier showed large discrepancies between expectations and evaluations of governmental performance by college-educated Negroes. Even higher gaps between expectations and evaluations of governmental performance, however, were noted for high school-educated Negroes. The high rate of dissatisfaction with governmental performance for high school-educated Negroes seems to have a depressing effect on conventional participation for this group. This raises the interesting question of whether conventional means of participation will be utilized by the high school-educated Negroes as they attempt to achieve an approximation of their expectations. Barring a massive increase in the opportunities for this group, and barring a narrowing of the gap between evaluation and expectation, it seems plausible to suggest that conventional means of political participation may be rejected in favor of more extreme types of activity.

Demonstrating and rioting may be considered forms of political participation, even though the literature of political science does not ordinarily do so. These kinds of activities are, of course, of a different order than activities such as discussing politics and voting. Presumably those who engage in extreme activities such as riots do so because they are more alienated and have less faith in normal political processes than those who participate in more conventional ways.

From a factor analysis of participation items, a "demonstrate and riot" factor emerged. The mean score for whites on this factor is 1.20; for Negroes, it is 1.51. (A score of 2 indicates "some responsibility" to perform the activity, and a score of 1 indicates "no responsibility" to do so.) Support for this dimension is low among whites regardless of education or age, with a slight tendency for those under 30 years of age to score higher than the mean. Among Negroes, however, the high school-educated and those under 30 years of age indicate "some responsibility" to engage in these unconventional forms of political participation. Although the mean scores for both whites and Negroes are low, the data show that Negroes are more responsive to a riot-demonstrate dimension than are whites, and that the most likely participants in such activities are young high school-educated Negroes and not necessarily the downtrodden.

While we have no systematic explanation for these findings, some speculations are in order. Negroes of the lowest socioeconomic status may have few aspirations beyond carving out a meager existence. They may not feel frustrated about being unable to achieve lofty goals because they have none. Negroes of the highest socioeconomic status, on the other hand, see some promise in life and have developed career aspirations which are possible to achieve. While it is probably true that some may feel the effects of discrimination (note their low evaluations of governmental performance in some areas), many find their race to be no hinderance, as they become beneficiaries of efforts by employers to hire "qualified" Negroes. The young Negro with a high school education is in no such enviable position. The school system, civil rights groups, the federal government, and other agencies of government have encouraged the young Negro to have high aspirations. The ability to "see how the other half lives" through the media of communication encourages the same thing. It is at this level of the socioeconomic ladder, however, that competition between Negroes and whites for scarce economic, social, and political resources is most keen. It is here that the effects of discrimination against Negroes may be most sharply felt, and frustration is strongest. Sensing relative deprivation more than other Negroes, this group seems more likely to strike out against society than are others.

CONCLUSIONS

This article has attempted to identify strains that contributed to the hostile outbursts in Buffalo in the summer of 1967. We have assumed that strain exists when people have a pessimistic view of their progress in life; when a considerable gap exists between what people expect from government and how they evaluate its performance; when sharp divisions exist in the community over political issues; and when some people express a willingness to go outside normal political channels to get things done. We do not find Negroes to be pessimistic about the future; on the contrary, we find high hopes for the future among them. Despite these high hopes, Negroes are not satisfied that government is doing all it should to alleviate their situation. Dissatisfaction is high among all Negroes, but it is especially high among the high school-educated. Dissatisfaction within this group finds expression in relatively low conventional political participation and a greater willingness to endorse more extreme ways for inducing change in the system. Clearly, these findings are indicators of strain within the Buffalo community. These strains are heightened by the fact that on significant issues such as integration, open housing, and school bussing, Negroes and whites differ sharply. Given the existence of these strains, it is not surprising that a hostile outburst took place last summer.

A good deal of hope is placed by many people in poverty programs, educational innovations, and related activities as means of reducing the likelihood of hostile outbursts. At present, however, various programs aimed at improving the condition of the Negro seem to have had the effect of making Negroes hopeful but not satisfied. To continue programs that make Negroes hopeful but not satisfied is to invite more strain and potentially more hostile outbursts. This is not to suggest the irrelevance of current programs, but only to emphasize that more needs to be done. Only when people feel that the *results* of programs are consistent with the *hopes* they have generated will the probability for urban violence be reduced.

NOTES

1. Neil J. Smelser, *Theory of Collective Behavior,* (N.Y.: Free Press, 1963).
2. For a full discussion of this technique, see Hadley Cantril, *The Pattern of Human Concerns,* (New Brunswick: Rutgers Univ. Press, 1965), pp. 22 ff.

Part IV
REACTIONS
TO VIOLENCE
Attitudes
and
Social Controls

Introduction

■ Perhaps of more long range significance than the occurrences of civil violence themselves are the responses and reactions of the individuals, groups and institutions, public and private, which comprise the community. How they respond, individually and collectively may well determine the potential for future violence. Thus, response and reaction to violence constitutes a critical factor in the analysis of civil disorder.

In this section we will deal with two types of riot response. The first is attitudinal, representing a mental response to the events themselves, reports of the events (including both accurate documentations and less accurate or even inaccurate rumors of the same events), and perceptions of the events which characterize civil disorders. Any disorder serious enough to affect the public order of the community is very likely to stir even the most apathetic citizen into thought (if not action), changing or reinforcing an attitude, or causing him to have an opinion on some aspect of the riot: cause, process, or consequence. The important questions for both analysis and action are concerned with the distribution of these attitudes, in time, in space and across the various segments of the population. Who reacts how to the riot, and perhaps of most significance, why?

In the first selection of Part IV, Dan Harper, a sociologist, reports his findings of the attitudes of white citizens towards the Rochester, New York, riot of July 1964. Professor Harper's survey sought to determine the extent to which hostile white opinions were a function of either the physical distance of whites from Negroes, in terms of where they lived (within a block, more than a block within the city, or in the suburbs), or the "sociometric" or social distance, in terms of whether whites were negatively affected by the riot (e.g., suffered property damage or were inconvenienced). He hypothesizes that those whites who lived closest to the Negroes and were sociometrically closest would be most hostile to them. In terms of the first variable, physical distance, he finds that the results are inconsistent, neither supporting nor rejecting the hypothesis. Those who lived closest to Negroes are less hostile than those who live more than one block, but more hostile than those who live in the suburbs, although the differences are not great. With regard to sociometric distance, to the extent that there are differences, those who were more distant were more likely to be hostile, i.e., those sociometrically closest to the riot were more likely to be sympathetic. Harper suggests two alternative explanations for his findings: (1) pre-existing attitudes, an explanation which argues that whites who live closest are more sociometrically closest to Negroes were basically less hostile to Negroes before the riot and continued to be so afterward; and (2) interaction, an explanation which argues that those who have contact with Negroes are more likely to understand their concerns and problems, and thus be more sympathetic. He concludes that the riot had relatively little influence on changing basic attitudes of whites towards Negroes.

In "Riot Reactions Among Clergymen" Jeffrey K. Hadden, another sociologist, reports on his interviews with seventy-seven clergymen in Grand Rapids, Michigan, immediately following the 1967 disturbances in that city. Hadden feels that reactions to the riot among the clergy was significant for two reasons: (1) the clergy has played an important role in the civil rights movement during its non-violent phase (1954-65); and (2) they have a "sustained opportunity to interpret and create understanding of the increasingly complex and deepening racial crisis in America" for a substantial segment of the American public.

Although the respondents included small numbers of Negro Protestants and Roman Catholics, Professor Hadden concentrates on the similarities and differences in opinions, attitudes and behaviors between "liberal" and "conservative" Protestants distinguished by the theological tradition of the clergymen's denomination. Although the range of opinions, attitudes and behaviors is considerable within both the liberal and conservative traditions (and amongst the Catholics), Hadden does find different patterns of response to the riot. Liberal ministers are more sensitive to conditions of frustration and injustice as a cause of the riot, were more likely to interpret the riot in terms of the failure of the white community and the church with regard to ghetto conditions, discussed the riot from the pulpit and with members of their congregation more often, and reported a more sympathetic and understanding attitude for the rioters' situation among their congregations. The conservative clergymen, on the other hand, tended to think of the riot in terms of agitators and lawlessness, and as a group seemed more concerned with re-establishing law and order than with understanding the causes of the riots. The reactions of their congregations were likely to be "fear, disgust and hostility." The level of knowledge about the mood of the ghetto and the degree of involvement in it was low among both groups, but again patterns emerged: liberals tended to see a role for the church but were unsure of what it should be, while the conservatives tended to feel that *if* the church had a role at all, it involved "preaching and teaching the Gospel."

A second kind of response to civil violence, which is treated in the last three selections of this section, is organizational and institutional. The first deals with the affect of the riots on the organizational structure of the Negro community, and the last two represent interpretations of the front line social control agency in riot situations: the police.

Enough time has elapsed after the first major riot of the 1960's, Watts (Los Angeles), to permit some analysis of its influence on the community. Harry Scoble, a political scientist and a member of the team of social and behavioral scientists involved in the Los Angeles Riot Study, offers an analysis of the influence of the Watts riot on Negro politics and the community organizational structure. In general, Scoble finds that since the Los Angeles riot, white political and civil leaders have tended to be more responsive to Negro demands, although their response has been more verbal and visible than substantive. The recommendations of the McCone Commission report represent a maximum program to most whites but only a minimum and largely symbolic program to Negroes. The potential danger of this gap in programmatic expectations is compounded by the fact that since Watts, the Negro leaders have articulated more demands on the white power structure. These demands have changed not only in intensity but in kind; they increasingly stressed welfare goals as opposed to the more symbolic, status equality goals of the traditional middle-class Negro leadership. This has led to a change in the style of leadership, which has become considerably more militant. Organizer and mass-agitator type leaders have emerged to contest with the prestige and token leaders for the sympathy and support of the Negro community. Scoble also suggests that the new Negro elective leaders, city and state, are probably not going to be able to produce public policies fast enough to satisfy demands of the militant leadership.

Scoble contends that the riots of 1965 in Los Angeles have produced three results which in combination have the potential for further violence. Negroes have used their electoral strength to elect Negroes to public office with unrealistic expectations about what they can do; the riots have also produced a supplementary Negro leadership aggregate which is much more militant and violent than the older, prestige, status leadership group; and there has been a significant increase in both the intensity and kind of demands made by the Negro community on the political structure. Scoble warns that "such increases in the Negro political leadership aggregate generate geometrically increasing expectations and demands compared with the achievement level currently permitted by American politics." This, he suggests, is the essence of a revolutionary situation.

The last two selections in this section deal with the police, the agency of social control which perforce must respond to civil violence. In "Cops in the Ghetto: A Problem of the Police System," Burton Levy, a law enforcement administrator close to the police situation, concludes that the urban police have too many systemic deficiencies to be effective in riot prevention or control. Levy challenges the traditional programs to improve police-ghetto relations, which are based on assumptions about the behavior of individual officers. He prefers to view the police department as a system with its own values, mores, and standards—and, all too often, "anti-black" is likely to be one of the more important values of the system. "The department recruits a sizeable number of people with racist attitudes, socializes them into a system with a strong racist element, and takes the officer who cannot advance and puts him in a ghetto where he has day-to-day contact with the black citizens." When the inherent conflict of a police-citizen encounter is compounded by law enforcement racism, it should not be surprising that most of the urban riots of the 1960's have started with a police encounter. Levy suggests that the issue of law enforcement racism has not been confronted with the same success as other civil rights issues because of an information gap (at least up until the publication of the reports of the President's Commissions on Law Enforcement and the Administration of Justice and on Civil Disorders), and the defensiveness and secrecy of the police themselves.

Levy's conclusion is not optimistic. Money alone—whether spent for higher police salaries or more police training—since it is directed at the individual officer, will do little or nothing to stop the pattern of police discrimination and brutality against minorities in the urban ghettos. The problem, he insists, is one of a set of values and attitudes and a pattern of anti-black behavior socialized within, and reinforced by, the police system.

The analysis presented by the late Dean Joseph Lohman suggests that the "crisis of the police" in the United States today is a structural deficiency which makes the interventions of the police "conducive to collective over-expressions of hostility in the society rather than the containment of individual expressions of hostility and/or violations of the law." He contends that the crisis has three dimensions: (1) the widespread disposition of the police to blame "troublemakers" and "agent provocateaurs," which when applied arbitrarily leads to a double standard of law enforcement (e.g., stop and frisk practices); (2) the absence of effective channels for expressing grievances by the ghetto sub-community— in the absence of real rather than symbolic grievance procedures, members of the minority community tend to perceive the police as "an instrument of an intransigent power holding group and irresponsive to their interests and well being"; and (3) the absence of machinery for effectively relating the police to the complexities of the urban community—in short, the inability and/or unwillingness of the police to adapt to their role as agents of social stability in a period of accelerated social change.

—L. H. M. and D. R. B.

WHITE REACTIONS
TO A RIOT

Dean Harper

■ It is clear that the urban riots by Negroes of the past three summers have evoked different responses in whites. Some have felt hostile toward Negroes—claiming that they ought to be sent back to the South from the northern cities, that they are ungrateful for what has been done for them, and the like; others have felt sympathy for Negroes, asserting that, although rioting is wrong, some may have good reason for their actions. Still others are more or less indifferent, or seemingly neutral—expressing mild support for Negroes in some respects and mild anger toward them in other respects. That any white expresses his feelings toward Negroes, *in general,* without differentiating among them, is, of course, an instance of stereotypic thinking and should not surprise us. But many whites can recognize the vast array of differences that exist in the motivations and behavior of Negroes as a group and yet be either hostile or sympathetic toward those who are caught up in rioting. Regardless of stereotypes, this variation in reaction is of interest to the social scientist and becomes a point of departure for further investigations.

The social scientist is interested in explaining the variation: why are some hostile and others sympathetic? Is it due to individual personality, varying as psychological factors vary? Or does it

AUTHOR'S NOTE: *This research was supported by the Group Relations Fund of the University of Rochester.*

depend on social conditions—varying with differing characteristics of the social structure? And the answer, of course, is that the hostility of whites toward Negroes is a function of both personality and social structure.[1]

One variable which has received less attention, but which many consider important, is "physical distance." Once negroes begin to move into a white neighborhood, send their children to what had been all-white schools, then many whites begin to get worried and hostile. The closeness of Negroes to whites generates hostility in whites which was dormant or did not exist before.

In the same fashion, the physical distance of Negroes from whites may influence the reactions of whites to rioting. Those who live close to the riot area might become anxious or apprehensive—fearful that rioting will spill into their neighborhoods—and then become hostile toward Negroes.

On the other hand, *perceived* physical distance may temper the effects of physical distance. That is, those who live close to the riot areas but who interact very little with Negroes may be less apprehensive than those who are somewhat farther from the riot area but who interact more with Negroes. For example, a family with children in a school which has increasing numbers of Negroes attending it may have become fearful of these changes in the school, may imagine that changes will occur on their street, and then may have their apprehensions exacerbated by the riots. Thus, interaction with Negroes through the schools may lead people to believe that they are physically close to Negroes and hence physically close to those who are rioting; consequently, they become hostile.

One influence on people's perceptions of physical distance consists of the civic or political boundaries that exist in every metropolitan area. These boundaries are, on the one hand, the lines which separate city from suburbs and, and on the other hand, the lines which separate one ward from another within the city. These are important because people think in terms of these sections of the community (suburb X or ward Y), locate each other within one of these sections, and may let their interactions be influenced by their relative positions in the ecological structure of the community. If a particular ward is defined as a Negro ward, those whites in it may feel closer to Negroes than whites in an adjacent ward which is not defined as Negro. Likewise, those in a suburb, though no farther from Negroes in actual distance than whites living in the city, may feel that the city-suburb boundaries provide them protection from Negroes.

The precise ways in which physical distance affects relations between whites and Negroes has seldom been investigated. To explore this matter was one of the purposes of the research reported here.

In the two weeks following the Rochester riots of July, 1964, a survey of white reactions to the riots was conducted.[2] Mail questionnaires were sent to 500 people in twenty different neighborhoods, with 228 returning completed questionnaires. In five other neighborhoods, another 108 individuals were interviewed. These two sets of individuals did not differ significantly from each other in age or sex. The average educational level of those who returned mail questionnaires was slightly higher than those who were interviewed: 12.3 years compared with 11.1 years. The results reported in this paper will be for the two sets of individuals combined into a non-random sample of 335 whites.

A number of different measures of hostility toward Negroes could be used. In this paper, however, answers to the following question, asked in the context of questions about the causes of the riots, were used as a measure of hostility.[3] (The first alternative was labeled a hostile response.)

Which of the following statements *best* represents your views?: For the most part, Negroes in Rochester have things pretty good, and they shouldn't complain.

For the most part, Negroes in Rochester have things pretty bad, and they have a right to complain.

Of the 336 respondents, 232, or 69%, chose the first alternative—the hostile response. That this percentage of people gave what is called a hostile answer may be of some interest; it suggests that most whites in Rochester had little sympathy with the problems of those living in the urban ghetto.

But of more interest is the effect of physical distance on reactions. The variable "physical distance" was created by classifying respondents into three categories—those living in the suburbs, those living in the city *more* than one block from Negroes, those living in the city *less* than one block from Negroes.[4] If the hypothesis that being physically close to Negroes generated hostility toward them is correct, then fewer of those in the suburbs should be hostile and more of those living within one block of Negroes should be hostile.

The frequency of hostile answers among these three categories of respondents is shown in Table 1.

TABLE 1

**Distance from Negro Families
and Hostility Toward Negroes**

	CITY		SUBURBS
	Distance from Negro Families		
	Within 1 block	**More than 1 block**	
Hostile answer	66.4%	75.7%	57.7%
Other answers	33.6	24.3	42.3
	(155)	(136)	(45)

Two things should be noted about this table. In the first place, the results are partly consistent and partly contradictory with our arguments given above. If we compare the city with the suburbs, we see that those whose homes are most distant from Negro families are least likely to be hostile. (However, 45 suburban residents is not a very large base from which to make such comparisons.) On the other hand, if we compare more distant city residents with less distant city residents, we see that those closer to Negroes are less likely to be hostile. This contradicts our hypothesis.

How can we explain this contradictory result? One possible explanation is that people's reactions to the riots are determined in large part by their more basic feelings toward Negroes—feelings that existed before the riots. As Negroes begin to move into a neighborhood, the more hostile whites move out and the less hostile whites remain. The respondents who live within one block of a Negro family are those who have chosen to remain because they are not so fearful of changes in the racial character of the neighborhood. Thus, our measure of hostility does not necessarily measure reactions to a riot; rather it may measure pre-existing hostility toward Negroes. The neighborhoods with Negro families have a smaller proportion of hostile whites than do all-white neighborhoods, because some of the hostile whites may have moved out of the former.

But another explanation can be offered. Those who live close to Negro families are more likely to have contacts with Negroes, and therefore see their concerns and worries more clearly, than are those whose homes are more distant. The contacts that come from living in a mixed neighborhood may make whites more sympathetic with the problems of the minority and able to recognize that they have grievances.[5]

Before examining another result which may enable us to select one of these two interpretations, a *second* thing should be noted about Table 1. The differences in percentages among the three columns are rather small; hence distance from Negroes, though it may have some effect on feelings toward Negroes, does not have a great effect. Again this contradicts the notions of many who feel that with increasing physical distance from Negroes, whites breathe easier and express less hostility.

The independent variable which has been used in Table 1 is "physical distance." Another independent variable which can be invoked in our search for an explanation is "sociometric distance." This refers to the "distance" of a respondent from someone who was directly affected by the riots. Some respondents may have been directly affected, e.g., the policeman who had to work long hours, the storeowner whose store was damaged by the rioting, and the like. On the other hand, some respondents may not have been directly affected by the riots but may have known others who were affected. Those who were affected or who knew someone who was affected will be referred to as "sociometrically" close to the riots. (This word is used because its most frequent use in the sociological literature is in reference to the choice of one individual by another; hence, it conveys the idea of social distance.)

Here one would argue that those who are sociometrically close will be more hostile; being affected by the riots—whether it be suffering several thousand dollars worth of damage, having to work long hours, or merely being inconvenienced—will make one angry, annoyed, and in turn, hostile toward Negroes. As with physical distance, this would seem to be a rather plausible hypothesis.

Table 2 arrays the data organized by the notion of sociometric distance. The respondents were categorized into one of four types—those who were both inconvenienced by the riots and knew someone who was, those who were not and did not know anyone who was inconvenienced, and those who either were inconvenienced or knew somebody who was, but not both. The kinds of inconveniences which people mentioned included being called to work, not being able to operate one's store or business (no one admitted any personal loss or damage), and the like.

TABLE 2

**Being Inconvenienced by Riots
and Hostility Toward Negroes**

	Did Respondent Know Somebody Who Was Inconvenienced?			
	YES		NO	
	Was Respondent Inconvenienced?		Was Respondent Inconvenienced?	
	YES	NO	YES	NO
Hostile answer	65.0%	63.9%	67.6%	75.9%
Other answers	35.0	36.1	32.4	24.1
	(93)	(58)	(65)	(120)

As before, we note that with respect to the proportion of hostile answers, these four categories of respondents differ but little. To the extent that there are differences, those who suffered no inconveniences nor knew anyone who did were more likely to be hostile. That is, those who were sociometrically close to the riots were more likely to be sympathetic.

Again one might explain these differences—small though they be—by the alternative mechanisms suggested above. On the one hand, it may be that those whose work brings them into contact with Negroes, and who would therefore be inconvenienced by riots, are those who are also relatively less hostile toward Negroes. If a policeman is quite hostile toward Negroes he may seek other employment, because he knows that he cannot work as a policeman and avoid contacts with Negroes. And a businessman who operates a store in an area which has increasing numbers of Negroes may decide to move his business to another area if he is sufficiently hostile toward Negroes. Plausible though these arguments are, they would seem to describe something which happens rarely. It is not that easy for a policeman to give up a job or for a businessman to move his place of business. Those who do would seem to be almost single-minded in their antagonism toward Negroes; the hatred would have to be so great that it would outweigh nearly all other considerations. This is not to say that it does not happen; it is to say that it happens only rarely.

The other explanation, that contact with Negroes lowers hostility and increases sympathy, would seem to be more fitting. Although many of those who have contacts may still be hostile, some become less hostile. Their contacts with Negroes make them aware of the grievances and problems which Negroes have.

In the final analysis, however, these are explanations for small differences. For these respondents, closeness to Negroes, whether it be physical or sociometric, has little effect on attitudes toward Negroes or reactions to a riot. This suggests that hostility or sympathy toward Negroes is created by events other than an outburst such as a riot. This study of Rochester provides little evidence for the hypothesis that the closer a white is to Negroes, the more likely he is to be hostile toward Negroes or to react with hostility toward rioting. Further, distance from Negroes seems to have little to do with hostility toward Negroes.

If closeness to Negroes has little influence on the hostility or sympathy which whites feel, then what does? It would seem that these feelings are basic personality characteristics that are created from birth by events that work slowly and over a long span of time; these are events which range from the way one is raised, as an "intra-familial" event, to the consequences of one's place in the social structure, as an "inter-societal" event.[6] Against these effects, a riot or similar disturbance appears to have relatively little impact.

NOTES

1. For the psychology of prejudice see, for example, T. W. Adorno et al., *The Authoritarian Personality* (N.Y.: Harper, 1950); for the sociology of prejudice see Robin M. Williams, Jr., *The Stranger Next Door* (Englewood Cliffs, N.J.: Prentice-Hall, 1964).
2. A longer report on this research, which incorporates this paper and proceeds from it, will be forthcoming. In it the methods of the study, and particularly of the analysis of the data, will be discussed in greater detail. For an earlier report of this research, see Dean Harper, "Aftermath of a Long, Hot Summer," *Trans-Action,* July-Aug., 1965, pp. 7-11.
3. The results reported here are essentially unchanged, regardless of which of several measures of hostility is used.
4. Several measures of physical distance can be used. The use of any one measure does not change the conclusions that could be made from the use of any other.
5. For other research that leads to the same conclusions, see Daniel M. Wilner, Rosabelle P. Walkley, and Stuart W. Cook, *Human Relations in Interracial Housing* (Minneapolis: Univ. of Minnesota Press, 1955) and the literature cited there. Also see Ernest Works, "The Prejudice-Interaction Hypothesis from the Point of View of the Negro Minority Group," *Am. J. Sociol.,* LXVII (July, 1961), 47-52.
6. In the longer forthcoming report of the research on Rochester, the role of some of these other factors will be discussed.

RIOT REACTIONS

AMONG CLERGYMEN

Jeffrey K. Hadden

■ The reaction of white Americans to the intensified rioting in Negro ghettos during the past four summers has been documented in a number of studies and public opinion polls.[1] The majority reaction has been confusion, disapproval, fear, and cries for law and order. Only a relatively small minority of whites seem to understand the underlying historical conditions which have produced these unprecedented eruptions of violence.

The purpose of this paper is to explore the reaction of a single occupational group—clergymen. The response of clergymen to riots is important for at least two reasons. First, between 1954 and 1965 clergymen played a role of increasing importance in the civil rights struggle in America. Empirical studies have found that the clergy are overwhelmingly committed to the achievement of civil rights and social justice for Negroes—probably more so than any other occupational group in American society.[2] But as the mood of the civil rights struggle has shifted from marches and other nonviolent strategies to cries of "Black Power" and intensified rioting, clergymen have become less visible. This declining visibility is easily understandable, but what has been the reaction of clergymen to the shifting mood of the ghetto? Have they been busy behind the scenes attempting to effect change, or have they, along with the majority of white Americans, become disillusioned with the changing mood and tactics?

A second reason for studying the clergy's response to riots is that they constitute the professional leadership of the largest institution in American society which professes to be centrally concerned with moral issues—the Church. As leaders they occupy a critical position vis a vis the pulpit, teaching, and personal counseling, to influence a broad cross-section of the American public. No other occupational group in our society has a more sustained opportunity to interpret and create understanding of the increasingly complex and deepening racial crisis in America. Thus, their understanding of and reactions to the rioting take on considerable significance.

This study reports the response of the clergy of Grand Rapids, Michigan, to the riots that occurred in that city on July 14-15, 1967. Limited financial resources and research staff played a significant role in determining the scope of the investigation, but Grand Rapids was selected for study because I felt it would be possible to conduct a more thorough study in a small city than in Detroit or Newark, where the rioting has been most intense.

BACKGROUND

Grand Rapids is a community of approximately 300,000 located in west-central Michigan. It is in many respects typical of other medium-sized cities in the Midwest. Population growth is moderate, although most of the growth is occurring in the suburban areas outside Grand Rapids, a pattern which has come to dominate most medium- and large-sized cities in America. Grand Rapids is one of the leading furniture manufacturing centers in America, but the industrial base is diversified and the level of employment has remained relatively high.

Like other cities in the Midwest, Grand Rapids faces problems. The exodus of upper- and middle-income families to the suburbs has created a squeeze on the central city government's financial operations. But like most other cities, Grand Rapids has turned to Washington for help. Through the efforts of Gerald Ford, House Minority Leader, whose home is Grand Rapids, the city has received a lion's share of federal monies.

Like most other cities in the North, Grand Rapids has experienced a rapid influx of Negroes from the South. Between

1950 and 1960, the proportion of Negroes increased from 3.9% to 8.3% of the central city population. City officials currently estimate the Negro population at approximately 20,000, which represents something less than a 50% increase during the current decade. Negroes have inherited the oldest and most deteriorated housing in the city. In 1960 the two census tracts with the highest concentration of Negroes had unemployment rates four times higher than unemployment in the rest of the city. Crime is higher in the Negro ghetto. The need for remedial education in this area is greater. In short, the ghetto conditions created by a large influx of Negro population are not significantly different from those in most other northern cities.

What is perhaps different about Grand Rapids is that its citizens thought they were aware of the problems and were taking constructive measures to improve race relations in the community. Most community leaders felt that Grand Rapids was a city that was working hard to eliminate the underlying causes of racial tension, and they were proud of their efforts.

And then on July 24, it happened. While Detroit was experiencing the worst violence that the nation has witnessed during this century, twelve Negro clergymen from Grand Rapids met on Monday morning and called for intensive meetings between city officials and Negroes to ward off the possibilities of rioting in the city. A spokesman for the group warned, "Time is short . . . . Unless something is done quickly, there could be trouble."

Little did community leaders realize the tragic propheticness of the clergyman's words. Late the same evening, Grand Rapids was rocked with fear and bewilderment as small bands of Negroes roamed through the streets of the south side breaking windows, looting, and burning.

Swift action by the police and state troopers brought the rioting under control early the second evening. Compared with Detroit and a dozen other cities that erupted during July, the civil disorder in Grand Rapids was relatively minor. There were 278 arrests and a dozen gun wounds, but no one was killed. Total estimated damage from 63 fires, looting, and window-smashing was initially set at $500,000, but was subsequently revised downward.

But as a whole nation has discovered during the past four

summers, the costs of rioting cannot be measured in terms of personal injury and physical damage. Grand Rapids, a community proud of its progress in race relations, was suddenly a stunned and confused city. In the pages which follow we will attempt to analyze the response of the clergy to the riots, and the role they played in the community during and immediately after the disturbances.

CLERGY SAMPLE

A total of 94 clergymen were interviewed during the week following the riot.[3] Several of the interviews took place in informal settings where it was not possible to take detailed notes with the interview schedule. While the impressions derived from these informal interviews are incorporated in the interpretations, the statistics reported here are based on 77 completed interviews with clergymen representing 20 different religious bodies. In order to deal with the data in a systematic way, the clergymen were classified into four groups: (1) Liberal Protestant, (2) Conservative Protestant, (3) Negro Protestant, and (4) Roman Catholic. The distinction between liberal and conservative Protestantism is, of course, a crude one. However, previous research, as well as the results of this study, indicate that the distinction is analytically meaningful. This breakdown resulted in the following number of clergymen in each group: Liberal Protestant, 33; Conservative Protestant, 31; Negro Protestant, 5; and Roman Catholic, 8.[4]

In a scientific sense, it cannot be claimed that the interviews represent a random sample of Grand Rapids clergymen. Numerous difficulties were encountered in the sampling. The large number of clergymen who were on vacation made it necessary to replace many names from the initial sample. Moreoever, the fact that many churches do not have secretaries did not make it possible for us to determine with certainty whether a clergyman was on vacation or simply out of his office. In spite of sampling difficulties, there is no compelling methodological reason to believe that the clergy interviewed do not constitute a fairly representative cross-section of Grand Rapids clergy.

CHURCH INTEGRATION

Racial integration varied significantly by the type of religious body. Only two of the thirty-one conservative ministers said that their church had any racial integration. A third of the liberal ministers reported that their churches were integrated. However, none of the liberal churches reported more than a token integration. In fact, the eleven integrated congregations have only about fifty Negro members in all, less than one-half of one percent of the members of these congregations. Three of the eight Catholic priests reported that their parishes were racially integrated. The integrated Protestant churches tended to be the large prestigeful downtown churches. In contrast, the integrated Catholic churches were either in or on the fringe of the Negro ghetto.

CLERGY ATTITUDES TOWARD VIOLENCE

As might be expected, the response of liberal and conservative clergymen to the riots was systematically different. When asked who they felt was responsible for the riots, three-quarters of the liberal ministers pointed to a sense of frustration and hopelessness which has been created by years of discrimination and the unfulfilled promises of whites. While more articulate than many, the following comment by a United Church of Christ minister reflects the mood of the majority of liberal clergymen:

> We are reaping the inevitable results of a hundred years of unfulfilled promises. Particularly in the past dozen years we have given Negroes reason to believe that their life chances are going to change. For a minority of Negroes, this change is occurring, but the tragic fact is that all of our efforts to eliminate discrimination and prejudice have failed. We have not basically altered the life chances of the vast majority of Negroes who are trapped and hopeless in the dark ghettos. The government has failed. The schools have failed. Business has failed. And most importantly, the churches have failed. We have preached the sweet niceties of brotherhood, but we have failed to call a spade a spade. We have let our congregation feel self-satisfied and pious because we have opened our doors to token integration, but we have not

communicated to them that this is but a small part of the task
which must be done. As clergy we have taken to the streets to
protest, which is all well and good, but we have abdicated our
responsibility to protest from the pulpit. As a clergyman, I am
shocked by the realization of how miserably we have failed to
respond to the injustices that have been perpetrated on the Negro
from the day he set foot on American soil.

Ministers from conservative denominations, on the whole,
seemed much less sensitive to the conditions of frustration and
injustice. About 10%, as compared with none of the liberal clergy,
felt that Communists were behind the riots, and an additional 30%
felt that the riots were the result of organized lawless hoodlums.
Another fifth of the conservative ministers said that they simply
did not know who was responsible for the riots.

But it would be unfair to portray all conservative clergymen as
passive and uninvolved. Many of them were quite concerned about
the riots. Several reflected considerable understanding. But several
others expressed the same overt bigotry and hostility that is
reflected in the following comment of a conservative Baptist
minister:

> The riots are just one more attempt on the part of Negroes to
> ram integration down the throats of the Americans. Forcible
> integration is un-American, unconstitutional, and un-Christian.
> Organizations like SNCC and the NAACP just help ferment
> trouble and arouse the Negro . . . . We must put down this law-
> lessness with force and stern punishment to let others know that
> we simply are not going to tolerate it.

It is important to emphasize that what is reported here are
general patterns. The range within both the liberal and
conservative traditions is considerable. Although it is more
difficult to draw generalizations among the Catholics, because of
our small number of interviews, the same wide range of response is
found in this tradition.

The immediate response of most clergymen to the riots was
shock and dismay. However, about 40% said that they were not
particularly surprised that it happened in Grand Rapids. This may
be a bit of Monday morning quarterbacking, but it is interesting
that the liberal and Negro ministers were more likely to report

that they were not surprised. This at least suggests that they were more aware of the ghetto conditions.

The deep concern that clergymen felt about the riots was reflected in the pulpits of Grand Rapids the following Sunday. Ninety percent of the clergy we interviewed who preached the following Sunday said that they mentioned the riots from the pulpit. Ninety-six percent of the liberals, compared with 85% of the conservatives and 81% of the Catholics, said that they discussed the riots.

The slight differences in the liberals' and the conservatives' mention of the riots is interesting in light of their feeling about discussing social issues from the pulpit. When we asked the ministers whether they felt it was appropriate to discuss social issues from the pulpit, 100% of the liberals, as compared with only 41% of the conservatives, gave an unequivocally affirmative response. Conservatives were much more likely to sidestep the question or give an affirmative answer with rather elaborate reservations. But in spite of their reservations, the riots touched a nerve, and conservatives were almost as likely to discuss them as the liberals were.

What the clergy of Grand Rapids had to say about the riots, however, varied enormously. Liberal and conservative clergy tended to vary systematically on several accounts. Conservative ministers were more likely to plead for law and order and abhor violence. And they were more likely to express thankfulness that the rioting was not worse and that it did not spill over into white communities.

Liberal ministers, on the other hand, were more likely to offer some interpretations of what had happened. They were much more likely to point out the failure of the white community and the church to respond to ghetto conditions. Many stated that if the white community continued to ignore the Negro ghetto, they could expect even more violent eruptions in the future.

Liberal ministers were somewhat more likely to report that they had conversations with members of their congregations about the riots (85%, as compared with 71% of the conservatives and 75% of the Catholics). We asked them how they felt most of the members of their congregations reacted to the riots. The differences between the liberal and conservative ministers were significant. Conservative ministers were more likely to report reactions of fear, disgust, and hostility (45%, compared with 21% for the liberals). They also were more likely to report that they didn't know how their congregations felt (23%, as compared with only 3% of the liberals). Liberal clergymen, on the other hand, were much more likely to report that the sentiment in their congregations traversed a wide range from sympathy and understanding to overt hostility (52%, compared with 10% of the conservatives).

The same general pattern appeared when we asked them how their congregations felt in general about the civil rights struggle. Liberals were somewhat more likely to report that their congregations were basically sympathetic (36%, compared with 26% for the conservatives). But again, liberals were much more likely to feel that their congregations were widely divided in their reactions to the civil rights struggle (45%, compared with 6% for conservatives). A third of the conservatives, compared with only one of the liberals, felt that the civil rights issues was not very salient for most of their members and that they were by and large apathetic or indifferent.

In summary, clergymen from liberal Protestant denominations were generally more concerned about the riots, and they tended to offer explanations that reflected some understanding of the underlying causes. Conservative clergymen, on the other hand, were not typically right-wing apologists, but they seem not to have thought about the problem as carefully as have many of the liberal clergy. As a group, they were more concerned with establishing law and order than understanding the causes of the riots.

Our interviewing attempted to move beyond how clergymen felt about the riots to what they were doing and what they felt ought to be done. Liberal clergymen were nearly unanimous in their belief that the churches should be involved in the civil rights struggle in this nation. But when asked for specific suggestions as to how the church should be involved, they talked mostly in terms of broad generalizations rather than specific programs.

We probed for more specific details by asking them how they had attempted to deal with the racial issue in their own ministry. Their answers became somewhat more specific; but for most, their involvement had not moved beyond talk. Virtually all of the liberal ministers report that they at least occasionally talk about the racial issue from the pulpit. About a fith report that they have held special adult education or study groups on the racial issue. A few are members of local committees that are working on some phase of the racial issue. But the predominant mood of their comments reflected a concern for what they should get their laity involved in various programs, but only a few reported that their laity *were* engaged in any kind of significant activity.

Many of the liberal clergy felt that preaching and teaching was about the only thing that they could do about the racial situation, but many others reflected a deep concern that this really wasn't enough. Among this latter group we sensed a considerable amount of frustration and uneasiness, as they responded to our questions about their involvement.

We asked Grand Rapids clergymen to describe the mood of the Negro ghetto. There were two distinct responses. One group felt that conditions were actually not as bad as they might seem to someone who came into the city right after rioting had occurred. To them, the rioting reflected the sentiments of no more than one-half of one percent of the ghetto. Furthermore, they didn't feel that the Grand Rapids rioters reflected the same angry mood of Detroit and Newark rioters. As one Methodist clergyman put it:

> We are making good progress on race relations in Grand Rapids, and most Negroes recognize this. Ninety-nine percent of the Negroes are just as disturbed and embarrassed by these disturbances as the white community. You can't really describe the mood of the Negro section as angry. This thing would never have happened if it hadn't been for Detroit.

While this was the most common response to our inquiry, particularly among conservative clergymen, those who seemed most concerned about the riots were generally candid in admitting that they really didn't know the mood of the ghetto. The following response from another Methodist minister is fairly typical of this group:

> I am ashamed to tell you that I don't really know what the mood of the ghetto is. The fact of the matter is that I know only by heresay. I have talked with a few of the middle-class leaders, but I haven't really spent much time in the ghetto.

While the honesty of those who were concerned can be admired, the fact remains that Grand Rapids clergy by and large have little firsthand knowledge of the ghetto. While our own contact with Negroes in Grand Rapids was limited, we got every indication that the mood of the ghetto was very tense and angry. This impression was reinforced during another visit to the city six months later.

There were no special meetings of the Grand Rapids clergy either during or after the riots to discuss what the churches should or could be doing. The director of the Grand Rapids Area Council of Churches seemed somewhat apologetic that the churches were not more involved, but he explained that it was virtually impossible for them to function in the summer months while so many clergymen were away on vacation.

During the riots, only a handful of white clergymen were out on the streets (our informants estimated between 11 and 20), but not in the Negro ghetto. They went into white neighborhoods and tried to ease the tension and fear among white residents.

If we had difficulty finding knowledgeable and involved clergymen among the liberal denominations, the search was even more futile in the conservative denominations. Only three or four of the conservative ministers reflected any real concern for the active involvement of the church in the civil rights issue. While the liberal ministers talked a great deal about what the church should be doing, this concern was largely absent among the conservatives. Most of them felt that if the church had a role to play at all, it involved preaching and teaching the Gospel.

The clergy of Grand Rapids, thus, present an interesting paradox. On the one hand, we found expressions of deep concern over the riots. On the other hand, our search for clergymen who were involved in any kind of constructive efforts to improve the racially tense conditions in the city turned up only a few rather feeble efforts. Even those who seemed the most deeply concerned admitted that their efforts had been inadequate, and furthermore that they were really out of touch with ghetto life.

INTERPRETATION

From a normative perspective, it would be easy for a strong advocate of civil rights to criticize the clergy for not doing more.

But this, it seems to me, misses the significance of the study. The clergy, perhaps more than any other group in white America, are concerned about civil rights. Why then are they not more involved?

Dozens of social science studies and riot commission reports in recent years have served to underscore two compelling considerations: (1) that white Americans are largely physically and emotionally removed from the ghetto, and as a result (2) they do not know or feel the depth and complexity of the problem.

This point is emphasized in the Six City Report of Brandeis University's Lemberg Center for the Study of Violence:

> ... the attitude of whites seems to be based on ignorance of or indifference to the factual basis of Negro resentment and bitterness.... If white populations generally had a fuller appreciation of the just grievances and overwhelming problems of Negroes in the ghetto, they would give stronger support ... to promote change.[5]

The same point is forcefully made in the *Report of the National Advisory Commission on Civil Disorders.* Commenting on the report, Senator Fred Harris, a commission member, said:

> ... What we have tried to do in this report is to let people see things through our eyes and to feel it in the pits of their stomachs like we do. And then if they can see what a terribly serious crisis this is for our country, there could be a great deal more willingness to try to do something about it.[6]

Clergymen may be more inclined than the general population to feel that civil rights is a serious moral issue, but, like the average white American, they are physically removed from the ghetto and hence fail to understand and feel the full weight of the problem.

The evidence would suggest that a large proportion of clergymen understand the problem of racial prejudice as a moral issue, but to use the words of Senator Harris, they have not felt the problem in "the pits of their stomachs." Those clergy who do understand the depth and complexity of the problem seem frustrated because they don't know what they can do.

But the clergyman who feels helpless and frustrated in the wake of a deepening racial crisis is essentially no different from other white Americans who are concerned with the racial crisis. This sense of helplessness to get at underlying causes is nowhere better illustrated than in the *Report of the National Advisory Commission on Civil Disorders.* Having labeled racism as the underlying cause of Negro unrest, the report fails to make a single recommendation explicitly directed toward the reduction of prejudice. Moreover, it fails to devote even a single page to exploring the nature and dynamics of prejudice.

Yet the problem of reducing prejudice is not so hopeless that it cannot be attacked. A good deal is already known about the nature of prejudice, as well as techniques for its reduction. Even modest research funding in this area could serve to further enhance our knowledge. If, however, the problem is to be attacked at its roots, the knowledge of social science must be translated into action programs.

Perhaps the most important implication of this investigation is that clergymen represent a potential source of manpower for action programs. At the present moment they seem to be relatively ineffective in this role.[7] Three critical steps would seem essential in the realization of this manpower potential. First, clergymen need to be made better aware of the realities of the racial crisis in America. Secondly, training programs need to be established which will educate the clergy as to the nature of prejudice and how it can be changed. Thirdly, strategies for action need to be developed.

Development of such a program would require broad cooperation between government, church bodies, and social scientists. Development of such cooperation is a formidable task in itself. But the more we understand about the racial crisis, the more aware we become of its seriousness and the urgency for action. The potential role that the clergy might play in this struggle should not be passed over lightly.

NOTES

1. For an excellent study of white reaction to rioting, see Richard T. Morris and Vincent Jeffries,

1. For an excellent study of white reaction to rioting, see Richard T. Morris and Vincent Jeffries, *The White Reaction Study,* Los Angeles Riot Study, Institute of Government and Public Affairs, Univ. of California, 1967.
2. For a detailed study of the role of clergymen in the civil rights

struggle, see Jeffrey K. Hadden, *A House Divided* (N.Y.: Doubleday, forthcoming).

3. The author wishes to acknowledge the following students who volunteered their assistance as interviewers for the study: Gene Calvert, Alan Chesney, Andrew Grossman, William Horvath, and Ed Silverman.

4. In this report, the terms "liberal" and "conservative" are used in reference to the theological tradition of the denomination of the clergyman. This is simply a shorthand to avoid continually repeating "ministers from liberal Protestant denominations," and "ministers from conservative Protestant denominations." While clergymen from liberal denominational traditions tend to be more liberal on both theological and social issues, this connotation is not implied in the use of the terms of this paper. The "liberal" denominations are: United Church of Christ, Methodist, Presbyterian, USA, Episcopalian, American Baptist, Disciples of Christ, and Evangelical United Brethren. The "conservative" denominations are: Reformed Church in America, Christian Reformed, Missouri Synod Lutheran, American Lutheran Church, Lutheran Church in America, Wisconsin Synod Lutheran, Regular Baptists, Assembly of God, Church of God, and the Salvation Army.

5. John Spiegel, "Six-City Study, A Survey of Racial Attitudes in six Northern Cities: Preliminary Findings," June 26,1967, Brandeis University, Lemberg Center for the Study of Violence, Boston, Mass.

6. C.B.S. Television Report on the Riot Commission Study, March 3, 1968.

7. Hadden, *op. cit.*, esp. chap. 6, "The Struggle for Involvement."

EFFECTS OF RIOTS ON

NEGRO LEADERSHIP

Harry Scoble

■ Following the devastating Watts riot or insurrection in August of 1965, a research team was organized at U.C.L.A. The present chapter, by a political scientist, is simply one piece of the total interdisciplinary research product. And the basic assumption of this piece is that the northern migration of the civil rights movement in 1963, the "War on Poverty"[1] of 1964, and the riots beginning in the summer of 1964 essentially make obsolete all that political scientists thought they knew concerning Negro political leadership, politics, and power prior to that time.[2]

AUTHOR'S NOTE: *The present research—on the impact of the Watts riots on Negro organizations, leadership, and politics—was supported under a contract from the Office of Economic Opportunity to the Institute of Government and Public Affairs at U.C.L.A. I wish to thank both agencies for their aid in permitting me to do the field research for this project. But at the same time, of course, only I am responsible for what is said here. I was also assisted by a grant from the Social Science Research Council of New York City, for which I gratefully acknowledge indebtedness and make the same disclaimer.*

Another aspect of my study is treated in "The McCone Commission and Social Science," PHYLON, forthcoming, 1968. Other reports from the Los Angeles Riot Study are available through the Institute of Government and Public Affairs at U.C.L.A.

But that basic assumption, or initial hypothesis, is grossly over-stated. In the case of the Watts riots, for example, one can observe important continuities with the past—as well as discontinuities. At one level, one can observe a "logical" historical progression from (1) the politics of order—e.g., lawsuits by the local NAACP, peaceful picketing, and other methods used in 1963 and earlier—through (2) the politics of disorder—e.g., a "sympathy-with-Selma" disruptive sit-in in the Los Angeles Federal Building by members of CORE and NVAC (the Non-Violent Action Commit-tee, a splinter-off from CORE) in March, 1965—to (3) the final politics of mass violence, in the 1965 revolutionary rioting of Watts.

Also, concerning Negro leadership, the riots represent a paradox: a continuous development from and with the past and, at the same time, a discontinuous rupture with that past. The continuity lies in the fact that the August riots constitute an accelerated completion of an age-generational revolution peace-fully initiated against "traditional" Negro leadership (during 1962-1963).[3] But, as many have remarked, the Watts riots indicate discontinuity as well, in the form of a nationwide lower-class revolution against middle-class political leadership, super-imposed upon the earlier age-revolution.[4]

Before stating the additional initial hypotheses of this study, it is necessary to indicate briefly the methodology employed. The two white daily newspapers and the two Negro weeklies were read thoroughly for a period extending from Janauary, 1963, to the field research period (April-June, 1966). But the main source of data was informal yet focused interviews[5] with key informants, selected judgmentally.[6] There were twenty-five interviews, ranging in time from one to four hours. Twenty-two informants were Negro. All were selected precisely because the newspaper analysis, or the cobwebbing technique,[7] suggested that they had the highest probability of having specialized knowledge in a particular deci-sional or issue area.

The first focus of my study was on limitations or constraints on Negro political power generally;[8] and the second was on changes resulting from the Watts riots, which is the subject here.

The initial findings of this study—which, because it is a single-city case study with a limited data-base, properly ought to be considered as tentative hypotheses for future comparative research—are as follows:

1. Since the Los Angeles riots, white political and civic leaders have apparently become more responsive to Negro demands.

   a. To this point, such responsiveness is more verbal and visible than it is substantive: the degree of public policy change that can be observed is minimal (as indicated below); there is furthermore a Machiavellian tendency within the political party system for Democratic political leadership to take the Negro as a traditional Democratic "given," and contrariwise for Republican leaders to write off the Negro vote.

   b. Prior to the riots, and evidenced by the 2-to-1 vote in favor of Proposition 14 to "repeal" California's open occupancy legislation, anti-Negro attitudes of whites appeared to harden (contrary to what was then occurring in other Northern cities and states, which continued to act upon Negro demands for legislation—at least until the general "hardening" evident by mid-1967); and this has produced a stalemate in the Los Angeles area.

   c. The open occupancy issue was what most inflamed most whites; while the intertwined issues of police brutality (i.e., unnecessary force), police harassment (i.e., subtle racism and "psychological warfare"), and under-protection were the issues which (from the Negro point of view) most nearly unified *all* Negro leadership and followership.

2. Since the Watts revolt, Negro leaders have articulated more demands.

   a. The organizer and mass-agitator types of leaders (as against both prestige and token leaders) have become more important in the Negro communities.

   b. The style of leadership, regardless of the base of power, has become more militant throughout.[9] (The denigration of ministerial moderate leadership—noted below—is by no means limited to Black Muslims or black nationalists.)

   c. The demands made have increasingly stressed welfare goals—i.e., those sought by the working-class Negroes—as distinct from the more symbolic, status-integration, equality goals of traditional middle-class Negro leadership.

   d. There is tension within Negro leadership, since it remains primarily middle-class and since its personal goals are primarily status-oriented.

   e. Negro political (elective) leaders are probably best able *personally* to overcome this tension between status and welfare orientation, because of the *generally* accommodative nature of American politics and the *presumed* flexibility of its political personalities. But they are exposed to systemic tensions and role conflicts because: (1) "with power comes responsibility" in American politics, meaning they must give up overt criticism of "The System"; and (2) the "white power structure" will try to coopt and/or "deal" with them, exposing them to competition for Negro leadership positions. In short, Negro elective leaders cannot now produce public policies fast enough to satisfy demands of civil rights and protest leadership, especially of the militants. As of early 1968 there were at least 154 Negro legislators in the 50 states, and also 215 Negro elective officials in the 11 ex-Confederate States (as against almost none in 1958)—but the point I am stressing is that such increases in the Negro political leadership-aggregate generate geometrically increasing expectations and demands compared with the achievement level currently permitted by American politics. And this, however trite it sounds, is the essence of a revolutionary situation.[10]

3. The McCone Commission Report recommendations represent a maximum program to most whites, including most white leaders, but only a minimum and largely symbolic program to Negro leadership and followership alike.

4. The structure of local government and politics make it especially hard to articulate Negro demands satisfactorily.

   a. In addition to the normal political discounts inhibiting political activity of all working-class individuals, the Negroes have special problems of apathy toward and fear about political participation, which further limits Negro electoral potential. There is also the traditional problem of the difficulty for Negro (or any other) leadership in suddenly shifting the normal party identification and voting behavior of the followership.

b. The Negro community lacks the reservoir of abundant organizational talents normally prevalent among a wide variety of white groups. For example, there is no full-time staff man in Los Angeles whose sole function is to plan strategy and tactics.

c. The division of authority and function between city and county, and also between state and federal levels of government, makes it difficult to pursue even a single goal (such as housing).

d. The weak mayor-council tradition, and the proliferation of independent boards and commissions of the city (i.e., the "civilian" Police Commission), also inhibit effective Negro political action.

e. Nonpartisan elections and—in the case of the Los Angeles Unified School Board—at-large districts may further operate to discount the voting-rate and/or numerical significance of Negro electoral power. Or, in the latter case, at least to force the selection of middle-class Negroes (such as the Reverend Jones) acceptable as coalition candidates to white middle-class potential allies.

f. The lack of an urban political machine and its tradition of service orientation and welfare for various low socio-economic status groups may also be a contributing factor, at least so far as induction into electoral politics goes.

g. Multiple power structures (or a fragmented power structure) will be reflected by—or will *tend* to be reflected by—power fragmentation within the Negro political and civic leadership structure. Multiple power structures may be just as difficult to deal with as a cohesive, opposed, and unified white power structure.

The above should be tested far more extensively before they are accepted as established findings. But following are further, more detailed conclusions about Negro politics in Los Angeles which are more firmly established.

With regard to implementation of the McCone Commission recommendations—minor though they were[11]—the basic conclusions of this study are that (1) white political leadership has proven exceedingly inept,[12] and (2) the few actual accomplishments are largely symbolic rather than substantive. On the latter point, for example: a Negro lawyer was designated as the president of the Police Commission; the City Council enacted as an ordinance, in 1966, the mid-1963 Executive Order of the Mayor forbidding discriminatory practices in municipal employment (while continuing to ignore all problems of promotion and upgrading); the creation of a "toothless" City Human Relations Commission, functioning primarily as a "PR" device; and minor adjustments of police procedures for investigating complaints by civilians and for behavior by the police in the ghetto.[13] Given these minor, largely symbolic accommodations, the analyst would predict that some political actors would come to appraise further violence as a more efficient tactic than available, conventional practices.[14]

An initial examination of the more than 250 incidents of Negro protest demonstrations in the Los Angeles area since approximately May, 1963, also leads to the conclusion that the protests have accomplished very, very little. One notes changes in leadership—some who were active in 1963 have in their frustration become alcoholics, or withdrawn into political apathy, or were disempowered by internal fractional battles (in one instance, at least, because of public incompetence). There are other observable changes in terms of the racial character of the crowds, the specific type of demonstration engaged in (pray-in, sleep-in, sit-in, etc.). But the major conclusion is the lack of public policy impact from such demonstrations. One is impressed, for example, that the major demands focused upon in mid-1963—with regard to de facto segregation of the public schools—remain the same sought-after goals today, with the exception that there has been the additional demand, superimposed on these, for quality educational facilities in the ghetto, and not just integration. In view of this, it is not at all surprising that the national debate on strategy and tactics engaged in in 1966-67 by Carmichael, McKissick, and the more traditional and middle-class national Negro leadership, is also reflected at the "local" level by such recent demands as that for "Freedom City"—for total separation of the Negro population as a self-contained municipal entity.[15]

There has been a very rapid expansion in the size of the Negro political class (or aggregate) in Los Angeles since 1961. Partly, this is the result of the growth of policy-and-issue oriented organizations such as UCRC (United Civil Rights Committee), the NVAC split-off from CORE, or TALO. Partly this is the result of changed hiring and promotion policies of governmental bureaucracies—e.g., Shaw as postmaster, or Murray as mayor's aide, or Hudson as president of the Police Commission. And partly this is the result of electoral change: population growth and apparently unexpected success in electing candidates—as in the three 1963 councilmanic victories—and the 1965 legislative reapportionment (which may, however, still systematically gerrymander against the Negro population in electoral interests). This expansion has been revolutionary rather than evolutionary. There has been no systematic Negro political leadership recruitment in Los Angeles. Congressman Augustus Hawkins, the one man who could have developed a second or third level of political leadership, grooming potential candidates and training protégés, did not choose to do so. Hawkins' following seems largely traditional rather than warmly personal, and his local organization seems amateurish and has a high turnover rate, especially in contrast with the professionalism and full-time devotion to politics of the younger Negro political leaders allied with Assembly Speaker Jesse Unruh—Mervin Dymally, Billy Mills, F. Douglas Farrell, Gilbert Lindsay, and now William Greene and Leon Ralph. It would not be surprising to see Hawkins directly challenged in the primary election two or four years hence by Mills or Dymally. On the other hand, enough "absentee political landlorship" may exist—of white elective officials representing substantial or even substantially Negro constituents, such as Supervisor Hahn and Councilman Gibson—to absorb the energies and ambitions of the more dynamic, younger politicos.

The contrast I wish to make is with that of the Chicago machine-within-the-machine model of Congressman William Dawson. There is a pattern of recruitment, long years of work, organizational loyalty, discipline and sanctions, training in machine skills—especially those of political accommodation and "brokerism"—and a quiet emphasis on tangible rewards for follower-voters in terms of welfare processes. In Los Angeles, on the contrary, wide-open recruitment and self-selection exist; these introduce a wide variety, and disunity, into the political leadership aggregate. These processes mean perhaps less and certainly different skill; and there has been an emphasis on intangible rewards to follower-voters in terms of rhetoric, ideology, and racial appeals. Under such conditions of recruitment, political—especially electoral—leaders would normally emphasize status (that is to say, the integration, equality, and symbolic racewide) goals. They probably would have done so but for the reorientation of the civil rights movement, the development of the poverty program, and the riots, which in combination forced a reorientation to welfare goals. At this writing, the best evidence of this was the split between the ACLU (American Civil Liberties Union) and the NAACP over endorsement of the June, 1966, Primary School Bond Issue of 189.5 million dollars; the NAACP stressed equality and the need for improvement within an obviously long-persisting ghetto, and so endorsed the bond referendum, whereas the ACLU remained attached to the white middle-class liberal goals of integration and could not recommend a favorable vote.

Elected Negro officials in the metropolitan area in 1965-66 developed their own political interest group known as CONEO, the Conference of Negro Elective Officials. Councilman Thomas Bradley was the head of this, and there was some indication of the development of unified strategy; but it is probably too early to assess the effect of this organization. From all evidence available, the organization was totally ignored by presidential assistant Benjamin Weinman in setting up the June, 1966, White House "Conference on Negro Rights." However, CONEO had in 1967 become the energy source of and model for a national organization of Negroes.

The Negro Political Action Association of California (NPAAC) was created in 1964 and is another recent, interesting, and potentially important development. If one refers only to Negro newspapers and periodicals, one gains the impression of widespread success and power of this organization—the first such in any state—but a more moderate and modest view should prevail. NPAAC is at variance with the black nationalist organizations because of its insistence that "to have an impact on the power structure, you have to operate within and through that power structure," and also because it officially encourages bi-racialism—although estimates available in 1966 put white membership at considerably less than 5% of the total, and there is no reason to believe it has increased since. In assessing the political significance of NPAAC endorsements in the 1966 Democratic Primary, it appears that NPAAC did not actively recruit candidates, it did not and could not supply "troops," and its financial contributions to endorsed candidates were quite trivial; rather, the organization waits until individuals self-select themselves, and then it endorses. It remains to be seen whether 1968 will prove different; meanwhile, if Congressman Hawkins were to take a greater interest in the organization, it might develop more effectively.

To a certain extent, NPAAC represents unification in the Negro political leadership aggregate—that is to say, institutionalization of the ad hoc unity originally achieved behind the Reverend James Jones when some 116 separate, and mainly Negro, organizations endorsed his candidacy for District 2 of the Board of Education in 1965. In another sense, however, NPAAC represents only partial unification at best, since the main forces in it have been independent Democrats, pro-Brown Democrats, and other allies of Hawkins. (Independents and Republicans are obviously excluded from NPAAC; Yorty Democrats are tolerated by it but ineffectual and inoperative within it; and Unruh Democrats seem to use it for their own purposes: they belong to the organization, just as many Negro political aspirants and leaders belong to the NAACP or the Urban League, because even though the organization cannot help them electorally, Negro political aspirants can be hurt by non-membership in it or by an organization blast against them.) Finally, thus far NPAAC has only partly been aimed at white Democratic politicians, in an effort, within the Democratic primary, to force the nomination of a Negro for statewide constitutional office. In the 1966 primary, for example, Negroes nominated a candidate for the position of Secretary of State—because it was then the only one held by a Republican, and a very elderly incumbent at that.

To be truly successful as a balance-of-power bloc vote, in both the primary and the general elections, the NPAAC would have to develop either into the specific modified-third-party form of the Liberal Party of New York City and State, or Negro political leadership would have to demonstrate that it can produce a significant amount of non-voting, or more desirably, of switch-voting and ticket-splitting among Negro Democratic voters than has been demonstrated to this point. An overtly Republican move has been made by a younger Negro leadership cadre identifying itself as the "New Breed" in Chicago in the past three years; it has not yet been reflected in Los Angeles, although certain black nationalists have been vocally critical of liberal Democrats as "the real enemy," and national reverberations of the Powell exclusion might well lead in this direction. The difficulty is that such national tendencies might well be checked by local Negro perceptions of the new Reagan Administration—for which no data yet exist. And prediction is made further difficult by uncertainty as to what, if anything, will be the domestic political impact—on Negroes and on Democrats—of the Vietnamese war.

The evidence available at this writing—especially from the 1966 Democratic primary State Assembly and Senatorial races, which centered on the Negro-population concentrations of Los Angeles—leads to the further general observation that good guys indeed finish last. The tentative finding is that Negro allies who are generally identified in the one remaining viable faction of the Democratic Party not only won, but that they tended consciously to violate the "rules of the game." For example, sample primary ballots were addressed to the voters in these particular districts. These ballots were identical in form, format, and print with the official sample ballot of the Registrar of Voters, but they also bore legible added check marks after the names of particular candidates and did *not* bear, as required by the state electoral code, the necessary boxed-identification of the fact that this facsimile ballot was not an official party publication but rather the propaganda of a non-party group. Other instances of apparently conscious attempts at manipulation, deceit, and fraud of the voters are readily available. Complaints must legally be addressed primarily to the County District Attorney, but the only statutory remedies are in the realm of civil action. Therefore, the sanctions are not particularly severe, even if one might win in court. Furthermore, fragmentary evidence on the often-admired "toughness" and indeed "brutality" accompanying the success of national political leaders—as in the careers of Presidents Kennedy and Johnson—may mean that "un-nice" tactics are always present, that the only unchanging rule of the political game is to win.

Another possible impact of the 1965 riots is that electoral Negro leaders may be displacing the older and more traditional

business/civic leader types; at least they are definitely supplement-
ing them. The proposition of displacement could be interpreted—perhaps overly interpreted—from the 1966 NPAAC
Convention's choice of Bill Williams, recently Congressman
Hawkins' field representative, over Norman B. Houston (insurance
executive and business leader) as the Negro statewide choice for
the nomination for Secretary of State. At the least, there is
evidence of supplementation, in the careers of Councilman
Bradley and Mills and in that of Senator Dymally.

The rise of the young militants—Ralph Reese and Tommy
Jacquette of SLANT (Self Leadership Among All Nationalities
Today); Robert Hall of NVAC; Jamal and Ron Karenga of US (a
black nationalist group); Ernie Smith of the Afro-American
Citizens Council, and others—is pushing and pulling the established
leadership into more extreme positions. (As of the time of writing,
the best evidence available suggests that the so-called RAM—the
Revolutionary Action Movement described by *Life Magazine* and
practically every other national periodical in mid-1966—exists,
primarily, at the level of heated and romantic discussion in Los
Angeles.) If there are para-military Black Panthers or Deacons of
Defense *organizations,* they are keeping out of sight. This is not to
say that there are not a minuscule number of "left-authoritarians,"
a small number of "new leftists" (or communal utopians, such as
SDS—Students for a Democratic Society), and of black nationalists—and these are potentially important as what might be
termed a proto-political or, more properly, para-political movement in the area. Perhaps more important is a group like SLANT,
which had in 1966-67 perhaps as many as 500 functioning
members, and it remains the only organization that currently has
effective contacts with both the proliferating black nationalist
groups and the established street gangs—the Slausons, the Businessmen, and others—whose members probably did participate actively
in the riots, who seem prone to violence, but who also are of the
age for political action and precinct organization.

This rise of militancy, which may be an age-generational split
within the Negro community along social class lines, is causing
tensions with traditional white liberal allies—Jews, the ACLU, etc.
For example, the United Civil Rights Committee (UCRC) was
originally hailed in 1963 as *the* unified, coordinating umbrella
agency (and also a metropolitan first). By mid-1966, even middle-

class moderates disparaged it: "There is increasing belief that decisions affecting Negroes ought to be made by Negroes"—i.e., alone. (And by 1968 the organization had disappeared.) The major thrust of black nationalism (in a secular sense: membership in religious Black Muslim groups seems restricted and relatively stagnant, yet admiration for these groups—and probably more generally for the late Malcolm X— seems widespread among all age and class levels of Negro leadership) is this emphasis on Negro decision-making. However, on the basis of United States political history, one would predict that the black nationalists are doomed to failure. Perhaps bi-racial proletarianism would be more effective. But, certainly, racist exclusionism leaves the "black power" leaders as "leaders" of the least politically mobilized or mobilizable sector of the electorate.

One would predict instead that black nationalism will go the classical course of traditional third-party protest movements and leaderships: to the extent that their demands seem both "legitimate" and especially threateningly attractive, the established, relatively moderate leaders will accommodate these demands—or pirate them, if you will. In this sense, division within the Negro leadership is probably not only inevitable but also politically desirable: Malcolm X in his posthumous autobiography asserted to Martin Luther King, "*You* need *me*." If there is indeed to be substantive rather than merely symbolic change in public policy, it would seem that this would be facilitated by a very aggressive and militant leadership among Negroes, who in effect can be denied recognition by "the white power structure" which turns instead to a more acceptable Negro leadership. In Los Angeles, as in the nation at large, however, no political analyst has the ability to predict that this state of affairs will come about; At the present time, white local political leadership could not be characterized by imagination, courage, or educational ability. If the personnel of the white leadership does not change substantially, if the present affluent stalemate remains, there seems no reason to predict that Negroes will continue to prefer the moderates to the militants, even though both may well be demanding precisely or almost wholly the same policy goals. The difference then would be in terms of strategy and tactics. Continued failure—"exhaustion of available political remedies"—could lead to increased violence, as previously suggested.

Public and private policy leadership is increasingly specialized and fragmented among Negroes. Influenced by the civil rights movement, the rediscovery of poverty, and especially by the riots of 1965, the number of policy arenas potentially affecting Negro demands and interests has risen far more rapidly than the reservoir of Negro leadership. In tracing Negro participation in various issues and decisions, the following sorts of specialization were revealed: (1) in the attempt to implement the recommendaton of the McCone Commission for a county hospital in Watts, the primary Negro participants were Negro doctors; (2) in the Hoover-U.S.C. Urban Renewal Project, the primary participants have been Dr. Christopher Taylor and area residents; (3) in the police brutality/harassment and reform of the Civilian Police Review Board issues, the early participants seem to have been Thomas Neusom and Thomas Bradley of the City Council, John Buggs of the County Human Relations Commission, and, to a lesser extent, Councilman Billy Mills—they were more recently supplemented by the major activists within the civil rights/ church/black nationalist coalition that was the Temporary Alliance of Local Organizations (TALO); (4) in the poverty program, primarily Representative Hawkins and Mrs. Opal Jones of NAPP (the Neighborhood Adult Participation Program); (5) the private training/employment, primarily the traditional-ministerial leadership plus others, such as the executive director of the Urban League; (6) in education, primarily Mrs. Tackett of UCRC; (7) in specific manifestations of the black power ideology, such as "Freedom City," primarily the younger militants of SNCC, CORE, and so on.

But there is a continuing search for a unified, effective all-embracing Negro leadership organization. Studies of Negro leadership in Southern communities—such as Hunter's study of Atlanta and Burgess' study of "Southern City"[16]—indicate that such organization normally exists in the South. Perhaps this is the mirror image among Negroes of the white wish, and the reality in the South, that there in fact be a unified, monolithic Negro power structure to deal with, and through which to control, all Negroes. In 1951, an abortive effort was made to create a Negro leadership organization in Los Angeles. It was initiated by Welsey Brazier, Executive Director of the Urban League, who attempted to enlist the support of the NAACP branch in the area, thus encompasing the only two then-visible Negro organizations. This unification attempt, known as ACO, the Association of Community Organizations, lasted some six or seven months and failed for a variety of reasons: Brazier suffered a heart attack; the white minister selected as the figurehead became much more interested in the "Draft Eisenhower" movement; the regional NAACP looked upon this as an empire-building strategy on the part of Brazier and the Urban League. This failure may illustrate the problem—the pervasive problem—of fear of sellout: no one leader or group is looked upon by the others as sufficiently neutral to be trusted with the power and prestige of a unified command. There are, however, other factors as well.

For example, E. Franklin Frazier's outspoken indictment of the Negro middle class, especially the professionals, in his *Black Bourgeoisie,* seems to apply to Los Angeles.[18] This belief was reinforced by the many Negro leaders interviewed. In addition, as indicated earlier, the Negro community lacks the resources—or initiative—for true bureaucratization. Financial contributions by Negroes to politics, at least until recently, have not been made by those able to make them with reference to any long-term and overall Negro strategy. But such financial resources are more lacking than for any other racial or ethnic group except, perhaps, Puerto Ricans or Mexicans. Related to this is the question of organizational talents per se. Those lucky few with the educational training and/or the occupational experience that give one organizational skills seem more inclined to go it alone in the business and professional world than to devote themselves to collective purposes—a devotion one finds in abundance in the middle class, the Jewish, and other white worlds. In the interviews, the respondents were encouraged to discuss the possiblity—advocated by John Buggs, Executive Director of the Human Relations Committee of Los Angeles County, among others—that the "Men of Tomorrow" might supply this long-missing need. This does not seem feasible. The roughly 150 members of this group tend to be younger men on the make, professionally and in business. This group meets, at most, once a week for the traditional luncheon. It has no staff. It has no continuing program. And these conditions, particularly the lack of qualified full-time staff directors, seem true of all available alternative organizations that might be considered for the "umbrella" role—the local NAACP, for example, has had no full-time staff director since 1956. The one possible recent exception is TALO—but this organization did not survive its mid-1966 police-surveillance campaign in the Negro ghetto, nor has a new coordinating agency yet arisen.

SUMMARY

Negro politics in Los Angeles seems now at a pause. The political/governmental aggregate has been expanded rather dramatically, but it is not providing particularly visible nor effective leadership for the reasons previously indicated. Since the riots, direct action demonstrations seem to have run their course. Some of the more demonstrative organizations have now retired to self-held programs in the ghetto (e.g., NVAC's "Operation Bootstrap" with its slogan of "Learn, Baby, Learn"); others seem directionless, demoralized, and disintegrating. This is not to say that direct action will not occasionally be employed and be effective, as in the 1966 protest demonstrations against the firing of Mrs. Opal Jones by the Economic and Youth Opportunities Agency. But Los Angeles' experience thus far seems to confirm James Q. Wilson's generalization that direct action (i.e., of *non-electoral* protest) will be limited to those infrequent situations in which success can be achieved. And the conditions of effectiveness are: (1) that there be an agreed-upon goal, the goal being specific rather than general, defensive rather than assertive, and welfare-oriented rather than status-oriented; and (2) that there be a definable target capable of acting, for which the costs of continuing demonstrations are real— the threat is credible—and from which the gains affect primarily Negroes, rather than constituting a "loss" for large numbers of whites.[19]

One is tempted to say that protests, if they are to be effective, must now be expanded in a more explicitly political direction (i.e., electoral reprisals). For the present, however, the prospects for such development are less than favorable. Inertia of party identifications weakens the power potential of Negroes, while the search for allies—who might themselves employ electoral reprisals—is hampered by the present weakness of potential candidates—i.e., the 1966 elections and evidence since suggest that locally labor unions remain weak, that the California Democratic Council has been faltering since 1964, that the Californians for Liberal Representation has not developed a membership base, that the "New Politics" effort has been unimpressive, and that the Mexican-American "community" remains split and distrustful of Negroes as of its own leaders. In an important sense, this raises a fundamental question as to the utility of conventional politics and of the capacity of the political system to *produce,* legal, economic, and social change rather than merely to reflect these. It raises a question posed by radicals which liberal democrats like not to face: "Must conditions become worse—with your deliberate aid—before they can become better?" Alternatively, unintended chaos resulting from the Vietnamese conflict may produce just that "grand, indeed revolutionary, realignment" of the party system which Bayard Rustin, among others, has called for, and which most political scientists insist is neither feasible nor desirable.[19]

Perhaps what one can therefore presently anticipate is simply a slow consolidation of Negro political power, continuing efforts toward organizational coordination, and probably also increasingly impatient waiting until political conditions change locally and create the possibility of a decisive election comparable to that of the national legislative "break-through" following the 1964 elections. One awaits accident and change, hoping to be prepared when they come.

NOTES

1. A political analysis of the Economic Opportunity Act of 1964 may be found in my chapter in Thomas Weaver and Alvin Magid (eds.), *Symposium on Poverty* (San Francisco: Chandler-Science Research Associates, forthcoming, 1968), and in a book-length treatment, *The Politics of Poverty* (Englewood Cliffs, N.J.: Prentice-Hall, forthcoming, 1969).
2. A good initial bibliography may be found in James Q. Wilson, *Negro Politics* (N.Y.: Free Press, 1960). The difficulty with Wilson's book—as with Robert A. Dahl's study of New Haven in *Who Governs?* (New Haven: Yale Univ. Press, 1961)—is that nothing much was happening in Chicago at the time of the field research. But, in both cases, the pace and intensity of politics changed rather dramatically *after* the book was written and published.
3. See Lerone Bennett, Jr., *The Negro Mood* (N.Y.: Ballantine Books, 1964); Louis E. Lomax, *The Negro Revolt* (N.Y.: New American Library, 1963); and Loren Miller, *The Petitioners* (N.Y.: Pantheon, 1966).
4. On intra-racial class segregation, see E. Franklin Frazier, *Black Bourgeoisie* (N.Y.: Collier Books, 1962), and also his chapter in

Arthur J. Vidich et al (eds.), *Reflections on Community Studies* (N.Y.: Wiley, 1964). And also see Ralph Ellison, *Invisible Man* (N.Y.: New American Library, 1952).

5. See Alexander Heard, "Interviewing Southern Politicians," *Am. Polit. Sci. Rev.,* SLIV (1950), pp. 886-896. A schedule of the total range of questions employed in this research may be obtained on request from the author.

6. For political reasons, I do not wish to enter the thicket of the still-continuing Hunter vs. Dahl et al exchange. It is relevant to note, however, that even the most hard-data political scientist inevitably ends up getting his data from other people reporting about themselves and other people. If this is "reputationalism," then let us make the most of it.

7. See Robert A. Dahl's essay in Stephen K. Bailey et al, *Research Frontiers in Government and Politics* (Washington: Brookings Institution, 1955).

8. For a treatment of general constraints on Negro political power, see my chapter on "Negro Politics: The Quest for Power" in Peter Orleans (ed.), *Reader in Sociology* (Boston: Allyn-Bacon, forthcoming, 1968).

9. By early 1968, because of rapid flux and change in Negro politics, the term "militant" may have lost any useful, unitary political meaning. At the present, militancy has at least three separate meanings—as to the rate of change, the substance of change, and strategy and tactics for change. (In future research, I hope to investigate just these.)

10. See Chalmers Johnson, *Revolutionary Change* (Boston: Little, Brown, 1967). The revolutionary potential is increased because of the resistance in the system to substantive change. This resistance arises in part because of fears of lower-income, often ethnic, whites frequently in public-service bureaucracies and unions. This inter-pretation—by Joseph Kraft, for example, in his 1968 columns on the New York City "garbage strike" by Italo-American sanitation workers—is a modern variant of the "Status Politics" interpretation to be found in Daniel Bell (ed.), *The Radical Right* (N.Y.: Double-day, 1962).

11. See my article cited in note 1, and the sources cited there.

12. In December, 1965, the McCone Commission Report was released. A major recommendation was the building of a 400-bed hospital in Watts proper. The county was to build the hospital, for 20 million dollars (of which the federal government would put up 8 million dollars). The five County Supervisors, in their wisdom, decided to have a referendum vote on the 12 million dollar bond issue—when, as a perfectly legal alternative, they could have directly voted such monies out of the general fund themselves. A bond referendum requires a minimum of 66.67% of the total vote cast. The vote in the spring of 1966 was only 64+%. As of this writing, a fine sign exists in Watts indicating where the new hospital will be when, or if, it is ever funded.

13. See *Los Angeles Times,* March 11, 1968, part 2, pp. 1 and 12; "(Chief) Reddin Opposes Heavy Weapons to Quell Riots" by Paul Houston. This is a lengthy article extensively quoting the chief of the Los Angeles Police Department as to how much and how well they have already met, or reasonably rejected, the fifteen recom-mendations on police/miniority relations made by the President's National Advisory Committee on Civil Disorders. For example, Chief Reddin reported that "roust-frisk" procedures were being reduced in the ghetto. But at least equally symbolic, the L.A.P.D.

had budgetary authorization to purchase two armored vehicles ("mainly for rescue"); it was equipped with buckshot (rather than less lethal birdshot); and it had stockpiled ("in dead storage") ten Thompson submachine guns and sixteen Reising semi-automatic rifles.

14. This is a prediction—but, as a social scientist, I am aware of the danger of self-fulfilling prophecies, and I do not make the statement lightly or happily.

15. The proposal was made by the head of SNCC in Los Angeles in late Spring, 1967. It is difficult to consider the proposal as seriously made in at least two cases. First, there was apparently no effort to anticipate negative reactions to the proposal, such as those centering on the tax-resource capability of the ghetto. And second, the legal procedures (e.g., petitions, referenda, etc.) for "disincorporation" and the "reincorporation" are horribly complex and time-consuming—and indeed, there was no visible evidence of any effort even to undertake the first step.

16. See M. Elaine Burgess, *Negro Leadership in a Southern City* (Chapel Hill: Univ. of North Carolina Press, 1960), and Floyd Hunter, *Community Power Structure* (Chapel Hill: Univ. of North Carolina Press, 1953).

17. Frazier, *op. cit.,* note 4. Frazier's view is primarily socioeconomic. Clark reaches a somewhat similar, but much more sympathetic, conclusion from this psychological perspective when, in discussing the incapacity for voluntary sacrifice by wealthier Negroes, he attributes it to the tremendous *in*voluntary sacrifice demanded of all in the ghetto. See Kennth B. Clark, *Dark Ghetto,* (N.Y.: Macmillan, 1965).

18. James Q. Wilson, "The Strategy of Protest," *J. Conflict Resolution,* V (1961), pp. 296-303.

19. Compare Bayard Rustin, "From Protest to Politics," *Commentary,* XXXIX (Feb., 1965), pp. 1-12, with James Q. Wilson, "Negro Politics in America," *Daedalus,* VC (Winter, 1965), pp. 940-986.

COPS IN THE GHETTO
A Problem of the Police System

Burton Levy

■ During the past five years, millions of dollars have been spent by police departments, much of it federally funded, for police-community relations programs (really "police-Negro relations"). The summer of 1967, with the destruction in Newark and Detroit, and the actual and threatened civil disorder in some thirty-five other urban communities, provides good reason to question the premises and assumptions on which these programs are based.

I have been a principal contributor to the notion that the gulf between the black community and the police in urban communities could be breached with lots of money spent for police recruitment and in-service training in human relations; for precinct police-citizen programs; and for generally upgrading and professionalizing the police service by raising salaries to retain current employees and to attract college-educated recruits. A short article I wrote two years ago brought requests for thousands of reprints.[1] I suppose I said what everybody wanted to hear: that is, that 96% of the Negro community are completely law-abiding; that 98% of the patrolmen never have complaints of brutality lodged against them; that whatever negative images now exist on both sides are a result of a history of brutal Southern police and the general stereotypes and prejudices that exist in America today. Therefore, I said, what is needed is a new dialogue based on fact, not myth, and a significant program of training for policemen.

My position is now completely reversed. Two more years of intensive experience with police in all parts of the nation, combined with results of other studies and statements by police officers themselves, have convinced me that the problem of police-Negro relations in the urban centers is one of patterns of values and practice within the *police system*. The new assumption is that the problem is not one of a few "bad eggs" in a police department of 1,000 or 10,000 men, but rather of a police system that recruits a significant number of bigots, reinforces the bigotry through the department's value system and socialization with older officers, and then takes the worst of the officers and puts them on duty in the ghetto, where the opportunity to act out the prejudice is always available.

This paper examines three major questions:

1. What is the present relationship between the urban police and the black community?
2. What, and how effective, are current efforts to improve police-Negro relations?
3. What is the nature of the police system as it relates to police-Negro relations?

BLACK ANGER

Every poll and survey of black-white attitudes toward the police produces the same results: Negroes believe that policemen are physically brutal, harsh, discourteous to them because they are black; that police do not respond to calls, enforce the law, or protect people who live in the ghetto because they are black. White people simply do not or are unwilling to believe that such racial discrimination by police officers actually happens. Louis Harris' national survey reported in August, 1967, that Negroes feel "two to one that police brutality is a major cause [of the civil disorder] — a proposition whites reject by eight to one. Only 16% of whites believe that there is any police brutality to Negroes."[2]

These attitudes in the Negro community are neither post-riot excuses to explain the disorder, nor new ideas planted by black power advocates Stokely Carmichael or Rap Brown.

In 1957, a poll of Detroiters showed that less than one white person in ten rated the police service as "not good"; while over four out of ten Negroes rated the police as "not good" or "definitely bad." Two-thirds of the Negro respondents referred to anti-Negro discrimination and mistreatment by police officers.[3] A 1965 poll in Detroit found 58% of the Negro community believing that law enforcement was not fair and equitable.[4] Similar results occur in other urban areas in the nation, North and South.[5]

Black hostility toward police is not confined to the poor or to those engaged in illicit activity in the Negro community. Black doctors, lawyers, and even police officers share the beliefs. For example, the Guardians, the New York City organization of Negro police officers, endorsed the establishment of a Civilian Review Board in opposition to the organizations of white officers. The Guardian president publicly stated that he had witnessed incidents of police brutality.

Finally, it is factually correct to note that virtually every incident of threatened or actual civil disorder in the urban ghetto began with an encounter between a police officer and a Negro citizen. Whatever the factual reality is — as contrasted to the belief systems — clearly the cops serve as the "flash point" for black anger, mob formation, and civil disorder.

To what extent does police brutality actually exist — i.e., the verbal insults and harassment, the negative selective enforcement and non-enforcement, the physical brutality? The problem here is that systematic evidence outlining patterns of behavior is difficult to obtain. It is equally difficult to gain evidence on individual cases.

The U. S. Civil Rights Commission's 1961 Report *Justice* noted that the U. S. Department of Justice received 1,328 complaints alleging police brutality in the two-and-a-half-year period from January, 1958, to June, 1960.[6] One-third of the complaints were from the South, and somewhat less than one-half of the complainants were Negro. Police officials note that few, if any, of these cases investigated by the F.B.I. have resulted in prosecution and certainly not conviction of any police officer. Still, the Civil Rights Commission concluded that "police brutality in the United States is a serious and continuing problem. . . ."[7]

The Michigan Civil Rights Commission, established by the new State Constitution of 1963, has the legal authority to accept, investigate, and settle complaints of discrimination by police because of race, religion, or national origin.[8] From January 1, 1964, to December, 1965, 103 complaints were filed against the Detroit Police Department. By December 30, 1965, the Commission found probable cause to credit the allegations in 31 cases, and the new Detroit Police Commissioner, in separate investigations by a newly established Citizen Complaint Bureau, had made similar findings. Citizens received apologies and medical expenses; officers were reprimanded and transferred.

Contrast the 1964–65 results of citizen complaints, when an independent review apparatus was available to Detroit citizens, with the departmental grievance procedure before the establishment of the Civil Rights Commission. The Detroit NAACP filed 51 complaints with the Detroit Police Commissioner from 1957 to 1960, and not one case was upheld. The police "investigated" 121 incidents involving Negro citizens during the first nine months of 1960, and according to the then Police Commissioner's sworn testimony to the U. S. Civil Rights Commission, in only one case was the officer "definitely at fault."[9]

The Detroit statistics cited above, both in terms of charges or findings of police brutality and the polls of attitudes, are not meant to make a special case for or against the Detroit Police Department. Today, Detroit is better in terms of police-Negro relations than some other comparable departments; it is worse than others. What is generally true of Detroit is also true of Boston, New York, Chicago, Los Angeles, and other urban communities. That is, the black community, from top to bottom, is angry about what they call mistreatment by police, from verbal abuse to physical brutality; further, that such mistreatment does occur — perhaps not to the high degree perceived by the Negro community, but certainly to a much greater degree than the police and municipal administrators have been willing to admit or correct. And when this relationship overlies the social and economic despair of the urban ghetto, it is little wonder that the cop on the beat (theoretically, the "foot soldier" of the Constitution)[10] becomes the "flash point" of urban disorder.

CURRENT REMEDIAL EFFORTS

Efforts to improve the relationship and increase attitudes of trust between the black community and the urban police move in three major directions: (1) professionalizing police by increasing education, training, and salaries; (2) establishing formal police-community relations programs for police dialogue with the Negro community; and (3) recruiting more Negro police officers.

The first program assumes that education and training — recruit, in-service, and off-duty — will produce policemen better able to cope with the general complexities in law enforcement and also be more understanding of their own conscious and unintended actions toward citizens. Higher salaries will enable departments to recruit more highly educated and talented men.

The following proposal by the Detroit Citizen's Committee for Equal Opportunity reflects the recommendations of most civil rights organizations and professional police societies:[11]

> We must support steps to further professionalize the police department. The increasing problems of our society are connected directly to the ever-increasing complexity of life itself. The dignity of the law enforcement profession and the demands we make of the police officers today call for a well-trained, well-educated, adequately staffed and adequately paid police department.

The problem with the education and salary approach to changing police attitudes and behavior is that there is no evidence that these kinds of efforts, while having other effects, do actually change attitudes and behavior. For example, public school teachers in urban communities, all with college degrees, have not shown any particularly positive attitudes or actions in their work with Negro children.

The second approach used to improve police-Negro relations is the organization of programs to facilitate communication, education, and understanding between the police and the citizenry. These programs usually involve face-to-face meetings between police and citizens. Some departments organize precinct meetings with adults and youths, or seek out local citizen programs where they may participate. Some police departments have regularly scheduled meetings with civil rights leaders; other departments have tried large and small community forums. Others concentrate on the youths — particularly in the inner-city.

No valid measurement of the precise effectiveness of programs of this sort has been undertaken anywhere in the nation. At best, it would be difficult to measure effectiveness over a short period of time. It is clear that police-community relations efforts in and of themselves are not sufficient to change police behavior. There are usually ten times more citizens than officers present at each meeting. Over the long run, it is difficult to sell a bad product — or to sell the product to people who are severely depressed socially and economically.

The recruitment of Negro officers assumes that the presence of a fairly representative number of Negro officers, at all levels within a police department, will serve to show the Negro community that the police department is not a white "occupation army," and that within the department the Negro officers will affect the attitudes and actions of their white counterparts.

The problem here is that the theory has not been tested because, with one or two minor exceptions, Negroes simply are not employed in any number in any department. Detroit has had between 200–250 Negro officers of a total force of 4,700 for the past decade, and significantly less before that. Officers who entered the Detroit department following World War II openly testify to the difficulties involved in getting a position at that time. Other cities are in similar situations. There are less than 60 Negro officers in all of the State Police departments in the nation.

THE POLICE SYSTEM

My challenge to the traditional programs to improve police-Negro relations is based upon analysis of the police department as a system, not the actions of individual officers — or, as they are sometimes referred to, as the few "bad eggs" in the department.

The systemic approach to the police has been undertaken by few social scientists. However, a recent study by Arthur Niederhoffer provides a record of trained, long-term observation and empirical evidence which strongly supports my own observation and limited research.

Niederhoffer is not an anti-police radical; he retired after 21 years in the New York City Police Department, earned a Ph.D. in Sociology at New York University, and now teaches police at the John Jay College of Criminal Justice. In Niederhoffer's introduction to *Behind The Shield,* he writes: "The great majority of policemen are men of integrity and good will. Yet it is a fact that a 'minority' goes wrong! Why this should occur even to the extent that it does among a body of men so carefully selected, is a mystery, but one that will, I hope, be less of an enigma by the time the reader comes to the last page of this book."[12] In fact, Niederhoffer's conclusions are a devastating indictment of the police system, particularly as it relates to police-Negro relations.

The police system, as described by Niederhoffer, comes out looking something like this: First, the police departments recruit from a population (the working class) whose numbers are more likely than the average population to hold anti-Negro attitudes; second, the recruits are given a basic classroom training program that is unlikely to change the anti-Negro sentiments; third, the recruit goes out on the street as a patrolman and is more likely than not to have his anti-Negro attitudes reinforced and hardened by the older officer; fourth, in the best departments, the most able officers are soon transferred to specialized administrative duties in training, recruitment, juvenile work, etc., or are promoted after three to five years to supervisory positions; fifth, after five years the patrolman on street duty significantly increases in levels of cynicism, authoritarianism, and generalized hostility to the non-police world. Finally, it is highly likely that the worse of the patrolmen will wind up patrolling the ghetto, because that tends to be the least-wanted assignment.

To put it more bluntly, the police system can be seen as one that is a closed society with its own values, mores, and standards. In urban communities, anti-black is likely to be one of a half-dozen primary and important values. The department recruits a sizeable number of people with racist attitudes, socializes them into a system with a strong racist element, and takes the officer who cannot advance and puts him in the ghetto where he has day-to-day contact with the black citizens. If this is an accurate description of the urban police system (and my personal observations over the past five years tell me this is so), then the reason is clear why every poll of black citizens shows the same high level of distrust and hostility against policemen.

Another nationally known law enforcement practitioner, now also teaching at John Jay College in New York, Donald J. MacNamara, said recently that "the police community is a closed society and it has its own customs, morals and taboos — and those who are not conforming to the police society, to its attitudes, to its customs and traditions, taboos and mores, are ostracized and then excluded . . . whatever prejudices and discrimination, whatever anti-minority attitudes he [the recruit] brought in with him, have been tremendously reinforced because they are part of the community attitudes of this police group of which he becomes a member."[13]

PROBLEMS IN CONFRONTATION

Civil rights organizations and the Negro community have successfully come to a confrontation with almost every other form of institutional racism, save law enforcement (albeit not all confrontations have been successful). Racism in voting, public accommodations, employment, housing, and education have been exposed and well documented, and have either been changed or the march (as in Milwaukee, as this is being written) goes on. Two major problems in confronting the police issue are the information gap and the defensiveness and secrecy of police.

White America's civic, religious, governmental leaders, reporters, and writers, have not dealt with the issue of racism in law enforcement as some have dealt with other civil rights issues. For example, Charles Silberman's best-selling *Crisis in Black and White* lists only two minor and passing references covering four pages to police, but has over twenty references to public schools, covering over one hundred pages.[14] Senator

Jacob Javits' excellent *Discrimination, USA,* with separate chapters on housing, education, and employment discrimination, has not one reference to law enforcement. The same is true of other civil rights studies. *The Speeches of John F. Kennedy in the Presidential Campaign of 1960* contains thousands of forthright words on civil rights in housing, employment, education, voting; however, not a mention of police or law enforcement.[15]

Police tend to be secretive and defensive, the good professional advocates as well as the old tough cops. The professional policemen — there are many, and particularly on the highest levels within the departments — seek education, training, and strict standards of professionalism within their departments. The professionals do not decry Supreme Court decisions or believe that local courts, newspapers or citizens are "out to get them," as the old-line tough non-professional "unleash the police" cop believes. The old cop clearly doesn't want interference or review because "only an officer knows how to handle the situation." The good professional doesn't want interference or outside review because the hallmark of a profession, they believe, is the ability to self-regulate the activities of those within the profession, as do doctors, lawyers, etc. Thus, while the professional and the old-line cop will split on most other issues, they do stand together on outside review or criticism.

Government officials and community leaders have another "hang-up" about confronting the harsh reality of the police system and its ability to withstand the minor effects of traditional "remedial" programs. That is, there are good, well-intentioned and intelligent law enforcement officials who recognize a serious problem and themselves are willing to say and often do say the "right" thing. And, in spite of the system, there are good cops actually working in the ghetto.

In November, 1967, a parade of Michigan's top police executives testified before a state legislative committee in opposition to "Stop and Frisk" legislation, an issue bitterly opposed by the organized and general black community, but favored by some politicians looking for a political issue and an anti-crime panacea. At the same hearing, a police chief

of a major Michigan city opposed across-the-board pay raises
for Michigan officers because "there are thousands [of officers]
who don't deserve their present salary."[16]

Other examples include the Michigan Association of Chiefs
of Police (MACP), which in 1966 became the first professional police organization to adopt a far-reaching civil rights
platform supporting equality not only in law enforcement, but
in education, housing, and employment.[17] And the MACP is
in 1967 engaged in a joint effort with the State Civil Rights
Commission (an incredible and unheard-of combination, say
some observers) to recruit hundreds of Negro and Latin-
American police officers.[18]

The problem, then, for government officials (like me) is to
retain a working and friendly relationship with the law enforcement professionals who sincerely seek change — and at
the same time tell them, and the community, that a basic,
long-entrenched part of the police system must be destroyed.

We must say that money alone — whether spent for higher
police salaries or more police training — will do little to stop
the pattern of police discrimination and brutality against
Negroes and other minorities in America's urban ghettoes.
Police dialogues with Puerto Ricans in Spanish Harlem, Negroes in Detroit, or Mexican-Americans in Texas, will not
significantly reduce the instances of police officers' abuse of
their power in those communities. The problem is one of a
set of values and attitudes and a pattern of anti-black behavior,
socialized within and reinforced by the police system.

CONCLUSION

This article does not propose specific actions to produce
systemic change in police-Negro relations in urban police
departments. Systemic change is possible. During the early
1960's, O. W. Wilson, Chicago's Superintendent of Police,
significantly destroyed police graft and corruption, attitudes
and practices that had become a primary part of Chicago's
police system. Secretary of Defense McNamara prevailed over
a defense establishment more entrenched and politically powerful than any fraternal order of police officers.

What is required are specific objectives by a mayor and police chief, committed to and strong enough to battle and prevail over the police system in their community — a Wilson or McNamara type. It will also require a political base, a sensitive power structure, and, if the white community is politically dominant, a change in the belief system of a large number of white citizens. The program of change will require internal and external controls over the behavior of cops in the ghetto.[19] A program and process must occur that will restructure the police system to exclude — or at least significantly minimize — the racism of cops in the ghetto.

NOTES

1. Burton Levy, "The Police and the Negro: Myth and Reality," *Newsletter,* The Archbishop's Committee on Human Relations (Detroit, Michigan), Aug.-Sept., 1965.

2. *Newsweek,* Aug. 21, 1967.

3. Arthur Kornhauser, *Detroit: As the People See It* (Detroit: Wayne State Univ. Press, 1957), pp. 123–124.

4. The *Detroit News,* Feb. 4, 1965.

5. The President's Commission on Law Enforcement and the Administration of Justice, *Task Force Report: The Police* (U. S. Govt. Printing Office, 1967), lists a dozen different attitude surveys confirming the discrepancy in black-white attitudes toward the police.

6. *Justice: 1961 Commission on Civil Rights Report* (U .S. Govt. Printing Office, 1961), pp. 26–27.

7. *Ibid.,* p. 26.

8. The new Michigan Constitution, Article I, Section 2, states that: "No person shall be denied the equal protection of the laws; nor shall any person be denied the enjoyment of his civil or political rights or be discriminated against in the exercise thereof because of religion, race, color or national origin. . . ." As a result of legislative action, opinions of the Attorney General, and court decisions, the Commission accepts complaints primarily, but not exclusively, in the areas of law enforcement, housing, employment, public accommodations, and education.

9. *Hearing Before the U. S. Civil Rights Commission: Detroit, Michigan, December 14-15, 1960* (U. S. Govt. Printing Office, 1961), p. 341.

10. Harold Norris, Professor of Law, Detroit College of Law, presents an intriguing concept of the unique role of the policeman on the street as the citizen's most important reference point toward the Constitution. "Constitutional Law Enforcement Is Effective Law Enforcement," Univ. of Detroit *Law Journal,* December, 1965.

11. Report of the Detroit Citizen's Committee for Equal Opportunity (1966) on Police-Community Relations (mineo.). This Committee is a blue-ribbon group of white and Negro business, labor, and civic leaders.

12. Arthur Niederhoffer, *Behind the Shield: The Police in Urban Society* (N. Y.: Doubleday, 1967), p. v.

13. Paper presented at the Pennsylvania State University Conference on Violence in American Society, May 26, 1967.

14. Charles Silberman, *Crisis in Black and White* (N. Y.: Doubleday, 1966).

15. This is a 1290-page document published by the U. S. Govt. Printing Office, 1962.

16. Statement of Dean Fox, Kalamazoo, Mich., Police Chief, before the Michigan Senate Crime Committee.

17. The Michigan Association of Chiefs of Police *Journal,* July, 1966.

18. The Police Recruitment Project of Michigan, Inc., is funded by a grant of $15,000 from the Office of Law Enforcement Assistance, U. S. Dept. of Justice, from June 1967–May 1968.

19. Herman Goldstein, "An Overview of Problems in the Control of Police Behavior," Univ. of Wisconsin Law School (mimeo.). Professor Goldstein was Executive Assistant to Supt. O. W. Wilson in Chicago from 1960–1964.

LAW ENFORCEMENT

AND THE POLICE

Joseph Lohman

■ American society has been developing a complex of subcultures which are driving it apart into local communities and groups, and in which the members are interacting among themselves and producing their own distinctive norms and values. These are the current subcultures of youth, of race, of suburbia, and of income (high and low). It is the reality of these subcultures which is so confounding to the established institutional structures and those who man them. It is not that there is a culture of crime. It is that there is such a plurality of subcultures that the problem of the individual's adjustment to commonly accepted norms is confounded, and that deviance and opposition to law and authority are generated as a matter of course. Crime, delinquency, and disrespect for law and the police are its logical accompaniment. We must develop means for modifying and preparing personnel to play quite new and meaningful roles.

In short, the demonstrations of the current day and the eruptions of a racial or youthful nature are evidence that we are at a critical juncture in the history of the United States and the development of its communities. That juncture is the emergence of a new kind of community—the metropolitan community which is not merely a bigger place, nor merely a change in population size nor in geographic location. The distinguishing feature of the new metropolitan concentration is that it represents a whole new set of human relations, and those human relations are the necessary conditions of action of the law and the police. The new community cannot be policed in terms which were appropriate to the village communities nor to the urban centers which preceded the new great and complex metropolitan centers.

The remedy of the failing respect for law is not a simple one, but the sober admonitions of Dicey and De Tocqueville point the way. In our democratic system, the power to make and implement the law lies in the majority of the community. It is to the majority that we must look in remedying the insensitivity and intransigence of the society in its relation to the groups who live marginal to the centers of power. The young, the poor, and the minority groups have frequently viewed the law as not of their making nor to their interest; the law is that of foreign power and the police is an army of occupation. Those who are thrice defined in their exclusion and deprivation—those who are young, those who are poor, and those who are of minority status—have been sharpest in their protest, most militant in their behavior, and least respectful of the law.

As Dicey observed, "Men come easily to believe that arrangements agreeable to themselves are beneficial to others." In going our separate ways and abandoning so many to an excluded life of deprivation, we are realizing the selfishness of individualism of which De Tocqueville forewarned the democracies. This individualism has its extreme expression in the polarities of youth and age, of black and white, of the affluent and the poor.

The remedy for disrespect for law is not to be found in merely admonishing the populace, and seeing all who oppose it as without reason. The remedy will require far-reaching structural changes in social institutions, one of which is necessarily the police. It will be found in a greater dialogue and a fuller participation in the counsels and decisions of the majority by those who have been and continue to be excluded from the making and implementation of the law. The remedy is in "just laws," democratically designed and sensitively enforced.

Recent incidents in cities through the United States, while undoubtedly an expression of complex relationships between minority groups marginal to and in considerable measure outside the mainstream of American life, emphasize the need for a re-examination of the traditional relationships between the police and the communities they are presumably serving. Subgroups of the society experiencing minority status and reflecting differentiated statuses of youth, low income, and ethnicity are actively expressing their concern over rights denied them and the deprivation attendant upon such denials.

The uniformed police, by the nature of their mission to maintain order and secure the established institutional arrangements, are, of course, conservative and supportive of the status quo. Correspondingly, they appear as the most visible and tangible representatives of an intransigent social order which insurgent groups seek to transform through their demand for equal opportunity and equal treatment. The police are, indeed, in extreme instances of alienated and estranged community life, likened to an "army of occupation." As a consequence, the police are today experiencing an unprecedented number of situations of confrontation. It becomes increasingly apparent that the police are confounded in their effort to enforce the law on the marginal few for lack of the confidence and support of the broader community in which the violater of law is situated.

The current hostility and the discouraging absence of confidence on the part of the general public have a serious and disabling effect upon the law enforcement function. The consequences are substantive in their effect and are reflected in the structure and organization of the police service. They attest to the need for structural and organizational changes as a necessary precondition of the effective engagement of the subculture of American society by the police, if the police are to be effective in the enforcement of law.

In summary, a number of the specific consequences and their attenuating efforts can be noted.[1]

1. Negative public attitudes are reflected in inadequate manpower supply. Able young men are reluctant to enter an occupation which, lacking the respect of the kinfolk and neighbors, enjoys low status among occupational alternatives. Nearly all of the major police departments are currently operating under their authorized limits of personnel strength, and difficulty in recruitment has become a chronic condition of the police service.

2. Hostility toward the police seriously impairs the morale of policemen. Correspondingly, the police are themselves disposed to become cynical and without enthusiasm for

entering or continuing in the service. Low morale is reflected in excessive turnover, personnel attrition, and early departure from the service.

3. A dissatisfied and disapproving public fails to support the police when issues such as salaries, number of officers, necessary equipment, and building construction and repair are pending before state legislatures, city councils, or executive authorities.

4. An antagonistic and cynical public is not likely to cooperate with the police in reporting violations, even when they are victimized. There is a general indisposition as well as reluctance to report suspicious persons or incidents, appear as witnesses, or generally provide information. Considering the relatively small proportion of all crimes known to the police, and the even smaller number which are cleared up through investigation, lack of citizen cooperation compounds the problem to critical proportions. The solution of an appreciable proportion of all crimes makes absolutely necessary the cooperation of the citizen public.

It becomes apparent that the police function may be so structured and so organized that the enforcment of the law upon the marginal few may be seriously impaired and even stymied. It follows that the immediate concern of the police in apprehending and suppressing law violators is contingent upon a deeper and more pervasive mission, namely, the activation of the whole public in the maintenance and securing of the peace of the community. It is, in my judgment, precisely because of this failure to define the basic and larger mission of law enforcement that so many law enforcement officials and other citizens of the society are bemused by the spectacle of so many of their fellow citizens in general, and of poor Negroes in particular, who have come to believe that law and authority are not *their* law and *their* authority.

It is apparent from the evidence of the "long, hot summers" that what troubles the police is not what confronts them in the streets, but the inappropriateness of their traditional posture in engaging these new and chronic situations of stress. There is, indeed, a crisis of the police. The crisis can be stated as follows:

There is a structural deficiency in the police systems of the

United States which makes the interventions of the police conducive to collective overexpressions of hostility in the society rather than the containment of individual expressions of hostility and/or violations of law. The forms of that structural deficiency are identifiable and can be made explicit. They are, first and foremost, the widespread disposition of the established authority and the police to blame troublemakers; to characterize situations of stress as brought on by "agent provocateurs." Correspondingly, that designation is extended to all who express their dissent in the form of "direct action." This structural deficiency is reflected in an extension of discretionary power to include the protection of the general community against such questionable elements, as well as the enforcement of the law in specific instances of overt law violation. A pattern of police practice is erected upon these premises which results in the familiar and questionable spectacle of the "double standard of law enforcement." Through the "double standard," there are promulgated a whole series of extra-legal, de facto, overreacting presuppositions which are taken for granted, like the air they breathe, by many police personnel.

1. The inevitability of vice or law violation when there are contacts between persons of differing racial extraction which are not a customary and accepted pattern of that community.

2. The necessity to enforce with police power the social customs and traditions of the community, apart from law (de facto segregation).

3. The necessity to invoke special action against minority groups which is not invoked against members of the majority group. Demonstrations, stop and frisk practices, and confinement of individuals to specific districts are instances of the double standard.

4. The necessity to regard all instance of civil disobedience as without any differentiating characteristics; hence the employment of uniform procedures in such differing conditions as:

   a. Challenging law for the purpose of testing the constitutionality of a law, and

   b. Challenging the morality of a given law.

A second structural deficiency is to be found in the absence of effective channels for expressing grievance. There is a widespread assumption that the attachment of marginal groups of the society has not been seriously affected by their actual or perceived experiences with the police and other authority figures. As a consequence, grievances, whatever their substantive merit, are regarded as a ploy of extremist and opportunist figures. Too often, complaints of police abuse or deprivation of civil rights fall upon deaf ears; and if there is an opportunity to register complaint, the channels for processing them are weak, ineffective, and unresponsive. If the opportunity for complaint is only a pale representation, no more than symbolic, the assertions and actions of the police in the "maintenance of peace and order" are viewed by those of minority status as evidence that the police are an instrument of an intransigent power-holding group and irresponsive to their interests and well-being.

A third and formidable deficiency, in the face of the current stresses in the relations between the police and the community, is the absence of a concept and/or machinery for effectively mediating the police to the present variations and complexities of the urban community. The newly emergent centers of power are an expression of the developing subcultures of metropolitan U.S. society. The ecology of the developing subcultures is in the pattern of the historic settlement of the low-income and ethnic groups of American society. Their aggrieved statuses and protestations take on collective significance, since they are in direct and intimate face-to-face contact and communication with one another. The distinguishing fact of racial identity, and the conditions of de facto segregation, have forged a new and unprecedented collective consciousness in the core cities of metropolitan United States. (If these newly developing collectivities are not effectively engaged and mediated to the official agencies of the community, they present an opposite potential of a new empirical condition of violence when confrontations occur.)

As Whitaker has observed, "It is not the changes in modern society which are affecting the morale of the police, but the failure of the police themselves to adapt to those changes."[2] I might further add, the failure of the society to provide the wherewithal by which the police can make those adaptations is the more basic consideration.

The rapidly accelerating "cops versus courts" controversy makes this point all too well.

On a Los Angeles freeway recently, I noted a car ahead of me with a bumper sticker—I suspect that all of you have seen some like it—"Support your local police." But I was surprised to find, "Support your Supreme Court." Now I suspect that the drivers of both those cars were motivated by the same high principle, concern with the integrity of law and order. Each in its different way was urging support of a common purpose. It is immediately apparent that each saw the objectives of law and order served differently by these two agencies of criminal justice. We must not be in the situation of the lion hunter whose gun jammed as he saw the lion coming at him. He went to his knees and shut his eyes in prayer. And when he opened them, he saw the lion was on his knees with his eyes shut in prayer. And so, in obvious relief, he said, "We both apparently believe in the same everliving God. We are praying to him, hence we can talk this thing over." The lion, as lions occasionally will, replied, "I don't know what you've been praying about, but I've been saying grace."

It's not clear, at the moment, who is praying for deliverance and who is saying grace, as between the proponents of the courts and the police. The public is alarmed by press reports, and by, indeed, police warnings of the rampaging upsurge in crime, especially violent crime in the city streets. The press reports the public concern. And the press and the public search for a scapegoat. There are bitter complaints from police and prosecutors about what they refer to as "bleeding heart" judges. Many private citizens, in widespread hysteria, are blaming their fear of walking night-time city streets on a succession of court decisions which seem, in the defensive cliches of many an observer of the passing scene, to handcuff, even strait-jacket, the police. They say that concern is for the criminal and not for his law-abiding victims. The sentiment is celebrated in even the highest and most authoritative places. But these court decisions are not, in any real sense, new. They are not the expression of new policy that is being promulgated, so much as they are an extension of the experience of law enforcement which is already the possession of those who are adequately informed and who have the economic resources with which to meet and treat of the law enforcement experience.

The better known of the controversial court decisions are those called the McNabb, the Mallory, the Gideon, the Escobido, and the Miranda cases. In general, they tend to extend into state courts restrictions already applicable to federal law enforcement agencies. This should be noted. Briefly stated, the Supreme Court outlawed the evidence obtained by illegal house search, without a warrant, in the McNabb case. The Court furthermore has thrown out confessions on a number of grounds: because police had not brought suspects before a magistrate without unnecessary delay, that the accused persons should have been instructed on their right to remain silent, and to consult an attorney. And in the famous Gideon case, the Court ruled that the accused person, though too poor to hire a lawyer, still had a right to counsel; and in the still more conflicted case of *Escobido vs. Illinois,* the Court voided a Chicagoan's murder confession because the police had refused to let him see his attorney, though the lawyer was in the very building where he was being held. As we read of these decisions, they ring loud of liberty and reason. In the Miranda decision, the Court set forth police guidelines designed to protect the rights of a suspect after arrest. Most Americans profess a love for liberty, for individual liberties, notwithstanding widespread confusion about the Bill of Rights. Many feel that the Court decisions are constitutionally and theoretically sound, but they are nevertheless uneasy. They are disturbed that a confessed rapist is turned loose on female society, merely because police, in protecting the community and in their determination to solve a heinous crime, have not questioned the confessor according to the rules of the game. The worst fears were confirmed when the same offender promptly committed two further offenses against women, and was sentenced to a penitentiary term in an eastern state. These are the troublesome facts and the doubtful mind in which the public regards the recent decisions.

It would be well to note at this point the equally disturbing opposite face of our coin, as Justice Arthur Goldberg observed in his opinion in the Escobido case, when he wrote, "We have learned the lesson of history, ancient and modern, that a system of criminal law enforcement which comes to depend on the confession will in the long run be less reliable, and more subject to abuses, than a system which depends on extensive evidence independently secured through skillful investigation." In other words, there is the suggestion that, in these cases, there are very profound and basic issues that affect us all; they are not to be evaluated only in relationship to the specifically accused one, who should or should not be in jail. The American public might well reflect on its long history of suspicion of police methods of interrogation and of general hostility to the uniformed police officer. It is not reassuring when a highly placed police official writes, as one did not so long ago, "It is hardly news that suspects of serious crimes often get worked over in the back rooms of station houses." Are the police handcuffed, and is this bringing on an unwitting assist to crime? Is the burgeoning crime rate brought on by a misguided solicitude of the courts; or are police persisting in methods which are a threat to the precious liberties of all, even if mounted in the best of intention and purpose? I suspect the truth lies somewhere between the two extremes.

We must see this problem, as the others, in the context of the changes which are taking place in our society generally, and with reference to the crime problem specifically. The development of our knowledge, through the behavioral sciences, is very rapidly changing the attitudes of the public, and is being reflected in the posture of the courts toward criminal behavior. The infusing of such knowledge, as a condition of judicial decision, has, of course, created much confusion and much misunderstanding between many law enforcement officials and other groups of persons in our community. It has led to a dangerous polarization of viewpoints concerning law enforcement objectives, on the one hand, and constitutional guarantees, on the other. If the current trend is allowed to continue unabated, it will not only affect the security of citizens, their persons and homes, but could have a serious effect on the administration of criminal justice itself.

There is in fact, in my judgment, no inconsistency between vigorous law enforcement and constitutional standards, providing the rules of the game—the legal rules concerning the constitutional guarantees— are not obscure and deficient. The courts, in addressing this problem in the aforementioned cases, are indicating their concern for the specifically limiting conditions which, in the society of our day, are at long last seen as denying what have been seen as the time-honored procedural rights of our constitutional system. The denial of the enjoyment of the procedural rights of our constitutional system is the central and overriding theme which has concerned the courts in these cases. In short, both state and federal appellate bodies, through these cases, have directed our attention to conditions of economic and minority status which have limited the representation of these individuals. Procedural rights are a sham and a mockery when the accused does not have at his side the necessary and indispensable procedural resources which are an automatic, almost immediate, possession of other elements of the community. In these respects, the current court decisions are, in my mind, only making certain and effective, for disadvantaged sections of the population, that which has been long available to those with knowledge and sophistication and those resources which are attendant upon adequate income. This, of course, creates difficulty for the police. But the answer is to be found not by insisting on the "right" to handle these groups marginal to the community in a different way and by standards different from those set forth under our Constitution, and which are best exemplified by the persons who have the resouces by which they can command those procedural rights and recognitions. It is to be found through bringing the police abreast of these developments in the community which give occasion to see and address those people who have lived in the shadow in the same way that they treat and relate to persons who have enjoyed their constitutional rights and privileges.

It is not the new decisions of the Court, nor the changes in our society on which they are based, which have affected, in this serious, challenging way, the work of the police. By and large, it is the absence of a program through which the police can adapt to these changes and to take note of this application of the general principles to all of their cases. This is the root of the problem, as I see it, for law enforcement in the United States today. The problem is further complicated by the fact that as agents of social stability, which is their basic and overriding task, they are finding their mission an exceedingly difficult and involved one in a society with an extraordinarily accelerated pace of change. How can stability be maintained when the society changes the rules of the game, almost hourly, which the police are called upon to enforce? Make no mistake about it, there are appearing before the courts of the United States today cases that have lain in the shadow since its inception as a nation, but are now celebrated in the white light of judicial review. We cannot take lightly the fact that there are sections of the community which answered to charges of law violation by invoking the constitutional privileges, while other sections of the community, living in the shadow, did not, could not, employ these protections. The police are being called upon to adapt their methods and procedures to the new situation. By and large, this is what they are doing, however difficult and distressing the task may be.

In my acquaintanceship with the police departments of the United States, I do not find that their modus operandi has been seriously modified as a result of the host of decisions which have come down from the high courts. They are not immobilized. They are not refraining from enforcing the law. As a matter of fact, I venture to say that in most cases they are continuing to enforce it, for the most part, in the traditional ways. But there are adjustments. The point is that the adjustments are themselves evidence that we are not confronted by a deteriorated model of law enforcement in the United States today. Police are not failing in their performance, as contrasted with some higher standard of the past. They are not deteriorating in their methods and procedures. Indeed, nearly everywhere, they are enormously improved. The police are better, they are not worse. It should be emphasized that the import of the court decisions is not that the police have become bad or brutal. It is that with respect to persons and places where the courts have been silent, they are now vocal. More often than not it is the public which has required a double standard by the police. The police have been encouraged and pressured to go beyond the limiting provisions of the law in those cases and situations which appeared to give security to the general community. They have been pressured to employ special measures where the community wanted crime stopped, especially in those places among those persons commonly associated with high rates of crime. Hence, the police took to themselves measures which stretch, in some degree, the power which the Constitution placed in them.

There is a way in which the courts and the police can join together, and face together as one, the new situation which the court is facing, and which the police in turn can face, rather than present the spectacle of an agency harassed in its mission. Courts can help clarify the present maze and jungle of case precedent which constitutes the way in which the problem is presented to the police.[3] The courts can and should formulate broad positive guidelines for the police in their daily work. There is precedent for this in the experience of the British courts and their police. Since as long ago as 1912, the nine Magistrates Rules of the British courts have resulted in a more successful record of criminal investigation and prosecution without infringing on individual rights than the parallel record in the United States. More recently, the rules have been modified and reconciled to the changing social scene. We are now familiar with the phrase "to caution" a person so that he might, so to speak, know that what he has to say might be held against him. The phrase has its origin in the rule of the British courts and the practice of their police. As we find in our police practice, we have in fact adopted it in response to rulings by the court. What the Magistrates Rules have provided by way of positive guidelines, we have attempted to approximate through negative case law. Hence, the police have been instructed by an assortment of negative decisions—by, so to speak, "slaps on the wrist"—rather than by positive guidelines to action. Make no mistake about it, policemen dispense more justice than all the courts combined, through such routine work as issuing traffic tickets and handling public complaints.

The steady stream of Supreme Court decisions and other appellate rulings on search and seizure, due process, and other aspects of police investigation, have often in the past created a situation of ambivalence and anarchy for much of their work. Most of these decisions have been limited to a particular set of circumstances judged after their occurrence. We must be concerned with the way in which the message of the Court is delivered to the police and to the society. Nearly all the highly publicized cases, the ones here mentioned, have presented the spectacle of a Kafka-like powerless individual, confronted by the overpowering majesty and the force of the law. These cases are, in fact, reflecting a fundamental change in the life of our society, for they represent classes of individuals that formerly were powerless. They are now receiving attention, because we live in a time when power is becoming shared in the society.

The growing controversy over police brutality does not mean that there is more of it; as a matter of fact, in a curious and unfortunate sense, the words "police brutality" have become one word. Wherever there are police, there is an assertion of brutality. Of course, this is not true. We know that there are some instances of brutal behavior by a policeman, but why do we have this common—this almost generic—reference to the conduct of the police as brutal? Because this term has come to note the way in which, in many respects, the police system is insensitive to changes that have taken place in the powerless centers, among the Negroes, among youth, among the aged. They are no longer disposed to be treated by the double standard which was, for the most part, operative in the past. The greater proportion of the community not only countenanced the double standard, but nudged the police in that direction because of what was presumed to be its own interest. The growing controversy over police brutality does not mean that there is more of it, but that individual instances of excess and abuse, where the public once was silent, are now cause for public concern. Police methods of operation are not a reflection of ignorance or insensitivity, but a collective formula for discharging responsibility, colored by the particular power complex of the society. Where professional criminals (you don't find them complaining about being beaten up by the police) and upper-class adults are aware of their legal rights, and have the resources to secure them, public defenders and legal staff, groups like the ACLU, the NAACP, CORE, and SNCC are currently providing a similar countervailing force for less advantaged groups.

Growing impersonality in society, individual detachment, and the flight of witnesses from involvement in legal matters have led police to focus their attention on individual suspects in attempting to solve crimes. Quick resort to effecting a confession is, under these circumstances, an easier way to approach the problem. Instead, law enforcement officers should rely on new scientific technology to help their work. The crime laboratories should be used not only in exotic cases, but as a condition of general police operation. The changes in the behavioral sciences have brought about changes in our attitudes toward criminal behavior, a condition which has not only complicated the life of the court, but is now complicating the life of the police. Therefore, the police too should develop, as the court developed. I see such a development as the means for composing the oppositions that now exist. For their work of prevention, the police will increasingly need to become agents of social welfare. To succeed in detection and investigation, they will have to become scientists, rather than persons involved in polemic discussion, or subtle extraction, of self-incriminating information.

Let me conclude by suggesting the sense in which we must look deeper than for a cliché, or a notion such as the fault of the courts or the brutality of the police. There must be developed a third force, a public force, that can provide a new kind of dialogue. From such a dialogue there can evolve a common understanding for the development of rules, and a common condition of its implementation. Let me quote to you a paragraph from an English treatise on the police, that applies equally well to the United States. In commenting on the future of the police, Ben Whitaker wrote:

> They are doing the difficult and dangerous job society demands, without any understanding by society of what their moral and professional problems are. The public use the police as a scapegoat for its neurotic attitude toward crime. Janus-like, we have always turned two faces toward a policeman; we expect him to be human, and yet, inhuman. We employ him to administer the law; yet ask him to waive it. We resent him when he enforces a law in our own case; yet demand his dismissal when he does not elsewhere. We offer him bribes; yet denounce his corruption; we expect him to be a member of society, yet not to share its values. We admire violence, even against society itself, but condemn force by the police on our behalf. We tell the police that they are entitled to information from the public; yet we ostracize informers. We ask for crime to be eradicated, but only by the use of 'sporting' methods. What. . .do we want the police for? Only by resolving the conflict in values between liberty and law enforcement, can we determine the paradox of the policeman's position in our future society.[4]

It is time for us, in terms of the current dangerous polarity, to think clearly and give our police a role and the condition of implementing that role which can help bring about a unity of purpose and action by the police and the courts, rather than accentuate the meaningless, purposeless, and self-defeating polarity which currently prevails.

NOTES

1. President's Commission on Law Enforcement and Administration of Justice, Task Force Report: *The Police* (Washington: U.S. Govt. Printing Office, 1967), chap. 6, pp. 144-215.
2. Ben Whitaker, *The Police* (London: Eyre and Spottiswood, Ltd., 1964), p. 166.
3. *Miranda vs. Arizona*, 384 U.S. 436 (1966).
4. Whitaker, *op. cit.*, p. 171.

Part V
CIVIL VIOLENCE AND THE POLITICAL SYSTEM

Introduction

■ The final set of readings in this volume on riots and rebellion is concerned with the relationship between civil violence and the political system within which they take place. The political system—i.e., the system responsible for the authoritative distribution of advantages and disadvantages in the community or the society—may be viewed from several perspectives in terms of this relationship. They range from the control function (which the polity assumes as the ultimate authority responsible for maintaining social order) to the preventive function (which includes the efforts made to assure that the laws of the community and the society are fairly, equitably and justly enforced in both letter and spirit). Much of the controversy over the interpretation of the urban civil disorders in recent years has, in fact, focused on the difference in emphasis placed on these two functions. It involves in part a confusion between legality and legitimacy, which we discussed in the introductory essay to this volume, and which Professor Neiburg comments on in this section. Those who challenge the legitimacy of the "system" and those who demand that it perform its law and order function effectively are both right in the sense that the political system includes both functions. Depending upon the depth of dissatisfaction and the degree of intensity of the rioters, the potential of the polity to restore order, however, may depend upon its ability to right the perceived and/ or actual wrongs. But where the capability and willingness to apply necessary force is available, enforced order may be restored without promising or implementing reforms in the political or socioeconomic systems.

Civil violence thus has political implications which may be more or less manifest. Most will recognize the Hobbesean obligation of the political system to maintain social order, but the level at which it is maintained and the manner in which this function is performed are continuing issues. The legitimacy issue is usually less clearly defined because more difficult to articulate, but it often has unambiguous political overtones when reforms in process and/or policy are made the *sine qua non* of community order.

In the first essay, political scientist H. L. Nieburg distinguishes between "frictional" violence and "political" violence. Although both involve what he calls "anti-social acts" by individuals and groups, the control of frictional violence at acceptable levels of cost and risk to maintain the norms of social relations is the legitimate function of the police power. Political violence, however, involves violent counter-escalation directed at the very system of social norms to be protected by the police power; it heightens the risk and increases the cost to society beyond acceptable levels. Adjustments and bargaining are required in order to maintain the legitimacy of the political system whereas the control of frictional violence relies on adequate legal authority and force.

The role of local politics in urban riots is examined in "Ghetto Revolts and City Politics" by political reporter E. S. Evans. He suggests that the dynamics of Negro politics in general, and the ghetto economy in particular, within the urban political system can be viewed as part of the riots' causal order. He analyzes the reasons for the relative powerlessness of the Negro community in terms of a poverty economy (which is reflected in such things as weak family structure, a status-oriented middle class which appears unrelated to a welfare-oriented lower class, and small, feeble Negro organizations) and the urban political structures which tend to stifle Negro influence in the political process (e.g., at-large, non-partisan elections).

He concludes that the riots of the 1960's are the result of the hope-filled but failure-ridden behavior of the urban Negro. Powerlessness and frustration in the orderly decision-making process of the urban polity has given rise to a form of "disorderly" politics. Local politics has failed in one of its major functions—conflict management—because Negro needs and demands have been accommodated without civil violence (not, many would argue, with it). The reforms being suggested in the aftermath of the riots should include political as well as social and economic reforms to allow Negroes in cities to participate significantly in the process of effecting social change.

T. M. Tomlinson, a psychologist and another member of the Los Angeles Riot Study (LARS) research team, draws on his Watts findings to discuss the implications of widespread ghetto support for rioting as justified, instrumental, political protest. "Intense political concern combined with felt impotence to exert influence on the political structure . . . is a cornerstone of social unrest" which pervades the urban centers. He suggests that a riot ideology has developed after Watts "took the lid off" by defining riots an "acceptable" response to the conditions of Negro life, making a large segment of the Negro population susceptible to the idea of violent protest. What produces riots, argues Tomlinson, is not related to political or economic difference among cities—a wide variety of them have had riots; it is rather the consensus within the Negro community that conditions are unacceptable, combined with the notion among a singificant minority that riots are "a legitimate and productive" form of protest. Tomlinson explores several alternative approaches to riot prevention and control (the War on Poverty, Negro self-restraint, and the police) and concludes that none of them will prevent the occurrence of popular riots "because the mood of many Negroes of the urban North demands them, because there is a quasi-political ideology which justifies them, and because there is no presently effective deterrent or antidote."

In the fourth paper in this section, Richard L. Meier, an urban planner, discusses how planning might be employed to avoid or control the causes of violence and disorder, which he views as "the last remaining mortal threat to urbanism as a way of life." He sees the maintenance of social order in massive urban centers as the fundamental task for society, both in the United States and abroad, and suggests that planners may be in a position to advise political decisions dealing with the stresses of urban life which feed civil violence. He concludes that the most promising alternative among several considered is the design of protections for the urbanite from environmental stresses and crowding which may lead to violence. Planners and political decision-makers should give attention to the development of varied opportunities for individual and group privacy.

The final selection on violence and the political system, by sociologist Martin Oppenheimer, addresses itself to the possibility of guerrilla warfare or armed insurrection directed at the immediate or ultimate overthrow of the established political order. Drawing on six historical analogous examples of urban ghetto insurrections in other societies (Paris, Dublin, Shanghai, Vienna, Warsaw, 1943 and Warsaw, 1944), he suggests that insurrection, or para-military activity ("any violent behavior of an organized sort directed either defensively or offensively against the military forces [including police] of the dominant power in society, by military elements associated with no regular or recognized government, that is, by irregular, partisan, volunteer, guerrilla and/or revolutionary forces"), can only be successful if two related conditions obtain: the support or neutrality of the vast majority of the population and the inability of the dominant power structure to suppress the insurrection or remedy the structural strains which give rise to it. Placing the potential for American urban black power insurrection in this framework, Oppenheimer concludes that such an effort would be doomed to failure. A black insurrection is not likely to have either majority population support or neutrality, and government forces seem well disposed to move in an organized and systematic fashion to put down such an attempt of force. His conclusion is straightforward: ". . . if White America is not to destroy Black America physically it must create the conditions for the success of the civil rights, rather than the Black Power, movement."

L. H. M. and D. R. B.

VIOLENCE, LAW, AND THE SOCIAL PROCESS

H. L. Nieburg

■ The function of the police power of the state is to maintain a threshold of force to deter and/or contain the ever-present margin of anti-social acts by individuals and groups. Some element of personal dislocation and anomie exists in the best-managed and most equitable societies. Even when isolated outbreaks contain germs of larger social issues (they usually do), they may be contained at acceptable costs by the appropriate and measured application of police power. This constitutes the normal function of the state in dealing with the private violence intrinsic to the social process. So long as this task be managed at acceptable risk and cost, police power protects most of the members of the community and enjoys general support. It need not raise political issues (i.e., concerning a change in the norms of social relations).

This kind of violence may be termed "frictional." To minimize and control it is the legitimate purpose of police power, which thereby maintains the norms of social relations. Such political grievances as frictional violence often implies are forced into peaceable channels, adjustment through debate, legislation, public policy, and private contract.

However, the characteristic pattern of contemporary riots afflicting American cities has shown a tendency toward violent counter-escalation against police action by elements of both Negro and white communities. While the white violence has been limited, the Negro violence escalates in response to police action, often with general support from the black community and with enhanced responsiveness, organization, and danger of future outbreaks. This phenomenon is different in kind from frictional violence. The capability of infinite escalation heightens the risk and increases the cost to society beyond acceptable levels; most important, it destroys the efficacy of normal methods of police power. This kind of violence is not merely frictional but must be termed "political." It addresses itself to changing the very system of social norms which the police power is designed to protect. It focuses grievances in recurring, deliberate, or spontaneous acts of violence, even at great risk and cost to the actors. The peaceable procedures of political adjustment fail to divert the escalation, whether because they are closed, discredited, halting, or simply untried.

The peculiar pattern of major social upheaval and political confrontation arises from the fact that the normal police security methods become counterproductive; they merely solidify the capability and likelihood of disruption by a group which is increasingly polarized and alienated. Efforts to deal with such outbreaks as merely frictional tend to create a vicious circle of violence and counter-violence which may discredit responsible leadership on both sides and make further disruption and alienation inevitable. It is these risks and costs that endow such violence with political efficacy and induce the general community to look for other remedies, not in escalation but in modification of access to the peaceable channels of adjustment and, eventually, of the norms of social relationships.

The urban riot as a political event can best be understood by Baron von Clausevitz' definition of war between sovereign states: the last resort of diplomatic bargaining and a continuation of diplomacy by other means. In the face of major political violence, the prevailing consensus of interest and power groups must choose between social-economic-political adjustments and the unpromising course of infinite escalation and counter-escalation of force. However spontaneous, isolated, and emotional the incidents that trigger the circle of disruption, the choice between infinite escalation and concessions confronts the prevailing social order, with no obvious third course in sight.

We witness more perilous phenomena each year, and imitative outbreaks proliferate not only in summer and by race but throughout the seasons and as models for disruptive action by other groups. Many observers of the dilemma recognize the distinctive element of political confrontation and crisis which the rioters themselves seem instinctively to feel. A Dutchman who rioted against Nazi occupation during World War II noted that, like the American Negroes, the Dutch rioters were "filled with elation by the fact that they were doing something." There was a community feeling that combined hope, impatience, and impulsiveness. They looted "to obtain trophies, not to get merchandise they could use profitably. Loot has to have symbolic value; strictly utilitarian goods are set on fire!"[1]

VIOLENCE AND THE SOCIAL PROCESS

Violence can be unambiguously defined as the most direct form of power in the physical sense. It is force-in-action. Its use is the continuation of bargaining by other means, whether employed by the state, by private groups, or by persons. Here all the attenuated politically-socialized forms of indirect power are brushed aside. The threat of force moves into action, inching from forms of demonstration and continued bargaining into a direct test of relative power by actual mutual destruction and defense. Bargaining, inducement, coercion, and possible accommodation now hang upon tactical and strategic weakness and advantage, the shifting tides of maneuver and battle. Yet short of the total collapse or destruction of the very means of struggle on one side, the element of bargaining, the continuous assessment of capabilities, risks, and costs is not suspended in the movement toward eventual accommodation. Power in the sense of raw violence, defense, and counter-violence is always in the process of measurement as the *ultima ratio* whose status at some point of respite becomes the provisional basis of political settlement.

A situation reduced to this extreme may well preserve survival values and future bargaining power, even for the weaker antagonist. As in all things, there is an economy of the use of force in terms of cost-risk/benefits, whether for nation-states or for disaffected minorities. The capability and determination to exact unacceptable cost from the enemy, even at greater cost to oneself, may be the only means available for a small nation or a minority group to seek to maintain some respect for its independence, values, demands, and political bargaining power, while avoiding the escalation of violence to extreme limits. The weak may lose, but they may also win by testing the cost-risk/benefit constraints of the strong; in any case, they may have no choice. The essence of risk for both sides arises from the loss of ability to limit, control, or predict the dynamics of a confrontation crisis. This is especially true of a domestic confrontation, where the groups involved have greater proximity and occasion for random actions, together with less formal organization and discipline.

The definitions of violence (uncontrolled and dysfunctional) and force (controlled and legitimate) which are common in the literature, we find inadequate and possibly tendentious.[2] The distinctions between capability, threat, and demonstration are more universally applicable and therefore useful: *force* = capability/threat of action; *violence* = demonstration of force tending toward counter-demonstration and escalation, or containment and settlement. Under this rubric, force and violence are two aspects that merge imperceptibly into each other. The actual demonstration (force-in-action) must occur from time to time to give credibility to its threatened use or outbreak; thereby gaining efficacy for the threat as an instrument of social and political change or control. If the capability of demonstration is not present, the threat will have little effect in inducing a willingness to bargain politically. In fact, such a threat may provoke "pre-emptive" counter-violence.

The "rational" goal of the threat of violence is an accommodation of interests, not the provocation of actual violence. Similarly, the "rational" goal of actual violence is demonstration of the will and capability for action, establishing a measure of the credibility of future threats, not the exhaustion of that capability in unlimited conflict.

By and large, all violence has a rational aspect for somebody, if not for the perpetrator. All acts of violence can be put to rational use, whether they are directed against others or oneself. This is true because those who wish to apply the threat of violence in order to achieve a social or political bargaining posture are reluctant to pay the costs or take the uncertain risks of an actual demonstration of that threat. Many incoherent acts of violence are exploited by insurgent elites as a means of improving their roles or imposing a larger part of their values upon a greater political system. Such acts tend to be identified as primarily political rather than frictional when the choice of targets, pretext, or victims (whether persons or property) symbolically represent social rather than personal relationships and categories,[3] whether *in fact* they do or not.

Violence and counter-violence tend, at least in initial stages, to be symbolic and to constitute domestically as well as internationally a threat by demonstration of latent but clearly potential escalation. In this sense, an aroused private group may control "force" and "forces" exactly as does the state. Conversely, the application of police power may escalate beyond the controlled function of symbolic counterforce. An individual policeman may use unnecessary and excessive means of pacification or constraint.

LEGITIMACY AND LEGALITY

This leads our attention to the underlying conflicts and power relationships of private groups within a nation, to analysis of the representativeness and vitality of the peaceable channels of social conflict, bargaining, and accommodation, and to consideration of methods to limit mindless escalation of destructiveness; to what might be called the *informal polity* or *social process* that underlies and gives vitality to the formal institutions of state and government.

The distinctive mark of the state as an institution is its unitary sovereignty which centralizes the authority of all the various normative systems of behavior that make possible the precarious balancing act of group life. This unitary sovereignty functions to protect private and public activity and bargaining which seek to achieve whatever values men contrive, to conduct and enhance their destinies. The nation contains a vast multiplicity of personal lives, energies, relationships, etc., which somehow must maintain order in the midst of change, and change in the midst of order; the group must endure and grow through continuous adaptation to the parameters of the human and physical environment. The power of the state gives authority to the institutions which mediate and bind diversity with unity, freedom, experimentation, and conflict, with social stability and institutional continuity. The state authority does not eliminate conflict but underpins the institutionalization of conflict and bargaining in ways which optimize consensus and values.

Political violence cannot be dealt with in terms of legality which views the formal system of law and order as sacred and inviolate and which tends to become a status quo ideology. This view has its place when it tends to justify the proper functions of internal security. Legality supports the prevailing consensus against erratic transitions, enabling the majority of citizens to conduct their private and public affairs within a more or less stable social environment. This doctrine may, however, become dysfunctional and inappropriate when applied to the violent events of present race relations. As political instruments, the threat or use of violence violate the procedural norms of legality. Private violence and forceful self-help are among the most important conditions which the norms are intended to replace. Yet this very value is what gives such methods efficacy as the last resort of political and personal bargaining. The bargaining process creates the substantive and procedural norms whose legitimacy alone makes enforcement feasible. The bargaining relationship of individuals and groups constitutes the heart of the informal policy of the social process. The threat or occurrence of violence has political efficacy either for or against the status quo when the danger and risks of state or private violence are credible. Thus violence is inescapably part of the bargaining process.

Legitimacy and legality of state authority are by no means synonomous. The procedural-structural aspect of social order generally bears the broadest consensus of values. However, the substantive norms of social relationships, the matters of rights and duties, leadership and policies, etc., carry no such broad agreement; they are the common grist of interest group politics. Legality is an attribute of sovereignty. It is an abstraction which confers the authority of the state upon the acts, records, elections, etc., of those who conduct the offices of state power, and upon the code of law which regulates behavior. Legality is the technicality of formal consistency and adequate authority.

Legitimacy, on the other hand, reflects the vitality of the underlying consensus which endows the state and its officers with whatever authority and power they actually possess, not by virtue of legality, but by the reality of the respect which the citizens pay to the institutions and behavior norms. Legitimacy is earned by the ability of those who conduct the power of the state to represent and reflect a broad consensus. This is the familiar doctrine enunciated in the Declaration of Independence. Legitimacy cannot be claimed or granted by mere technicality of law; it must be won by the success of state institutions in cultivating and meeting expectations, in mediating interests and aiding the process by which the values of individuals and groups are allocated in the making, enforcement, adjudication, and general observance of law. Not all law is legitimate in this sense, whether because it is unenforced, unenforceable, or responsive to sporadic and arbitrary enforcement here or there by this or that police chief, policeman, judge, or jury. Once-legitimate laws may still retain legality after losing legitimacy. Legitimacy refers to public consensus and actual practice, while legality refers to technical consistency of doctrine and prescribed practice.

Whatever the historical, logical, or illogical culmination of events that unifies a population under central state sovereignty, the subjective aspect of its unity, i.e., its legitimacy, makes it a nation. The objective aspects, territorial boundaries, the letter of the law, the monopoly of legality and police power, in themselves do not make a nation, in fact they may generate more violence than collaboration, unless a nation is built in the minds and hearts of the people. It is the consensus that supports the informal polity that constitutes the nation. Those who occupy the offices of state power face each day the continuing task of validating the legitimacy of the state by the way they manage and shape the life of the nation. This must be done because and in spite of divided regional loyalties, economic rivalry, ideological and religious conflict, cultural variety, etc.

The tension of normal social life is aswarm with ambivalence, conflicting loyalties, and shifting alliances of convenience. Competition and cooperation are by no means incompatible, but are poles of a continuum in a working social order. All individuals and groups are torn between conflicting interests and impulses and struggles for leadership and influence. Hostility and cooperation are intertwined in a complex maze whose day-to-day movements and adjustments are never conclusive. Ambivalence is the nature of the bargaining relationship, and all social relationships in the informal polity are a form of exchange of values, accommodation out of conflict, and conflict out of old accommodation. New states concocted from old colonies face a more difficult task, but even well-established advanced nations must continuously prove and improve their legitimacy. The means of force are essential to maintain a threshold of deterrence against separatism. But such means are not a substitute for the positive achievement of values which can win legitimacy and create self-enforcing informal as well as formal processes of unity. There is always a danger that the prevailing minority that holds power will use legality and force to deter all pluralistic politics and opposition. In the swiftly changing human and physical environments of modern technology and international relations, such a mood can quickly dislocate a nation, destroy legitimacy, escalate violence, and endanger the peace of the world. A state system whose central and primary values become the negative and costly ones of internal security and repression is soon riddled by subversions, interventions, assassinations, and extremist fits and seizures of all sorts.

All state systems must integrate into the power structure at least those groups which are self-conscious, organized, interested, and able to exercise private power in the streets if barred from the magic circle. The notion of a totalitarian regime able to ride roughshod over practically all its citizens, enforcing from above dogmatic ideological norms and/or preventing political change from below, is a myth. No system can long operate without legitimacy. Every regime, whatever its narrow base, means of access to power, or ideology, must set about building consensus somewhere, and most importantly among those who themselves retain the capability of imposing high cost and risks of concerted action if too obtusely neglected.

NOTES

1. Jan Boeke, Letter to the Editor, *Science,* III (Nov., 1967), p. 577.
2. See Georges Sorel, *Reflections on Violence* (Paris, 1905; N .Y.: Collier, 1961).
3. See Allen D. Grimshaw, "Changing Patterns of Racial Violence in the United States," *Notre Dame Lawyer,* LX, 5, (1965), 534–548.

GHETTO REVOLTS AND CITY POLITICS

E. S. Evans

■ America's new urban explosion continued into its fifth successive year in 1968 with riots in the Negro ghettoes of more than
150 cities. Explanations have been offered by police officials and
others concerned with the breakdown of law and order and by
civil rights leaders more concerned with injustice and disorder.
Social scientists have explored the social, cultural, and psychological forces involved and argue strongly that ghetto economics—
poverty amid affluence—are the roots of the twisted vines of
inner-city pathologies that ensnare the Negro today. Few,
however, have examined the role of local politics in these
outbreaks of violence. Do the riots have political causes, too?

That the local political system is a factor was suggested a year
before the first long, hot summer of 1964 by political scientists
Edward C. Banfield and James Q. Wilson in their book *City
Politics.*[1] They pointed out that the government has two purposes: the service and political functions. In the service function,
city hall provides police protection, public parks, and the like. In
the political function, the governmental processes "manage
conflict in matters of public importance."[2] To repress the political
function, as municipal reform movements have more or less done
in many cities, is to prevent some groups from asserting their
needs, wants, and interests, the authors said.

> Where there exists conflict that threatens the existence or the
> good health of society, the political function should certainly

take precedence over the service one. In some cities, race and
class conflict has this dangerous character. To govern New York,
Chicago, or Los Angeles, for example, by the canons of efficiency—of efficiency *simply*—might lead to an accumulation of
restlessness and tension that would eventually erupt in meaningless individual acts of violence, in some irrational mass movement,
or perhaps in the slow and imperceptible weakening of the social
bonds. Politics is, among other things, a way of converting the
restless, hostile impulses of individuals into a fairly stable social
product (albeit perhaps a revolution!) and, in so doing, of giving
these impulses moral significance.[3]

New York, Chicago, and Los Angeles—each has suffered a
violent, if not irrational or meaningless, mass movement, if not
such individual acts. New York has had three riots in three years.
Chicago had a minor outburst in 1966, and a major riot in 1968.
The 1965 Watts riots in Los Angeles was the worst eruption until
Detroit exploded two years later.

A look at the political system and Negro communities of such
cities might show how the riots came about and the part politics
played in the outbreaks. Detroit is particularly interesting. It was
reputed to have one of the nation's best anti-poverty programs.
The city administration there was noted for its liberal racial
policies, especially in police-Negro relations. Also, with half the
Negro membership of the U.S. House of Representatives, Detroit
Negroes apparently enjoyed considerable political influence.

Negro politics may be seen as part of the riots' causal order.
The weight of social science evidence indicates that the ghetto
economy is the primary cause of the riots. From poverty stems a
volatile web of slum pathologies complicated by racial discrimination—social ills like crime and drug addiction, poor physical
health, slum mentality including alienation—the culture of
poverty. When set off by some inflammatory incident, typically a
police effort at social control, the complex burns furiously.
Political links are found in each part of this chain: (1) the political
condition of the Negro community results largely from the same
economic causes; (2) this condition is part of the web of ghetto
illnesses, reinforced by the other forces and reinforcing them; and
(3) this disorderly politics of poverty plays a central role in the
outbreak of the riot. The voices of the ghettoes, as recorded in
press reports of the Detroit riots and more rigorous studies,
support these points.

POVERTY POLITICS

The fact that the Negro community, especially its lower stratum, is politically uninfluential, or even powerless, should be apparent. If the Negro had the power, he most likely would have improved his physical and social conditions. Moreover, evidence of Negro helplessness is not hard to obtain. Studies of the Detroit political system and issues involving the race disclose only limited Negro influence.[5] At the very least, Negro power is not proportional to the size of the Negro community anywhere, and is not increasing according to population growth.[6] The inner-city Negro is just not heard, even by the more enlightened officials. As Frank Ditto, a community organizer in Detroit, said shortly after the riot there:

> The thing is these Goddamned people know what to do on a certain level, and these folks can tell them of the central things, the basic things. These are the folks that they refuse to negotiate with or even talk with. This is the problem. They won't listen to Ralph Bunche. They won't listen to Roy Wilkins. They won't listen.

Why is the Negro so powerless? "The anomaly of the Negro's numerical strength and political weakness," Banfield and Wilson wrote, "can be explained largely in terms of two interrelated factors: the class structure of Negro society and the character of urban political systems."[7] They discussed six features of the society that, although they did not go this far, seem to have economic causes for the most part.

SOCIAL FACTORS

First, "the predominance of a lower class lacking a strong sense of community" is a result of weak family structure in Negro society. The breakdown of the family partly results from historic economic factors, as the Moynihan report showed, and its continuation today is a result of Negro fathers' inability to earn sufficient income[8]. The economic forces that make the family weak also cripple the Negro's sense of community attachments, because there is no strong family to inculcate them. These attachments may be less weak today because of increasing race consciousness, but they are still difficult to develop. Ditto tried to instill his brand of "black pride" in two Detroit rioters. Here is the conversation:

Ditto:	Do you realize that just 150 years ago when black babies were born and brought forth onto the earth, the slave masters would snatch them away from their mother and send them off to some other plantation as soon as they were old enough? They separated brothers and sisters and fathers and sons.
First rioter:	We've come a long way since then, though.
Ditto:	Just let me finish the point. I'm trying to tell you Negroes don't stick together because of the psychological brainwashing they've been through for hundreds of years. You understand what I'm saying, brother? Ask him. He's your friend. He can articulate it to you better than I can.
Second rioter:	I'll tell you one thing. It would have been hard to get me over here if my brother hadn't been selling me for some beads.

Second, "the inability, or unwillingness, of the Negro middle class to identify with the lower class and provide leadership for it" stems partly from the lower class's poverty-inspired style of life. This factor also has important economic implications for Negro politics. The middle-class Negro is more interested in social-status improvement than in economic improvement, which he already enjoys. The lower-class Negro, on the contrary, seeks economic welfare. The middle-class Negro may even oppose welfare benefits when he feels his social or economic security threatened. Although these divergent interests may converge at some point, like support for a riot, the difference remains important.

Third, the fact that "a fairly large and growing number of young Negroes have more education than the job market enables them to use" is hard to trace to economics, but it may be less significant than it was a few years ago—before hiring a Negro secretary who looks like Lena Horne and casting Negro actors in television commercials became so popular. Nevertheless, it still has serious implications for such Negroes' political attitudes, because to the extent that it occurs, it enhances the frustrations of those who are potentially the most politically active.

Fourth, "the relative fewness of Negro entrepreneurs and the consequent importance of professionals in the Negro community" means that those with the income and time for political activity are often apolitical doctors, teachers, and civil servants. The lack of a more politically oriented middle class, of Negro business owners and independent operators, stems directly from the lack of financial resources. A budding Detroit businessman turned rioter expressed it this way:

> "I talked to one of the leading town citizens. Me and him were very good friends. I said, 'Well, why don't you invest some money and I'll invest some—I have a little money—and we could start a car-wash set-up here, an automatic for a quarter.' He told me to go to the bank and see what the bank said. I talked to the banker. You know what he told me? He told me to get four Negroes to invest. I said, 'Well, I heard that you all didn't want Negroes holding businesses.' He said, 'No, we didn't say nothing about Negroes holding businesses. If you can get enough people and collateral behind you, we'll take care of the rest of it.' You think I could? I couldn't get nothing. I couldn't get a quarter."

Fifth, "many important individuals and institutions have a vested interest in the maintenance of segregation." The Negro consumer's economic weakness allows him to be exploited by merchants; he cannot shop elsewhere. More fairly run ghetto businesses depend on economies of scale and other advantages of slum market concentration.

Sixth, the fact that "Negro organizations are small and feeble" is largely explained by the other five factors, especially their economic causes or aspects. The smallness stems from lack of community ties among the poor, and the weakness from lack of financial resources.

Thus, most of the social features limiting Negro political effectiveness can be traced to the economics of poverty. Also, many have direct economic effects on ghetto politics, like lack of economic control, as well as social implications, like deep cleavages in the class structure. As for the political factors limiting Negro influence, Banfield and Wilson analyze four types of political systems.[9] These restrictions are economic as well as legal or formal.

POLITICAL FACTORS

First, in the ward-based machine system, the Negro political machine must be built, like earlier ethnic machines, on an economic base of patronage jobs and other material benefits. Indeed, because the lower-class Negro would respond reliably to specific material benefits, a machine would appear to promise strong Negro influence in city politics. However, because civil service regulations and other reform measures have greatly undermined the political machine's economic foundations, the opportunities for building such organizations are few.

Moreover, where Negro machines do operate, they are parts of citywide, white-dominated machines, and therefore operate on the white bosses' terms. Congressman William L. Dawson's Negro sub-machine in Chicago is an example. Although it has placed eight Negro aldermen on the 50-member City Council, the eight are more responsive to Mayor Richard J. Daley's demands than to the Negro voters' needs. This relationship may change as the Negro population comes to comprise a larger, and maybe eventually a dominant, proportion of the city. However, the maintenance needs of such organizations, reflecting both internal and external demands, will continue to limit Negro power in local politics. Nevertheless, the small-ward system of councilmanic election undoubtedly facilitates symbolic Negro influence, irrespective of economics. Chicago got its first Negro alderman in 1915; Detroit, where countil members are elected at large, did not elect a Negro councilman until 1957.

Second, the ward-based but weak organization, one without much in the way of economic benefits to dispense and therefore built instead on personal followings or factional alliances, would give Negroes more independent influence. However, these systems often weaken Negro power by further splitting the Negro community along socioeconomic lines. New York, Cleveland, and St. Louis are examples. In these cities, both middle- and lower-class Negro representatives sit on the council and glare at each other. Also, personal followings and issue-oriented organizations are difficult to build among alienated low-income groups.

Third, under the rare proportional representation system of election, at least as it once operated in Cincinnati, the Negro leader has to depend on developing a personal following, because there is no economic base for a machine. Cincinnati abandoned P.R., it should be noted, in 1957, after a militant Negro was elected to the City Council.[10]

Fourth, the much-used nonpartisan, at-large electoral system is probably the last conducive to Negro influence. The reason lies at least partly in economics; it takes more money to conduct a city-wide campaign for office. Also, to be acceptable to the electorate-at-large and the white civic groups whose support is needed, the Negro candidate must be a moderate with middle-class means and manners. Machine-less Detroit is an illustration.

Nonpartisanship by itself, as in Los Angeles, where candidates run from large electoral districts without party labels, similarly blocks Negro influence. Again the reason is at least partly economic; without partisan support, the Negro candidates cannot draw on party chests for campaign funds. Thus, in most urban political systems, economics plays a significant role in reducing Negro influence.

OTHER FACTORS

According to Wilson's analysis in *Negro Politics,* the differential rate at which Negroes lag in politics in Northern cities depends on three other factors: the in-migration rate, the density of Negro sections of the city, and the size of the city's basic electoral unit.[11] The economics of poverty lie at the root of these, too.

The rate at which Negroes move into Northern cities depends, of course, on the state of the national economy and even more on rural economic conditions in the South. Although the push forces for urbanization of American Negroes are greater than the pull forces, economic opportunities in the cities, at least to the migrant's thinking, are a significant force. The density of Negro ghettoes, which determines how easily the Negro vote can be gerrymandered, also depends on the Negro's economic ability to move out. The electoral unit size, which is also a factor in gerrymandering, stems partly from campaign finances, as shown for at-large elections.

All these economic factors make effective most of the social and political restrictions on Negro influence. But another reason for the Negro's political weakness must be considered. This is a direct economic cause. If money is a source of political power, it follows that poverty makes low-income Negroes politically deprived as well as culturally, socially, and psychologically under-privileged. The pattern of proportionally few and small Negro businesses, low wages, joblessness, and other economic lacks means that most money spent in the ghetto goes right through the community to more affluent groups. Negroes do not control the ghetto economy to which they are restricted. Without that control, they cannot control their own political community, either. Lacking such parochial power, they cannot influence the city political decision-making process on which they depend for progress.

In addition, low levels of political participation are usually associated with low-income groups.[12] So the ballot, another source of political power, is crippled in Negro precincts, and this weakness is related to economic factors. Ghetto economics could likewise be expected to undermine reliance on knowledge or information or whatever other bases of power there may be. However, in the absence of economic strength, such resources as services and other cheap benefits may be the best available means for building a Negro political organization that will get out the ghetto vote in controllable blocks.

Thus, the Negro's political poverty must be considered an effect of ghetto economics rather than a primary cause of the conditions leading to the urban riots. The result may be summed up by HUD psychiatrist Leonard Duhl's political definition of poverty: "The inability to command the events that affect you."[13]

FRUSTRATION POLITICS

Negro politics is tied up in the complex of ghetto pathologies. It is fragmented by the internal social, cultural, and psychological consequences of poverty, as well as other forces both within the ghetto and imposed from without. Besides the direct and indirect economic constraints, two major internal conditions which hamper political activity are cultural alienation and social cleavage. These are reinforced by political weakness, and result in psychological and political frustration, which feeds back to the original factors in a cycle of accumulating advantages.

ALIENATION

Estrangement means not only that large percentages of lower-class Negroes fail to vote, but also that they do not participate in political activity of other sorts. Consequently, grass-root organizations do not sprout on ghetto pavement.

Non-machine types of political organizations built on non-material benefits or ideology do not readily appeal to the lower-class Negro life style. The personal following appears to be the most prevalent method of political organization among lower-class Negroes. But it is probably impossible to collect a following except where the politician can confine his pitch to a district or ward dominated by Negroes. Such is the case with former Congressman Adam Clayton Powell of Harlem and Congressman Charles Diggs, Jr., of Detroit.[14] Although both are elected from relatively large Congressional districts, Powell's is homogeneous, and Diggs' has the added advantage of being smaller than the councilmanic electorate in Detroit.

Alienation also prevents the formation among the lower classes of interest groups which might operate effectively in some factional-alliance system, such as Detroit's. The strongest of such groups are dominated either by middle-class Negroes, like the National Association for the Advancement of Colored People in Detroit and elsewhere, or by whites, like the United Auto Workers Union in Detroit. Similarly, the political club is a middle-class institution based on the satisfactions of membership, but for lower-class individuals, discipline would be difficult to maintain without material inducements.

Along with political alienation goes lack of empathy, even for other members of the same social or ethnic group. Here is more of that conversation between Frank Ditto and the rioters he was trying to get organized:

Ditto:	All right, you've got it. What about all those cats that haven't?
Rioter:	Well, let them help theirself. That's the way I look at it.
Ditto:	Do you think they're helping theirselves? Do you think they're going to stir the white people up?
Rioter:	What's that old saying? God bless the child who's got his own. They've just got to get out and get theirs, because I'm getting mine.

CLEAVAGE

Class conflict is an overwhelming, though often overlooked, fact in Negro society. Much evidence indicates that the middle-class Negro does not identify with his lower-class "brothers." Where Negroes have obtained political influence, it has usually been exercised by the middle class, as in the UAW in Detroit. "The Negro community," David Greenstone said in a 1961 report on Detroit politics, "is rapidly developing a significant middle class that wishes to disassociate itself from the lower class, sometimes even at the cost of race ends."[15] The lower-class Negroes "are beginning to resent the Negro middle-class leadership," E. U. Essien-Udom wrote at about the same time in his *Black Nationalism.*[16]

Ample evidence of this split showed up in the Detroit riot. Congressman John Conyers, one of the city's two Negro Congressmen, tried unsuccessfully to calm the rioters on the first day and was driven back by stones. "They are a leaderless community," he told Timothy Bleck of the *St. Louis Post-Dispatch.*[17] "They're alienated from us. We don't speak their language. We give $100 dinners, and some of these people don't see $100 a month. They [the NAACP and Urban League] have not been geared to the economic have-nots in the Negro community. They are middle-class organizations geared to middle-class people."

However, Rennie Freeman, another Detroit community organizer, saw no such class division in the riot. Asked whether there was hostility toward the middle-class Negro, he replied:

> Not as much as you would think. All Negroes in the community, man, belong to what's happening. They were there from the first. These people are standing out there, you know, giving the black power sign. You know, women in housedresses outside on the porch doing this shit, you know. And it is clear that this thing about the lower class being disenchanted is bullshit, man. *Negroes* are disenchanted with this shit. It's not just the cats that have got the guts to throw bricks that's pissed off. Everybody is pissed off. For the simple reason that you don't see no significant attack launched by any part of the Negro community or just the black nationalists. So what you have is fast support. It's clear. I mean you go to a community and, you know, these chicks that sit on the P.T.A. boards give the black power sign when there's smoke all around. You can forget that other shit, man.

The accuracy of this estimation of middle-class support for the riot may be challenged, though. These "P.T.A. chicks" were on the porch, not the street. Their clenched fists may have been protective gestures. However, hard evidence of such support was turned up by Raymond J. Murphy and James M. Watson, U.C.L.A. sociologists, in their study of Watts attitudes, "The Structure of Discontent."[18] Their interviews showed high levels of support for the riot among the higher socioeconomic groups of Negroes. Apparently the lower-class' welfare demands and the middle-class' status interests converge in the crisis.

FRUSTRATION

The political failures of the lower-class organizations have further alienated Negro masses. (Nothing fails like failure.) Frustration is likewise the product of all these forces, and at the same time embroiled with them. It is the product of outside political and economic factors as well.

The frustration stems on one hand from Negro helplessness. One of the Detroit rioters, asked whether he thought he could do anything to improve his life without white aid, put it this way:

> Naw, I ain't going to say that, because I can't get along without Whitey. You can't get along without him. He pays your check. He's got all the money. He's got all the commodities. He's got the gas company, the light company. You can't do without him.

On the other hand, frustration results from all kinds of obstacles erected by white society. These include all the measures of racial discrimination—"All we want is what our money can buy," a Detroit anti-poverty worker told reporter Bleck[19]—and political barriers.

The frustration also comes from internal political constraints. Greenstone reported that in Detroit, political issue development reflects the frustrations of a socially dispossessed group struggling for recognition in nonpartisan government and the growing power of the Negro community, apparently referring to the expanding middle class.[20] But the sources of political frustration are numerous. Racism, one aspect of it, often results from a political position outside the power establishment, Wilson has observed; and conservatism, from inclusion.[21]

Further, if a machine will give the Negro maximum political influence, it will also subject Negro needs to white organizational demands. If a small-ward electoral system will increase Negro influence, it may also increase representation of anti-Negro groups. If partisan local elections will facilitate Negro power, they will also aid opponents with the same party designations. If militancy will solidify potential Negro power, it will also prevent the alliances the Negro politician needs. If mass violence expresses the Negro's demands, it also repels most Americans. Even if the Negro comes to dominate city politics, the cities still lack the resources and state authority to serve his great needs. The name of the Negro political game is, in fact, frustration.

But success, at least at the polls, is possible. Carl B. Stokes' recent election as the first Negro mayor of Cleveland could set a new style for Negro politics. He succeeded in attracting the white support he needed to win, up to 30 percent of the vote in a few white wards, without losing mass Negro support.[22] Whether he can maintain this uneasy alliance while preventing another outbreak of violence in Cleveland remains to be seen.

DISORDERLY POLITICS

Just as national frustrations contribute to the rise of nationalistic states in international society, Negro frustration in America has given rise to black nationalism. With this new political consciousness rose the "black power" movement. Its leaders clearly advocate political force. For example, Floyd B. McKissick, national director of the Congress of Racial Equality, told the July 1967 Black Power Conference in Newark, N.J.: "The white man is the judge, jury, and the executioner in his system, and he first made the law so as to control us. His concept of 'law and order' means legal methods of exploiting blacks. We object, and we resist."[23]

Political frustration seems to be a central factor in the total pathology of ghetto life and to play a forceful role in the outbreak of the riots. Harold D. Lasswell traced such a path to violence in *Democracy Through Public Opinion*:

> If the level of insecurity is high—owing perhaps to economic breakdown—opinion may not become unified in time on behalf of measures for the alleviation of distress. . . . Insecurity may continue to pile; the flame of propaganda burns ever more brightly on behalf of every conceivable proposal; yet nothing but frustration results. Young and old grow disillusioned with the process of democratic discussion, and hotspurs demand action, even violence.[24]

As the Negro's insecurity, demands and frustrations pile on one another while America remains uncommitted to solution of the race-poverty-urban problem, Negro political activity has followed a similar pattern. Weak in the usual and legally acceptable processes, he has turned increasingly to what might be called, as is war, politics by other means. Inititally, he found the established local and state political arenas closed to him or unresponsive. Then he went to court, a less representative but more high-minded arena. He was successful there, but litigation is slow and implementation disappointed him. He also pressed the national political parties and the Federal Government, with fitful success and many failures.

When the decision-makers continued to lag in meeting his demands, especially at the local level, he resorted to uninstitutionalized means, "direct political action" aimed at both a power structure and popular opinion. He boycotted buses, picketed employers, sat in lunchrooms, demonstrated at city hall, and marched on Washington. The forms of such disorderly politics included all sorts of disruptive activities, but for years remained nonviolent, even when whites reacted violently. With each little contest won, hopes soared, but the next goal proved more difficult. Each move involved more people, rose in tempo, and became harder to control. Finally, in the largest cities outside the South, where goals are both nearest and most elusive, some spark from police-Negro friction sets the ghettoes ablaze. Thus, the riots climax hope-filled, failure-ridden political behavior. In Ditto's words:

> I see the charred buildings as nothing but frustration. American black people have suffered for over 400 years to gain the right of a good education in order to obtain good employment, in order to obtain some values and so forth, and they have been

denied this. Why? They have been asking white America why. White America has said nothing. 'Just have patience,' you know. And after 400 years the patience has worn out, and people are beginning to say this by sniping, looting, burning, or any other way they can. They have used every method that I can think of, from petitioning to demonstrating, marching and going to Washington and marching from Selma to Montgomery—all in a nonviolent way. They have died in Germany. They've died in Korea. They've died in Vietnam. They've died here in America for this same thing. And nothing has happened.

Saying that powerlessness in orderly decision-making underlies the riots is not saying it is the sole or paramount cause of the riots. Too many other potentially violent attitudes and frustrations are involved to single out any one as directly or immediately responsible. Poverty is probably the prime cause, but it is both economic and political poorness, and complex intermediate factors intervene between the roots and bitter fruit. Poverty politics is a central factor, however, because the explosion has largely political characteristics. Violence can be a primitive form of protest; and mass protest, whatever its cause or nature, is a political act.

Reports and studies of the recent riots show, of course, that the ghetto people voiced many different reasons for rioting. For some, it was economic retaliation; for others, social rebellion. Some claimed catharsis; still others, other motives or mixed motivation. Said one Detroit reveler:

> This is not a riot. A lot of people have a misconception of it. This is nothing but—like the man said—pure lawlessness. People was trying to get what they could get. The police was letting them take it. They wasn't stopping it, so I said it was time for me to get some of these diamonds and watches and rings. It wasn't that I was mad at anyone or angry or trying to get back at the white man. They was having a good time, really enjoying themselves until them fools started shooting.

He too may have been enjoying himself, but others acted more purposefully. Louis E. Lomax quoted a Detroit woman who owed $912 on an original debt of $285. "Yes, I burned that store down," she said. "That's one bill I will never have to pay. I made sure the office and the records went up in flames first!"[25]

Political motives showed up frequently in press reports of the Detroit riot. "You all still haven't got the message," a leader of public housing tenants told reporters after a "frustrating" meeting with city officials. "Nobody wants to listen to it like it is. Just because you are high and I am low, that doesn't mean you are supposed to call the shots and I am supposed to bend. The upper class is going to have to take a chance on the poor man."[26]

Indirect evidence that power relationships played a part in the riots appeared in a study by Brandeis University's Lemberg Center for the Study of Violence of the levels of Negro dissatisfaction in three riot-torn and three riotless cities. Most Negroes told the interviewers they felt their city governments were doing too little for racial integration. This attitude was the only one found corresponding highly with the overall degrees of Negro dissatisfaction. "Negroes are shifting to the opinion that only intense forms of social protest can bring relief from social injustice," the "Six-City Study" concluded.[27]

Murphy and Watson found compelling evidence of similar motivations in their Watts survey. Their sample agreed almost unanimously that the main targets of attack were white merchants and police—symbols of white domination, especially of economic exploitation and social control.

"Those who are better off seem to evidence considerable anti-white sentiment which is significantly related to their participation in violence," their preliminary report said. "Those less fortunate rebel against discrimination and appear to be motivated mainly by economic discontent. Mistreatment or exploitation by whites (merchants and police) seems to be a source of riot support for all levels in the ghetto."[28] Thus, middle- and lower-class Negro demands appear to unite in the violent attempt to grasp power.

"A significant number of Negroes, successful or unsuccessful, are emotionally prepared for violence as a strategy or solution to end the problems of segregation, exploitation, and subordination," the U.C.L.A. study concluded.[29]

If the riots were partly political revolts, why was the violence not aimed at more obviously political targets? Why did the rioters not burn some symbol of white power—city hall, say—rather than their own neighborhoods and homes? "It is this self-destructive character that lends justification to describing such acts of violence as revolts rather than as riots," Murphy and Watson noted unconvincingly.[30]

The suicidal bent may stem in part from some self-despising twist of ghetto mentality, or other factor. The answer lies also in the limited aims of the Negro's political demands. He does not want to upset or replace the American social, economic, or political systems. He just wants in, a share of the benefits, equitable influence in political decisions—in his expression, "a piece of the action." So he does not attack the whole system, or its larger symbols, but only the smaller parts which concern him daily, the merchants and police. Lacking a scalpel, he uses a firebrand.

POLITICAL REFORM

The three points borne out here—over-simply, that the violence results from disorderly politics, which results from frustration, which results from political poverty—mean that local governmental processes are failing to convert the restless Negroes' demands and hostility into an orderly revolution. The cities are failing to manage conflict as they are supposed to. Today's violent revolts constitute a frustrated revolution.

Robert Conot says in *Rivers of Blood, Years of Darkness* that Watts "placed on record that the Negro has become a power in the cities, a power that civil authorities do not have the strength to cope with."[31] But Richard M. Elman feels Canot is wrong: "Once the rioters had brought violence to a point where police restraint was no longer an option, the white show of force was unequivocal, even, at times, indiscriminate. It was Negro—not white—power which was brought to its knees."[32]

Elman is right that the riots were crushed. However, Conot is right, in the larger sense, that the revolt cannot be stopped. The greater question is whether the impending cycle of riots and repression, which could culminate in open racial warfare, will be avoided. This requires profound reform.

Any new program which would try to change only one or a few factors in the knot of ghetto ills, the intermediate causes of riots, would probably have little impact, as the current war on poverty has shown. More basic reforms are needed in the distribution of urban political power and economic wealth.

The importance of local political forms might be illustrated by comparing Los Angeles or Detroit to St. Louis. The first two appear to have the political systems least favorable to Negro power, and have had the worst riots. St. Louis, which has not so far suffered such violence, may have a political system most advantageous for Negroes. Its small-ward representation, partisan elections, several patronage departments, and weak-machine politics all increase Negro influence; while other features, such as civil service, provide balance. However, because other riot-rocked cities may have similar political structures, this suggestion is tentative—maybe even foolhardy.

A better explanation of why some cities have had riots while others with equally bad social and economic conditions have not may lie in the origins of the ghetto dwellers. A survey of Detroit Negroes following the riot, which was sponsored by the *Detroit Free Press* and Urban League, showed that the Northern-born or -reared Negroes were three times more likely to riot than were Southern immigrants.[33] Similar findings are contained in the Watts study and other reports. Civil rights activists in St. Louis say the city has avoided such violence because there is no militancy among its Negro masses. Many black St. Louisans came from Mississippi in the last ten to twenty years. Their passive, Southern attitudes—compared to a possibly more aggressive frustration expressed by children of earlier migrants—could be the reason for the lack of rebellion, or at least its delay.

The reforms, in any event, should be political as much as economic and social. For a redistribution of economic wealth is not likely without a redistribution of political power. Power reform is fundamental because it would enable the Negro to get the social and economic changes he needs. But moreover, since so much of the Negro's political weakness stems from money poverty, and since the cities also lack resources, economic reform is even more basic.

"We're getting close to the economic power distribution and the wherewithal and the affluence of society," Mel Ravitz, a Detroit councilman and Wayne State University sociologist, said following the riot, "and we're not willing to share it. I don't know how to do it, but it must be done. Even if money, more money, is not the answer, my whole training as a sociologist has biased me towards looking to the economic situation as a road to the answer."

Ravitz was echoed by a Detroit ghetto dweller quoted by J. Anthony Lukas of the *New York Times.* "Yeah, sure, Jerry Baby's done a few things for us," he said referring to Mayor Jerome P. Cavanaugh. "But that's just it, man. We're tired of having people do things for us. We want to do things ourselves. They're willing to share some of the corn and 'taters with us, but when it gets right down to the nitty-gritty of power, they aren't sharing anything."[34]

NOTES

1. Edmund C. Banfield and James Q. Wilson, *City Politics* (N.Y.: Vintage, 1963); esp. chap. 2.
2. *Ibid.,* p. 18.
3. *Ibid.,* p. 21.
4. The quotations from Detroit, unless otherwise attributed, are from interviews conducted by Tom Paramenter of *Trans-Action* magazine, who made his transcript available to me. For other use made of these interviews see his "Breakdown of Law and Order," *Trans-Action,* Sept., 1967, pp. 13-22.
5. See David Greenstone, *A Report on the Politics of Detroit* (Cambridge, Mass.: Joint Center for Urban Studies, 1961; mimeo.); and Robert J. Mowitz and Deil S. Wright, *Profile of a Metropolis: A Case Book* (Detroit: Wayne State Univ. Press, 1962), esp. chaps. 2, 6, 8, and 10.
6. James Q. Wilson, *Negro Politics: The Search for Leadership* (N.Y.: Free Press, 1961), p. 22.
7. Banfield and Wilson, *op. cit.,* note 1, p.294. The following discussion is based on chap. 20, esp. pp. 291-303.
8. See Lee Rainwater and William L. Yancey, *The Moynihan Report and the Politics of Controversy* (Cambridge, Mass.: M.I.T. Press, 1967).
9. Banfield and Wilson, *op. cit.,* note 1, pp. 303-308.
10. See Ralph A. Straetz, *PR Politics in Cincinnati* (N.Y.: New York Univ. Press, 1958), esp. chap. 8; and Kenneth Gray, *A Report on City Politics in Cincinnati* (Cambridge, Mass.: Joint Center for Urban Studies, 1959, mimeo.).
11. Wilson, *op. cit.,* note 6, chap. 2, esp. p. 23.
12. See Robert E. Lane, "Why Lower-Status People Participate Less than Upper-Status People" in Lewis Lipsit (ed.), *American Government: Behavior and Controversy* (Boston: Allyn and Bacon, 1967), pp. 87-97, which is an extract from Lane's *Political Life* (Glencoe, Ill.: Free Press, 1959).
13. Lecture Nov. 20, 1967, at Webster College, St. Louis.
14. See Greenstone, *op cit.,* note 5, p. V-26.
15. *Ibid.,* p. V-35.
16. E. U. Essien-Udom, *Black Nationalism* (Chicago: Univ. of Chicago Press, 1962), p. 304.
17. Timothy Bleck, interview in *St. Louis Post-Dispatch,* July 30, 1967.
18. Raymond J. Murphy and James M. Watson, "The Structure of Discontent: The Relationship between Social Structure, Grievance, and Support for the Los Angeles Riot" (Los Angeles Riot Study, Institute of Government and Public Affairs, University of

California at Los Angeles, MR-92, June 1, 1967, mimeo.), p. 114.

19. Bleck, *op. cit.,* note 17,July 31, 1967.

20. Greenstone *op. cit.,* note 5, p. V-38.

21. *Ibid.,* p. 34.

22. See Jeffery K. Hadden, Louis H. Masotti, and Victor Thiessen, "The Making of the Negro Mayors, 1967," *Trans-Action,* Jan.-Feb., 1968, pp. 21-30.

23. Extracts from McKissick's speech were published in the *New York Times,* July 30, 1967.

24. Harold D. Lasswell, *Democracy Through Public Opinion* (Menasha, Wis.: George Banta Publishing Co., 1941), p. 21.

25. Louis Lomax in *St. Louis Globe-Democrat,* Aug. 20, 1967.

26. Quoted by Bleck in *St. Louis Post-Dispatch,* July 31, 1967.

27. "Six-City Study: A Survey of Racial Attitudes in Six Northern Cities: Preliminary Findings," a report of the Lemberg Center for the Study of Violence (Waltham, Mass.: Brandeis University, June 1967, mimeo.), p. 22.

28. Murphy and Watson, *op. cit.,* note 18, p. 114.

29. *Ibid.,* p. 115.

30. *Ibid.,* p. 43.

31. Quoted by Richard M. Elman in "Victims of Hate which Hate Produced" *New York Times Book Review,* Aug. 20, 1967, p. 3.

32. *Ibid.*

33. Philip Meyer, "The People beyond 12th Street: A Survey of Attitudes of Detroit Negroes after the Riot of 1967," *Detroit Free Press,* Aug. 20-22, 1967; the three-part survey report has been reprinted as a booklet distributed by the Detroit Urban League.

34. "Postscript on Detroit: 'Whitey Hasn't Got the Message,' " *New York Times Magazine,* Aug. 27, 1967, p. 51.

VIOLENCE
The Last Urban Epidemic

Richard Meier

■ Violence and disorder are the last remaining mortal threats to urbanism as a way of life. Cities, it must be reiterated, have been death traps from the beginning of history until the start of this century. If cities were not pillaged, raped, and burned, or torn asunder by revolutionaries, they would be reduced by famine, or visited by one pestilence after another, so that deaths on the average remained significantly more numerous than births. Urban centers grew primarily because the overflow from the countryside had to find some outlet, and powerful cities occupied the more accessible sites between the landless farmers and the available new land. Thus urban settlement could be renewed and expanded only by infusions of new blood (much more male than female) from the farms. Only when the expanding industries of the nineteenth century in some favored locales began to exert a strong attraction for immigrants did the outlook change.

Spectacular shifts in expectations originated in northwest Europe at that time which have now diffused to cities everywhere in the world. Smallpox, cholera, typhoid, typhus, mea-

AUTHOR'S NOTE: *The author is indebted to the U. S. Public Health Service for use of material from a larger manuscript produced for contract No. PH–86–66–120 ("Resource Materials on Health and Social Well-being in the Environment of the Metropolitan Region") with the Center for Planning Research and Development, University of California, Berkeley.*

sles, malaria, and dysentery were rather speedily brought under control via quarantine, immunization, and sanitation, and eventually with a series of potent drugs.[1] It was not until the decade of the 1960's that ravages of "galloping consumption" (tuberculosis) in the most densely populated sectors of the poorest cities came under control, and admissions to the sanitaria began to decline. Thus only the behavioral epidemic of violence remains a serious threat to life in cities.

The problems of disorderly behavior experienced in American cities are still trivial as compared to those yet to come elsewhere in the world. Our difficulties can be traced to a spurt of rural immigrants two to three decades back that added 20–50% to the population of a metropolis. Virtually all of this influx originated from subsistence-level farms or low-income rural communities. Much larger numbers will be impelled to move in Asia, Latin America, and Africa, if any success in economic development is achieved. The urban-rural disparity is even greater there, so tribal, linguistic, racial, and religious ghettos must come into existence on a scale that is sometimes ten to a hundred times that experienced in America. The potentials for the amplification of protest and the escalation of violence are so great that a policy of keeping the immigrants at home is often considered. Yet without the growth of cities, and the institutions that only large-scale urbanism can support, economic development as we know it today would come to a stumbling halt. Thus the propagation of social order in the course of massive urban settlement is a fundamental task. Any approaches to solution found in the United States may well be refitted to meet the more urgent needs overseas. Preventive techniques, as worked out by public health practitioners, have been found to be transferable across cultural barriers in most instances; therefore similar modes of thinking about the prevention of violence may find an equal degree of acceptance. This direction promises to be by far the most effective in our search for indirect controls.

Epidemic prevention has already been extraordinarily influential in changing the rules for city building. Urban planners take the latest water supply, sewage treatment, and air purification standards into account as prerequisites for the design of cities. Are there any other policies that should be added which would minimize the civil disorders that threaten the future?

More specifically, to what portions of the city should the stream of migrants be directed — the central areas, the fringe, an industrial sector, or where? How shall they be grouped into neighborhoods and communities? What kinds of assistance are most needed? What physical arrangements seem to be conducive to peace and security? What institutions need to be built up as rapidly as possible? To what alarms and alerts should authorities be sensitive? The answers to these questions require a search for information, but the need to make political and planning decisions soon forces us to adopt a decision focus rather than an inquiry into fundamental causes.

PLANNING CONTROLS AND CIVIL DISORDER

Because so many of the disorders of a metropolis have been traced to the presence of a poorly acculturated population of new settlers, the long-range planners often attempt to reduce the stresses by restricting immigration. Since the "freedom to move" is a fundamental civil right within the United States, indirect measures have sometimes been taken, such as forbidding the extension of Welfare payments during the first three years of residence, as in some of the most attractive counties of California, in order to transmit a harsh image to the landless indigents "out there."[2] Restriction of flow has been possible wherever an international boundary existed, but no nation has been able to implement a policy of internal migration control, even with the strictest regulations, without bringing economic development to a halt. The evasions become too numerous to prosecute.[3]

Once people had arrived, the typical administrative solution was to build public housing for the needy group, even though many recognized that state subsidies of this sort do not buy peace. Isolation in a standardized administration-produced ghetto (e.g., the public housing project) corks up the sentiments for a very short period, only to see them unloosed by an apparently trivial incident thereafter.

The transportation system can be organized so that the locus of the damage due to a hostile outburst is specified. In a city highly dependent upon electrified mass transit, the buildings in the recalcitrant neighborhoods are most subject to damage, but rolling stock and stations may be partially destroyed. Where buses form the main basis for movement in and about the city, the threat is again experienced primarily by neighbors in the vicinity of the outburst, although a few buses may be burned. The automobile-oriented metropolis that is beginning to dominate America poses extra difficulties, because the poor feel deprived of "automobile gratifications." Auto theft therefore becomes the most common crime on the police register, but its incidence will be spread over an area many times larger than the ghetto, because the best prizes are to be found in the high-income communities. Youths living in an auto age feel compelled to obtain vehicles with their first earnings, the main justification being that jobs are almost always at a distance and will require commutation by expressway. Thus the gangs become mobile, and virtually any area in the metropolitan region becomes vulnerable to disorders initiated by delinquents. Built-in forms of quarantine may be expected to work less well in any future that allocates a vehicle to a young male upon coming of age.

In newly developing countries living close to subsistence, the planning strategy must be different. There the new immigrants are best settled on the metropolitan periphery, where they build their own homes and gardens and organize their own community. Such levels of cooperation can only be achieved in relatively homogeneous settlements where the settlers have a common origin and a similar social status. The urban form that results is a series of corridors radiating from a center. Buses, later replaced by electric railways, provide the most economical source of transport. Epidemic violence arises either from population pressures, breaking out when a neighboring community believes its territory is threatened by invasion, or is inherited from a long history of feuds and injustices in the countryside from which the immigrants came.

Planners can help most by foreseeing which feuds may be brought to a boiling point by proximity, and then allocating land in advance so as to minimize the likelihood of guerrilla warfare in the city. They can designate sites for new colonies early enough to prevent overpopulation of the first reception centers. They can assist the rapid growth of community controls over the socialization of the youth by giving high priority to the useful local facilities. However, plans by themselves are puny instruments at best, and can do little more than set the stage for other kinds of action aimed at violence prevention. What new policies are indicated?

THE SEARCH FOR SPECIFIC CURES

Medicine has made a great deal of progress in the past century by drawing upon the biological and physical sciences for specific drugs or extracts that create either an immunity to the contagion or a quick cure. Alvin Weinberg has dubbed this remedial use of science "the technological fix."[4] The fact that certain drugs are commonly called "tranquilizers" alerts us to the immediacy of this possibility.[5] However, a quick investigation reveals that the tranquilizing drugs with discernible effects cannot be freely administered to a population, because they cause complications which require prescription in advance and constant supervision by physicians while under the influence of the drug. Management of violence with drugs is most advanced within institutions which house acutely ill mental patients, but there the principal ethical justification remains that of the personal safety of nurses and caretakers, and not the control of violence for the sake of peace and quiet.

A review of experimental literature intended to enlighten us on the properties of violence that could eventually lead to the preparation of therapeutic agents requires a translation of *macro* terms of community scale into the *micro* observables in the controlled laboratory study. Violence, or illegal coercion as defined by political scientists, becomes *physical aggression* among the social psychologists. Among social biologists, it is *aggressive behavior*. At the physiological and biochemical levels, the key terms become *rage, arousal,* and *anger*.

They ask themselves, "What are the alternatives for aggression?" A model of Feshbach's outlines the regulation of a response sequence, one branch of which involves violence followed by catharsis.[6] Berkowitz carried out a number of experiments on incitement to aggression which caused him to become most concerned about conditions in the environment preceding the immediate occasion for frustration.[7] If cues in the environment suggest that aggression is an expected response, then violence is more likely to occur.

A persuasive explanation of the Los Angeles Watts riot of 1965 can be constructed from these theories. The frustrations within the Negro community were built up by a decline in employment and income when everyone else in the region was visibly prospering. Negroes could do very little for themselves. Enhanced striving rarely was rewarded, due to race prejudice; the reform program that had been backed in the past yielded no visible results. The response of the majority was to withdraw or to "turn the other cheek." A youthful minority, however, had had few lessons in the repression of aggressive feelings. It was also following closely the events of the civil rights movements in the South over television. The programs had been purveying fictional violence ever since they could remember, but violence in the news featuring their own people as both victims and victors was more persuasive.[8] Add to this background the history of hostility to the police, so that uniforms and patrol cars become intensely threatening images. In such a context, the "repeated aggression sequence" becomes possible, but far from certain. The therapy indicated at that stage seemed to be jobs, rather than tranquilizers.

Communities other than Watts are likely to suffer from *overcrowding*, a condition which has a direct connection with violence in biosocial systems. It leads to enhanced fighting to defend territory and to maintain status in the community. The costs of fighting are severe, not only due to injuries and destruction but also from the drain upon the adrenalin supply and its effects upon the organs.[9] Crowding in animals results eventually in infertility and a reduced reproduction rate, but the history of man has not given us a clear indication of consequences, because these conditions of *anomie* were interrupted by wars and famines. The experiments in the laboratory suggest that *privacy*, as much as it can exist for animals, seems to be essential for community peace.

Privacy implies that a choice is available to an individual (or a primary group) to use an aspect of the environment (e.g., a door, a hiding place, etc.) to protect himself from "others." The complete control of the situation would allow those transactions which have been initiated to be completed without interruption. With wealth and authority a person may manipulate *spaces, walls, lighting, rituals, schedules, calendars, and uniforms* so as to keep interruptions, along with the errors they cause, down to a tolerable frequency. The greater the density of signals demanding attention, the greater is the need for privacy. Thus the design of arrangements for privacy is the best tranquilizer society has yet invented: it reduces the strains that set the stage for violence and lowers the rate of spread as well.

CONCLUSIONS

The last great plague of the cities can be productively viewed as an epidemic of violence. The current state of knowledge suggests that alarms can be set and alerts organized which reduce the likelihood of precipitating incidents. The halting of immigration is not feasible, even in authoritarian political systems, but a policy of forming self-help communities seems to work in most parts of the world. Such communities are admittedly ghettos, and will certainly create problems at a later date, but they permit immediate escape from the under-employment of overpopulated rural communities. Once the contagion begins, it may be possible to intervene via the communications network and interrupt the spread. Distribution of anti-violence drugs to the relevant populations thus far does not appear to be promising because, if they are effective, continuous medical supervision is indicated. How-ever, hope should not be abandoned, because a number of interesting leads should be explored. The design of protection from environmental stresses and crowding that may lead to violence opens up many more alternatives. For better defenses against violence, the most attention should be given to the provision of varied opportunities for privacy.

NOTES

1. Kingsley Davis and Hilda H. Golden, "Urbanization and the Development of Pre-Industrial Areas," *Economic Development and Cultural Change,* III (Oct., 1954), 6–26. The demographic histories reviewing the history of mortality in various world pop-

ulations are somewhat more balanced on assessing the hazards than the histories of public health. The latter emphasize the diffusion of the innovations from urban centers, mainly seaports, to towns and villages.

2. Unemployed girls can beat this stratagem by becoming pregnant and applying for Aid to Dependent Children, which depends upon Federal funds. The social costs of illegitimate children are rarely taken into account by county authorities.

3. Recent reports from Moscow suggest that its authorities are ready to abandon past attempts to limit migration by decree and a system of internal passports; cf. Peter Hall, *The World Cities* (N. Y.: McGraw-Hill, 1966).

4. Alvin M. Weinberg, "Can Technology Replace Social Engineering?" *The Bulletin of the Atomic Scientists,* XXII (Dec., 1966), 4–8.

5. A relatively comprehensive bibliography is to be found in John M. Davis, "Efficacy of Tranquilizing and Anti-Depressant Drugs," *Archives of General Psychiatry,* XIII (Dec., 1965), 552–572.

6. Seymour Feshbach, "The Function of Aggression and the Regulation of Aggressive Drive," *Psychol. Rev.,* LXXI (1964), 257–272.

7. Leonard Berkowitz, "Aggressive Cues in Aggression Behavior and Hostility Catharsis," *Psychol. Rev.,* LXXI (1964), 104–122. Cf. also his paper in this issue.

8. Leonard Berkowitz, Ronald Corwin, and Mark Hieronymous, "Film Violence and Subsequent Aggressive Tendencies," *Public Opinion Q.,* XXVII (1963), 217–229.

9. J. D. Carthy and F. J. Ebling, *Natural History of Aggression* (London: Academic Press, 1964); Konrad Lorenz, *On Aggression* (New York: Harcourt, Brace, and World, 1966).

RIOT IDEOLOGY
AMONG
URBAN NEGROES

T. M. Tomlinson

■ For the most part, the substance of this paper uses a data base provided by a study of the Los Angeles riot of 1965.[1] In that study a random probability sample of 585 Negro respondents residing in the 182 census tracts which made up the riot curfew area were interviewed by indigenous Negro interviewers. The interviewers were located, hired, and trained by the survey staff; they entered the field within two months after the riot and completed their task six months later. A sample of whites, similar in size and stratified by socioeconomic status and racial composition of the area (integrated and non-integrated) was gathered at the same time from areas outside of the curfew zone.

An omnibus interview schedule was used which covered mainly social-economic-political issues and riot participation and evaluation. The interviews were generally long, usually requiring upwards of two hours for completion. Respondent cooperation was very good, and interviews were seldom halted for reasons of attention or cooperation lag. The coded data were analyzed by conventional cross-tabulation and statistical techniques (chi square) in the computer facilities available at U.C.L.A.

AUTHOR'S NOTE: *None of the data or information reported in this paper, except where expressly noted, was accumulated while the author was in an official capacity with the Office of Economic Opportunity. The views expressed here are those of the author, not his agency.*

Although the conclusions of this paper are based on data drawn principally from the Los Angeles riot survey, it should be kept in mind that the sense of those data are essentially duplicated by several other riot studies, most notably the one following the Detroit riot of last summer, 1967.[2] Thus the generalizations which follow will describe a set of conditions that exists in most Northern urban centers.

SOME CONCLUSIONS FROM THE LOS ANGELES STUDY

In describing the results of the Los Angeles study, it is appropriate to begin by dispelling some myths that persist in the minds of many whites, a large number of public officials, and some Negroes.

1. It is a myth that only a tiny fraction (3% to 5%) of the Negroes living in the riot zone(s) participated in the riots. The best estimates indicate that upwards of 15% of the people interviewed in Los Angeles (and Detroit) claim to have participated actively in the riot. Measures of participation were made from a number of direct and indirect angles, and each time the rate was about the same.

2. It is a myth that an overwhelming majority of the Negro community disapproves of those who supported the riots. At least 34%, and perhaps as high as 50%, express a sympathetic understanding of the views of the supporters.

3. It is a myth that most of the Negro community views the riot as a haphazard, meaningless event whose thrust was a disregard for law and order. On the contrary, 62% saw it as a Negro protest, 56% thought it had a purpose, and 38% described the riot in revolutionary rhetoric (revolt, insurrection, revolution). And it was a justifiable protest; 64% of the respondents who said it was a protest also said the victims deserved attack.

4. It is a myth that Negroes expect and are afraid of white retaliation and a decline in the quality of the relations between blacks and whites. Favorable effects were expected by 58% of the Negroes; unfavorable effects by only 18%.

In sum, participation in and support of riots in the Negro community is by no means the position of a tiny minority of malcontents. The riot in the eyes of a large proportion of Negro citizens was a legitimate protest against the actions of whites, and the outcomes are expected to produce an improvement in the lot of Negroes and their relations with whites.

Departing from the context of mythology, two additional important points should be noted:

✗1. Of the 56% who claimed the riot had a purpose, each cited one or another of the following goals of the riot: (a) to call attention to Negro problems, (b) to express Negro hostility to whites, or (c) to serve an instrumental purpose of improving conditions, ending discrimination, or communicating with the "power structure." In all cases these responses give justification to the action, whether it be a release of pent-up frustration or simply making the point that injustice and inequality exists among Negroes in America. The riot has been assigned the purpose of letting whites know "how it is" for Negroes in this country.

2. The second point deals with the fear the events of the riot generated in a majority of the Negro citizens. No matter the justifications for the event, 71% of the sample expressed dismay about the burning, destruction, and killing. Louis Harris' riot survey also describes the stark terror in his respondents as they talked about the fires. Thus political purpose is involved in riots, but the response of the citizens is also one of fear following its events.

MILITANCE AND RIOT IDEOLOGY

Probing the data a bit deeper, an atempt was made to find out who was most likely to view the riot positively, to participate, and to expect a positive outcome. To this end the sample was divided into three parts based on the respondents' expressed sympathy with radical militant Negro organizations. Those who sympathized were called *militant,* those who were antagonistic were called *conservatives* (or more precisely, counter-radicals) and those who had no pro or con position were called *uncommitted.* Rather than describe each of these groups, this paper will concern itself with the characteristics of the militants compared to the non-militants. The data support the following conclusions:

1. Militants, those identified by their radical sympathies, make up 30% of the sample and are most often found among male youth. They are more likely to be brought up in an urban setting, somewhat better educated, relatively long-term residents of the city (over ten years), more involved in religion, and in possession of a more positive self-image. They are equally likely to be working and tend to be more sophisticated politically than the non-militants. In short, they are the cream of urban Negro youth in particular and urban Negro citizens in general.[3]

2. Militants are the most deeply aggrieved and claim to have had higher rates of personal contact with the police under conditions usually described as police misbehavior.

3. They are more likely to view the communication media as unfair in their portrayal of Negro problems.

4. They are not markedly anti-white, but they are considerably more disenchanted with whites than the non-militants.

5. They are the most active politically and more likely to endorse the advancement of the Negro cause by any method necessary; they will endorse all conventional civil rights activity and in addition will lend disproportionate approval to use of violent means. Three times as many militants (30%) as non-militants endorse the use of violence as a legitimate last resort.

6. In describing the riot, the militants are more likely to use a term from the revolutionary lexicon than the non-militants. By a ratio of almost 2 to 1 they claim to have participated actively in the riot. They were much more favorable to the riot and its events than the non-militants, and project that view onto the community at large by claiming larger rates of community support for the riot. They are more hopeful of positive change in race relationships (which undercuts the notion that they are clearly antagonistic to whites, i.e., they look for and apparently desire an improvement in Negro-white relations).

7. Finally the militants place the responsibility for change clearly in the laps of the whites. The non-militants tend to take the view that both races must change to achieve rapprochement, but the militants clearly ascribe the locus of change to whites.

Now where does this take us? In the first place, it must be continuously kept in mind that the differences between militants and non-militants are those of proportions; more militants feel this way than non-militants. It does not mean that only militants feel this way. A majority of the entire Negro population is aggrieved, angry and disaffected. For example, compared to whites in the Los Angeles study, there are dramatic differences in the level of trust of elected officials and police by Negroes. Negroes express far higher rates of perceived political disenfranchisement and impotence to bring about change. At the same time, however, they appear to be deeply committed to bringing about change. The picture therefore is one of intense political concern combined with felt impotence to exert influence on the political structure, and that is a cornerstone of social unrest.

Thus the climate which fosters riots is endemic in American society and in the Northern urban centers particularly. The Los Angeles riot took the lid off by disinhibiting a riot response to the conditions of Negro life that had always existed. The response of the Negro ghetto residents suggests that a sufficient proportion of them view the riot as a justifiable protest form to account for the absence of counter-riot behavior on the part of the Negro community. What then are the implications of these data for the future of urban violence in America?

It would seem that a sort of simplified riot ideology has taken form, and riots today have assumed the shape of a popular movement. Support, or at least sympathetic understanding of the purpose of riots, characterizes a large segment of the Negro population. Within this segment are imbedded a group of sophisticated, activist young people who have provided the riot with political interpretations of purpose. They have created a riot ideology, and this ideology has infected the thinking of other less sophisticated but equally disaffected individuals. It should be emphasized that this is not a description of a conspiracy. It is a description of a portion of the population that for a variety of historical and current reasons are susceptible to the idea of violent protest. That idea emerged in its clearest form in the aftermath of the Los Angeles riot and has been blown across the country on the winds of pervasive Negro discontent.

The creation of a riot ideology has a number of implications for this society. In the first place, it does not allow one to cite the actions of "agents provocateurs" in accounting for the occurrence of a riot, unless one is willing to call the mood of the people by that name. Second, and most important, it implies that nothing can be done to stem the tide of urban disorders until they run their course. There are no immediate responses within the repertoire of any agency or person which are sufficient to expunge the outrage that gives birth to Negro violence, except the Negroes' own fear of the burning and killing, and that comes only after the riot has occurred.

Now it may appear that this is an unnecessarily pessimistic position (assuming the foregoing is interpreted as being pessimistic), or perhaps it appears that other equally plausible, but less stark, views of the underpinnings of urban riots have been overlooked. Let us examine some of the other possibilities.

Riots have occurred in cities with every type of administrative structure. They have occurred in model cities; indeed Detroit was a well known "model city." They have occurred in cities receiving relatively large sums of poverty money, and ones receiving relatively small amounts. They have occurred in cities with compact ghetto enclaves and in cities in which the ghetto was distributed over a large area. They have occurred in cities with relatively high Negro employment and wage rates (Detroit) and in cities with relatively low rates (Watts). They have occurred in cities with relatively large proportions of Negroes and cities with relatively low proportions.[4]

Clearly what produces riots is not related to the political or economic differences between cities. What produces riots is the shared agreement by most Negro Americans that their lot in life is unacceptable, coupled with the view by a significant minority that riots are a legitimate and productive mode of protest. What is unacceptable about Negro life does not vary much from city to city, and the differences in Negro life from city to city are irrelevant. The unifying feature is the consensus that Negroes have been misused by whites, and this perception exists in every city in America.

Thus it is the thesis of this paper that urban riots in the North will continue until the well of available cities runs dry. They will continue because the mood of many Negroes in the urban North demands them, because there is a quasi-political ideology which justifies them, and because there is no presently effective deterrent or antidote.

DETERRENTS AND ANTIDOTES

There are a number of approaches to riot control, none of which, if the thesis of this paper is correct, will serve to foreclose the occurrence of popular riots. Nevertheless it is worth discussing some of these approaches simply to buttress the assertion that even if something could be done, nothing will be.

Taking non-repressive methods first, where does the War on Poverty fit into all of this? The answer at present is nowhere. Congress, reflecting the attitudes of whites in this country, has already made it clear that it is not interested in using the poverty program to ameliorate, not to mention remove, the causes of urban violence. It recently appropriated a mere 1.733 billion dollars for the poverty program, and that only after a long and desperate fight by the proponents of the program. Little has been said by the political leaders of this country which would indicate both an awareness of the social causes of riots and a receptive position to costly but non-repressive methods of riot control. Quite the contrary, the public position of the Administration is that riots, for whatever reason or whatever legitimate grievance, will not be tolerated. But this position simply reflects the position of the bulk of white Americans, and represents an awareness by the country's leaders that it is impolitic to suggest that whites be prepared to make a personal sacrifice to remove the conditions of Negro life that generate a riot response.

The problem of white reaction must also be faced by any investigative body appointed by the Administration, in this particular case the National Advisory Commission on Civil Disorders. Advisory commissions are appointed by the President (or in a given state by a Governor), and their recommendations, so far as they are taken seriously, require the support and advocacy of the appointing figure. When the issue is explosive and when the truth is unlikely to be palatable to the majority of the electorate, the commission is in a ticklish political position. Its recommendations must not offend the negatively disposed (white) majority or their elected representatives (Congress), because then the appointing leader (President or Governor) is placed in the embarrassing and impolitic position of having to support recommendations which almost surely will alienate him from a significant proportion of the electorate and the legislature. Thus the real disaster of the situation is that the Commission on Civil Disorders must be responsive to the political realities of the country. In this context it cannot deliver the full message if by doing so it stands to antagonize a substantial number of voters. The voters, directly or through their representatives, would reject both the recommendations and those who made them and support them — i.e., the recommendations would not be implemented, and of those who advocate them, the politically vulnerable would be turned out of office.

Thus it is inconceivable that politically sensitive appointed groups would be free to issue a report which had the potential to be violently unpopular among a majority of the citizenry. The Commission is bound by political considerations to make a statement which, from the viewpoint of practical politics, is both feasible and palatable. And when the issue is Negro behavior in the face of white racism, and when historically such commissions have dealt with this issue by making "half-a-loaf" recommendations, it seems unlikely that a full loaf will be requested by a politically sensitive body faced with overt white hostility to precisely the measures the Commission must advocate. Thus it seems implausible to expect the Civil Disorders Commission to recommend much beyond what it is feasible for the Administration to expect the country to tolerate.[5] The recommendations may go beyond others, e.g., the McCone report following the Los Angeles riot, but once again whites will be able to avoid a confrontation with the realities of Negro life and once again the Negro ghetto dweller will feel that he has been sold out. And if this analysis is correct and predictive, it will indeed be sad, because for the first time in the history of this country, a national body of powerful men has been given access to the full body of data, and they will have once again concluded that the data and the required response to that data are, for reasons of political security, too explosive to allow an official utterance which truly "tells it like it is."

What then is left? Restraints imposed by the Negro community on its brothers? Not likely in consideration of the mood among the Negro populace which justifies riots, and since the conditions which provide that justification will continue in the absence of the economic wherewithal to change them.

The police? They might, but even at best that will involve incredible cost in lives and property. Furthermore, it is already clear that counter-violence by the police, national guard, or any other agency of public control exerts no deterrent effect; it helps to stop or control riots once they have ignited, but it does not apparently deter the impulse to ignition. It appears, however, that it is the police that the society is banking on. The police are the "patsies" for a country which seems to feel that it is cheaper to kill Negroes for burning and looting than it is to spend the money and create the climate which might produce a life situation which obviates these responses.

And so popular riots will continue and nothing of significant worth will have been done to relieve their causes. And because nothing is being done now to alter the situation, the future is bleak. Popular riots may run their course, but what will take place after they cease if the country decides against the necessary sacrifices now on behalf of the Negro ghetto dweller?

After the popular riots are over, probably in a few years, there will be a retrenchment by both blacks and whites. If at this point nothing is done by way of relieving the conditions which foster the riots, then it is entirely conceivable that politically-motivated black organizations, both public and clandestine, may indeed actively foment civil disturbance. Then there may well be the generation of formal revolutionary groups, whose cause, however futile, will be violent harassment, if not outright destruction of the urban centers, and with them the character of American society.

How can this be averted? A number of events occurring
in concert might serve to reduce the possibilities of this outcome. Among them are the following:

Belatedly, and perhaps too late, *massive infusions of money
and industrial resources into the ghetto*. White society simply
must realize that this is essential for their own and the total
society's welfare. They must realize that, if for no other
reason, it is in their best long-range economic interests to
make a relatively short-range financial sacrifice.

*The unification of the street militant and the Negro middle
class in the common cause of Negro development*. This may
or may not happen. If money is available to make it possible
and justifiable for the Negro middle class and the militants to
coalesce and to work toward a common goal, then the development of the entire Negro population toward equality and
affluence may take place. But as of now, there is no reason
for the two to combine forces, and even if they should want
to (and the burgeoning ethnic character of the Negro movement united by the commonality of skin color suggests that
such an event is possible) they have no point or purpose around
which to unite except protest. Therefore the Negro must be
given the chance to organize around economic and political
projects that provide for the unification of the factions within
the Negro community and allow all Negroes to pursue the
constructive goals of political and economic power.

*White society must demonstrate faith in the concept of
Negro equality,* i.e., that the whites are truly willing for the
Negroes, as Negroes, to enter into a society which is black
and white. Negroes have lost faith. They no longer believe
that whites will allow them to take their place in this society
regardless of what they might do by way of "proving themselves." For the Negro to stop rioting, he must first feel that
honest and legitimate action will be sufficient to gain entry.
Thus, the country must enact open housing laws, open its
trade unions, provide equal access to the courts, and strike all
the other overt and covert devices which serve to keep the
Negro "in his place." The point of this is twofold: there is
no place in a democracy for institutionalized discrimination,
and there is no other way to restore the faith of the Negro
American in this quasi-democracy save by demonstrating

to him that it is a true democracy worthy of his commitment. The Negro must believe it is worth his time to work and achieve. He must believe that the trappings of affluence will come as easily to him as they appear to come to most whites. He must be able to say to himself, "Nothing is keeping me here except the absence of my own industry." As it stands now, there are ample real justifications for erratic behavior and disaffection. The myths and stereotypes still exist in the minds of whites about the nature of the Negro American. Whites still justify their actions in terms of fanciful or superficial beliefs about Negro behavior. These beliefs are prevalent in white liberals, and they are endemic in the average citizen. Neither Negro behavior nor white attitudes will change until the Negro is given a true chance to develop his potential. And he will not have a true chance until the country decides to make available to him the climate and resources which lead to the outcome which we all presumably want. Self-help projects in the face of legal and extra-legal restraints are a lie. If the Negro is to do what whites evidently want him to do — namely, become like them — they must accept two facts: the Negro is black and will remain so, and self-help is a hypocritical sham if whites refuse to provide equal access to "their" society.

NOTES

1. The survey was carried out by several members of the sociology and psychology faculty at U.C.L.A., but the data reported here stem mainly from reports prepared by Vincent Jeffries, Richard Morris, David O. Sears, and T. M. Tomlinson. Copies of the reports may be obtained by writing to the project coordinator, Nathan Cohen, Institute of Government and Public Affairs, University of California, Los Angeles.
2. I am indebted to Phillip Meyer of Knight Publications (Detroit Free Press) and the Detroit Urban League for the materials from the Detroit riot study of 1967.
3. These conclusions are similar to those of Gary Marx in *Protest and Prejudice* (N. Y.: Harper & Row, 1967).
4. They have not occurred in the modern South or in cities with Negro mayors. The data aren't in on the latter category yet, but they may provide a clue about the necessary change in the

political structure, i.e., the assumption of political power by Negroes may be necessary to foreclose riots. Time will tell.

Two things operate to maintain calm in the South: a history of repression and high out-migration of young Negroes. What this means is that there is a low frequency of the population which is most susceptible to the riot mood of the Northern Negro. Such of that mood as exists takes form in a context which has traditionally and violently inhibited its expression.

There is, however, one important characteristic of Northern cities which serves to distinguish them from those in the South; they have received the influx of Negroes who have migrated out of the South and now live in the urban ghettos of the North. But the data from the riot surveys lend no credence to the notion that riots are a product of recent migrants from the South who have failed to adjust to city ways or who have brought traditions of violence. Quite the contrary, the average rioter has lived ten years or more in the city of his choice. As Barbara Williams has pointed out ("Riots and the Second Generation," unpublished report prepared for the Office of Economic Opportunity, 1967), the riot data do however support the hypothesis that rioters are typical second-generation youth. Traditional behavior by other migrant groups has seen crime rates of the first generation holding at the rate of the country of origin, but sharply increasing with the second generation. About one generation of "new" Negroes exists in this country. By "new" I mean the generation which has seen the development of the Negro drive for equality. We are now faced with the second generation who, unlike their parents, are unwilling to settle for the luxury of being an American and the token gestures of gradualism. And so instead of, or perhaps in addition to, high crime rates, we see high riot rates, and the participants are most likely to be those late teen and early twenties youth who are the second-generation offspring of those migrants from the South. But discontented youth per se are not the cause of riots, nor do they account for all Negroes who are discontented.

5. At the time this paper was being drafted (December, 1967), the Advisory Commission on Civil Disorders was engaged in behavior which was difficult to interpret except in terms of the half-a-loaf, classic-politics-of-compromise, thesis. For example, it seems at this date than an interim report, part of which was entitled the "Harvest of American Racism," has been put aside. But by the time this paper is published the "facts" about the internal workings of the Commission will only be of historical concern (although the ramifications of its action may reverberate for some time to come). What will stand over time is the realization of the half-a-loaf argument, and that this investigative body, as others with political liabilities, will not have faced up to the hard reality of stating the truth about the racism embedded in the character of this country or the economic and attitudinal sacrifice that whites must make to restore the health of the society.

PARA-MILITARY ACTIVITIES IN URBAN AREAS

Martin Oppenheimer

■ Considerable attention has recently been drawn to the possibility of guerrilla warfare in the urban Black ghetto. A recent Harlem handbill, for example, states, "There is but one way to end this suffering and that is by Black Revolution. Our Revolution is a unity of the Black Man wherever he may be. . . . When we unite we can end our suffering. Don't riot, join the revolution!" On the other extreme, as it were, are warnings emanating from police and army circles. *The New Republic* quotes a Colonel Robert B. Rigg, writing in the January, 1968, issue of *Army* magazine, as predicting "scenes of destruction approaching those of Stalingrad in World War II." Colonel Rigg's viewpoint seems to be that ". . . in the next decade at least one major metropolitan area could be faced with guerrilla warfare requiring sizable United States army elements. . . . "[1]

If we then proceed on the widely-held assumption that there will in fact be further disorders in our urban ghettoes in the years to come, given the general failure of society to solve the problems of the poor and of the Black community, and that some of these disorders may well take the form of para-military outbursts, what is the prognosis for such insurrectionary attempts?

AUTHOR'S NOTE: *This article is a revised version of a paper read at the Meetings of the Eastern Sociological Society, Boston, Massachusetts, April 5-7, 1968.*

In order to arrive at an answer to this question, this paper will first attempt to define more accurately some of the variables involved, within the broad context of collective behavior theory; it will then compare the military perspective of the urban ghetto in the United States to historically analogous situations; and finally, it will suggest some interpretations as to why so much attention is now being paid to this phenomenon.

Collective behavior theory suggests that insurrection is one kind of social movement among many, involving a collective effort to create a more satisfying culture, or to maintain a satisfying culture against perceived threats to it. In the literature of social movements, strain of one kind or another is what motivates individuals to participate in movements. Thus the prerequisite to insurrectionary attempts, too, is that the status quo is perceived as inadequate to the solution of certain problems as they are perceived by some group in population. In addition, the group must feel that only a military strategy is feasible to bring about, or to prevent, the desired, or undesired, change. The issue then becomes whether that group is capable of accomplishing its goal, especially in the face of the armed opposition of the "establishment."

It is an axiom that warfare is conducted like a zero-sum game, in which the successes of one side are more or less proportional to the defeats of the other; thus an insurrection, too, succeeds to the degree that the government is unable to contain and defeat it, and the government in turn maintains itself or is toppled, depending on the weaknesses and strengths of its antagonists. At the same time, these two elements interact within a wider social and historical context which tends largely to predetermine the characteristics of the antagonists, and therefore the outcome.

An insurrection, then, can be successful only if (1) it has achieved the support *or neutrality* of the vast majority of the population; (2) the dominant power structure is unwilling or unable, for whatever reason, to function in a coordinated manner either to suppress it or to solve the structural strains which give rise to it. The two points are related—inability of government to function to suppress para-military activity or to solve problems is closely related to government's failure to obtain the support or cooperation of certain functionally essential elements in the population, specifically in the military forces, the economic structure, the political decision-making area, and, less importantly, in the area of the ideological rationale for the status quo—e.g., religion, philosophy, the academy, etc. The reverse is also true: as government (possibly due to economic crisis, war, corruption, etc.) is unable to function, the population increasingly refuses to take risks for it. As support diminishes, government fails to function; as it fails to function, support diminishes.

Armed insurrection can obviously be distinguished from other kinds of collective behavior because it involves violence which is directed toward the immediate or ultimate overthrow of the established order, or, in the case of defensive insurrections, toward the defense of a subgroup of society against the further encroachments of the established order—two rather different types. But once violence as a posture vis-a-vis the status quo has been established, there are a host of other dimensions which enter into the picture.

Three dimensions are of particular interest: a historical dimension involving the rural-urban continuum, an associated dimension involving degree of political consciousness, and an independent dimension related to the distinction between a coup or putsch and a revolution—that of numbers of people involved. The following table suggests eight different types of collective behavior involving violence.

Since these are ideal types, they rarely exist in pure form. For example, an urban rebellion is almost invariably accompanied by a variety of individualistic, nonpolitical acts such as hoodlumism, and is frequently also accompanied by what can be called mass proto-political behavior, such as ethnic pogroms and looting. Again, individualistic political violence such as terrorism is often integrated into what are destined to become mass revolutionary movements. An intermediate phenomenon, the "resistance movement," typical of the occupied European nations of World War II, also involves individualistic acts carefully integrated into a master plan involving, ultimately, masses of people.

Historically, it would seem, urban mobs and riots are proto-political forms which give way to more sophisticated political forms such as modern social movements (trade unionism, for example, or working-class political parties) where such forms are feasible. In an authoritarian society (e.g., Czarism), where sufficient reforms are not forthcoming, or where democratic outlets are excluded, riots are replaced by revolutionary movements, involving armed insurrection.

TABLE 1

	Rural-Urban	Political Consciousness	Masses
Social banditry, vendettas, early Mafia	+	-	-
Peasant uprisings, re-vitalization movements	+	-	+
Guerrilla "foco" (Debray)	+	+	-
Guerrilla liberation army (Zapata, FLN, NLf, etc.)	+	+	+
Gangsterism, contemp. Mafia, jacket clubs, hooliganism	+	-	-
Riots, vandalism, looting	+	-	+
Terrorism (selective assassination, possibly leading to a coup)	+	+	-
Rebellion or "rising"	+	+	+

Similarly, the proto-political formation of the countryside, social banditry, in turn tends to be replaced by modern guerrilla warfare. In the developing nations, the analogy is that of the re-vitalization movement, which often develops into the modern national independence movement.

Para-military activity, then, can be defined as any violent behavior of an organized sort directed either defensively or offensively against the military forces (including police) of the dominant power in society, by military elements associated with no regular or recognized government—that is, by irregular, partisan, volunteer, guerrilla, and/or revolutionary forces. It implies at least a minimal political goal (as distinct from banditry or gangsterism), and a long-range terminal point, that of the actual attempt to seize control of a society or community. It is distinct, as well, from conventional reformist politics, and from nonviolent direct action (which may be either reform or revolution-oriented). Included in the definition is violence along a continuum of numerical involvement, from terrorism, the resistance movement or underground, and the coup or putsch, to the rebellion—which, if successful, is a revolution.

Urban para-military, guerrilla, or insurrectionary activity, either preceded by a resistance (underground) movement or a more spontaneous type, can be successful in overturning the established order only where, in effect, that order is already so decayed that a mere push will suffice—e.g., Petrograd in 1917, or Caracas in 1958. That is, it will normally be part of a nationwide revolutionary movement, often equal partners with or even subordinate to rural guerrilla activity. In cases where it is not related to rural activity, the countryside will have been neutralized, at least, and the communications system which normally would function to bring aid to a beleagured government has been disrupted or struck.[2]

Interestingly enough, no reputable analyst gives any revolution a chance unless the general population has been at minimum neutralized by the structural strain mentioned earlier; no writer, from conservative to radical, anywhere suggests that urban insurrection can be anything but wiped out once the rebels are geographically isolated and the government is in a position to move against them in a unified fashion. Relating this discussion now to the question of Black insurrection or guerrilla warfare in our urban centers, it would seem apparent that the American population on the whole has not been neutralized by structural strain, nor is it likely to be sufficiently disaffected within the next decade so as to refuse to support the government in suppressing urban uprisings; all evidence is to the contrary. Further, it is also apparent that the rebels would be isolated in our cities, and that for some time to come the government would be in a position to move in a unified fashion against them. Debray's description of South American miners holds for North American urban workers: they are "bound to their place of work, together with the women . . . and the children; exposed to all kinds of reprisals . . . without the material possibility of turning themselves into a mobile force (they) are simply condemned to slaughter."

SIX URBAN UPRISINGS

Some data were gathered covering six cases of urban uprisings, all but one in this century. Excluded from these cases were short-lived urban uprisings in which the population was mainly unarmed. For example, the Berlin and Poznan uprisings of June, 1953, and June, 1956, respectively, were both scarcely more than general strikes with attendant marches and rioting. Also excluded were general rebellions in which the urban uprising was only one part—for example, the Hungarian Rebellion of 1956, or the Irish Revolution of 1919—and urban uprisings such as Petrograd and Caracas, where the rebels were neither isolated nor confronted with a unified power structure.

The following criteria were used in this selection, which, it should be emphasized, is only illustrative. First, each was at minimum a guerrilla outbreak, backed by a significant sector of the local urban population—that is, each was a genuinely popular "rising." Second, in each case the rising developed into a rebellion which successfully took over an entire city or a large sector thereof. Third, the rebels in each case were isolated, either acting totally alone, or effectively separated from support from the outside. Fourth, in these illustrations, the government continued to function effectively at least outside the city, and managed to move in a unified fashion against the rebels within a short time. These are, in fact, the conditions confronting Black Power strategists when they talk about para-military activity.

Specifically, the assumption is that a Black rebellion would be a genuinely popular uprising and might militarily be able to take over the Black ghetto areas of some cities. However, if this were to take place, it would be without any logistical base in the countryside or abroad, and the urban uprisings would be territorially and logistically isolated. The further assumption is that of the Black Power advocates themselves—namely, that we live in a basically racist society which would not hesitate at a reactionary, completely military solution to Black para-military activity. The six cases supply us with ample data to support the idea that even racially and culturally similar groups do not hesitate at such measures. The Black guerrilla, then, would be in a worse position than almost any of the rebels in the cases studies. He is completely isolated from logistical support, and entirely at the mercy of modern technology, from electronic snooping to aerial bombardment. In short, historical evidence would seem to doom the Black guerrilla.

TABLE 2

CASE	DATES	CASUALTIES
The Paris Commune	Mar. 28-May 28, 1871	20,000 to 30,000 dead vs. 83 officers & 794 men of the Versaillese
The Easter Rising (Dublin)	Apr. 24-29, 1916	No figures available
Shanghai	Feb. 21-Apr. 13,1927	About 5,000 dead
Vienna	Feb. 12-17, 1934	1,500 to 2,000 dead vs. 102 Heimwehr
Warsaw Ghetto	Apr. 19-May 15, 1943	Several thousand killed, 56,000 deported, vs. about 20 Germans
Warsaw Uprising	July 31-Oct. 2, 1944	100,000 to 250,000 dead

The conclusion must be agreement with Feliks Gross: "Without outside support, without an internal crisis . . . social unrest, or international defeat, [a resistance movement] does not succeed. . . . "[3] Nor is this conclusion particularly new, for as long ago as 1885, in an introduction to Marx' *Class Struggles in France, 1848 to 1850,* Engels pointed out that as early as 1849 street fighting had become obsolete. "In future," he reminded revolutionaries, "street fighting can be victorious only if this disadvantageous situation is compensated by other factors." In the main, he argued, armed insurrection simply plays into the hands of the ruling class, giving its military agents an opportunity for wiping out revolutionary forces.

It is well to note, by the way, that in most of the cases studies, the rebels had no illusions about winning without aid—in three of the cases (the Easter Rising, Shanghai, and the Warsaw Uprising) the rebels moved on the (false) assumption that aid would be forthcoming, and in two others they moved only in desperation, with no real hope of success. Even the Paris Commune hoped that the remainder of the country would organize communes to go to its aid. Furthermore, in five of the six cases, the actual outbreak of the rebellion took place only as a defensive measure against severe encroachments by the dominant power, and would probably not have taken place at all without this repression. The Warsaw Uprising is the only exception to this. And in three of the cases, the rebellion was against foreigners, although perhaps it can be argued that the American white power structure is basically foreign to the Black population—but having a clearly foreign oppressor does seem to make it easier to unite the population for a rebellion.

An alternative form of para-military activity in the urban context that has been mentioned in "militant" Black Power circles is that of the longer-range "Resistance Movement," which involves a wider network of activity including propaganda, espionage, and terrorism (including sniping), but which does not assume an actual attempt to seize power within the immediate future. The prognosis for such movements, given modern technology, is not much rosier than for the urban rising. Again, without outside support (from rural areas or from abroad, or both) or internal crisis, it cannot hope to last indefinitely. Sooner or later it must face failure, or attempt the revolution. To remain "contained" over a long period of time is equivalent to failure—"La Guerre" becomes "Fini."

Several longer-range conclusions can also be drawn from this data. The dominant power structure can cope with para-military activity in a combination of two ways, similar in most respects to its strategy in any insurgency war. It can move radically to solve the problems of the population, thus cutting off the guerrilla's base of support in the populace; or it can move to suppress military activity through counter-insurgency warfare and other military means, including aerial bombardment. This seems far more typical. The "liberal" solution, that of attempting to combine these two strategies, is inherently inconsistent, because the use of military means is almost inevitably bound (in an urban situation, particularly) to injure the innocent, and if the guerrillas truly stem from the local population, to win them more support and thereby undermine "the other war," that of reform measures. In the long run, then, the Establishment must choose either a radical or a reactionary course. The liberal course gives the *appearance* of inconsistency, thus creating an illusion (at least among some Black revolutionists in this country) that the power structure is *inherently* incapable of dealing with para-military activity, and that therefore the Black revolution can win.

These six cases indicate that such a view is indeed an illusion. Given the dismal evidence, why has so much attention been given by the white press, and by some spokesmen in Black Power circles, to para-military phenomena? There are three possible interpretations for this fascination, apart from the view that attributes these cries of alarm to the military's need to justify its existence and its appropriations: First, para-military affairs in the Black ghetto may constitute a pseudo-event, that is, may largely be a creation of press attention which in turn results in a self-fulfilling prophecy. For the white community, which seems to require a rationale for ditching the civil rights movement and refusing significant aid to urban areas, the urban riot has become highly functional. The attention given by the white press to what is really the peripheral phenomenon of para-military Black Power is equally if not more functional. For lower middle-class people in the suburbs who have long repressed itchy trigger fingers, talk of urban risings by the Black population has afforded an outlet for repressed hostilities which can only be exceeded by an actual outbreak of warfare. For the Black community, the large-scale attention focussed upon a small minority of para-military types seems evidence that such activity may in fact be dangerous, hence effective as a threat with which to coerce concessions from white power structures; such a view is functional to recruitment for paramilitarism.

Related to this is a second interpretation, that of *machismo.* As Horowitz recently pointed out, the guerrilla mystique is the incarnation of "virility in speech, action, and dress, virility expressed by bravado, courage, and ruthlessness,"[4] precisely those qualities denied the hitherto castrated Negro male by American society. For the young ghetto Black, para-military posturing may replace the juvenile fighting gang as his way of finding the masculinity denied him by white society.

A third related interpretation is more mundane, perhaps to the point of now being a cliché in social science as well as informed lay circles. With the failure of the integrationist civil rights movement to achieve the promise of the "American Dream," Black leaders have been forced to the conclusion that for the time being, at least, the civil rights movement is dead, and that Black Power—that is, the cultural, social, and economic autonomy (and, given the ghettoized character of the Black population, territorial autonomy) of the Negro—is the only viable strategy. We are familiar with the fact that Black Power is a vague concept; this should not be surprising, because it is a new movement. Nor should it be surprising that on its periphery there will be some who advocate violent or military tactics. We must remember that it took from about 1825 to the publication, in 1896, of Herzl's *The Jewish State* for Jews to even begin to clarify the issue of "Jewish Power," and that, almost from the beginnings of Zionism, many Zionist as well as non-Zionist ghetto groups had para-military auxiliaries.

If our society is to avoid the tragic consequences of a reactionary or military response to "Power" movements, such as those described in this paper, the alternative must be a radical response, one which will obviate the necessity for autonomist strategies. In other words, if White America is not to destroy Black America physically, it must create the conditions for the success of the civil rights, rather than the Black Power, movement.

NOTES

1. *New Republic,* Jan. 27, 1968.
2. The precise relationship of urban (working-class) and rural (peasant) revolutionary movements has been the subject of a continuous debate involving such names as Lenin, Trotsky, Guevara, Fanon, and Debray. See, for example, Debray's *Revolution in the Revolution?* (N.Y.: Grove Press, 1967).
3. Feliks Gross, *The Seizure of Power* (N.Y.: Philosophical Library, 1958).
4. Irving L. Horowitz, "Cuban Communism," *Trans-action* 12 (Oct., 1967), pp. 7-15 ff.

BIBLIOGRAPHY

GHETTO REVOLTS
AND CITY POLITICS

John R. Krause, Jr.

■ This bibliography lists works pertaining to a number of forms of internal political violence. The emphasis, however, is on material relevant to racial turmoil in the United States. Five major classes of material are listed.

Part I, "Theories and General Descriptions," contains, in section A, general works which discuss such concepts as aggression, collective behavior, conflict, rebellion, revolution, and violence. Similar works which focus on racial violence in the United States are presented in section B. The material in Part I varies from broad, journalistic, descriptive explanations to some of the more advanced theoretical work found in the behavioral sciences.

The works in Part II are examinations of specific historical incidents of violence. These are divided into four sections based on the type of violence and its geographic location. Sections A and B contain references to incidents of racial violence in the American South and the American urban community, respectively. Sections C and D present material on non-racial violence in the United States, and on political violence in other parts of the world. The emphasis in C and D is on works useful, for comparative purposes, in the study of American racial violence.

Part III focuses on the majority response to racial violence. Works which analyze the attitudes of white citizens, the role of the police, and the actions of governmental bodies are presented. Many of the reports listed in III C are among the best reports of urban racial violence available. They are listed here rather than in Part II because, unlike reports of private persons, they are in themselves acts of public policy with direct political consequences, and also because they tend to have certain "blind spots" which reflect their origins.

Part IV, "The Voice of the Ghetto Violent," focuses on the attitudes of Negroes in the urban community. The first section lists works by or about specific individuals or groups. The material in Section B presents findings about the aggregate feelings of the ghetto Negro.

Part V lists special issues of journals which have been devoted to the problem of racial violence in the United States.

Part VI presents a selection of unpublished but available materials on political violence.

I. THEORIES AND GENERAL DESCRIPTIONS

A. General

ADAMS, BROOKS. The Theory of Social Revolution. (N.Y.: Macmillan, 1914).

AKE, CLAUDE. A Theory of Political Integration. (Homewood, Ill.: Dorsey Press, 1967).

AMANN, PETER. "Revolution: A Redefinition." Polit. Sci. Q. (March, 1962).

ARDREY, ROBERT. The Territorial Imperative. (N.Y.: Antheneum, 1966).

ARENDT, HANNAH. On Revolution. (N.Y.: Viking, 1965).

BERKOWITZ, LEONARD. Aggression. (N.Y.: McGraw-Hill, 1962).

BLOCH, HERBERT, and GEIS, GILBERT. Man, Crime, and Society: The Forms of Criminal Behavior. (N.Y.: Random House, 1962).

BOULDING, KENNETH. Conflict and Defense: A General Theory. (N.Y.: Harper-Row, 1962).

BRINTON, CRANE. The Anatomy of Revolution. (N.Y.: Vintage, 1956).

BUSS, ARNOLD H. The Psychology of Aggression. (N.Y.: Wiley, 1961).

CALVERT, P. B. A. "Revolution: The Politics of Violence." Polit. Studies (Feb., 1967).

CANETTI, ELIAS. Crowds and Power. (N.Y.: Viking, 1963).

CANTRIL, HADLEY. The Psychology of Social Movements. (N.Y.: Wiley, 1941).

COSER, LEWIS. The Functions of Social Conflict. (Glencoe, Ill.: Free Press, 1956).

DAHLKE, H. OTTO. "Race and Minority Riots—A Study in the Typology of Violence." Social Forces, XXX (May, 1952).

DAHRENDORF, ROLF. "Toward a Theory of Social Conflict." J. Conflict Resolution, II (June, 1958).

DAVIES, JAMES C. "Toward of Theory of Revolution." Am. Social. Rev., XXVII (Feb., 1962).

DAVIS, ALLISON. "Caste, Economy, and Violence." Am. J. Sociol., LI (July, 1945).

De REUCH, A. V. S., ed. Conflict and Society. (Boston: Little, Brown; 1966).

DEWEY, JOHN. "Force and Coercion." Internatl. J. Ethics, XXVI (April, 1916).

———. "Force, Violence, and Law." John Dewey's Philosophy. (N.Y.: Random House Modern Library, 1939).

DOLLARD, JOHN, et al. Frustration and Aggression. (New Haven: Yale Univ. Press, 1939).

ECKSTEIN, HARRY, ed. Internal War. (N.Y.: Free Press, 1964).

———. "On the Etiology of Internal War." History and Theory, IV, 2 (1965).

FEIERABEND, IVO K. and ROSALIND L. "Aggressive Behaviors Within Polities, 1948-1962: A Cross-National Study." J. Conflict Resolution, X (Sept., 1966).

FROMM, ERICH. "Different Forms of Violence." Fellowship (March, 1965).

GAMSON, WILLIAM A. "Rancorous Conflict in Community Politics." Am. Sociol. Rev., XXXI (1966).

GRIMSHAW, ALLEN. "Factors Contributing to Colour Violence in the U. S. and Britain." Race, III (May, 1962).

GURR, TED. The Conditions of Civil Violence—First Tests of a Causal Model.

(Princeton, N.J.: Princeton Univ. Center of Internatl. Studies, 1967).

———. "Psychological Factors in Civil Violence." World Politics (Jan., 1968).

HATTO, ARTHUR. "Revolution: An Enquiry into the Usefulness of an Historical Term." Mind, XLIII (Oct., 1949).

HOBSBAWN, ERIC. Primitive Rebels: Studies in Archaic Forms of Social Movements. (N.Y.: Norton, 1959).

HOFFER, ERIC. The True Believer. (N.Y.: Harper, 1951).

HOPPER, REX D. "The Revolutionary Process." Social Forces (March, 1950).

HOROWITZ, IRVING LOUIS. Radicalism and the Revolt Against Reason. (N.Y.: Humanities Press, 1961).

JOHNSON, CHALMERS. Revolutionary Change. (Boston: Little, Brown; 1966).

———. Revolution and the Social System. (Stanford, Calif.: Hoover Institution Studies, 1964).

LANG, KURT and GLADYS. Collective Dynamics. (N.Y.: Crowell, 1961).

LaPIERRE, RICHARD T. Collective Behavior. (N.Y.: McGraw-Hill, 1938).

LeBON, GUSTAVE. The Crowd. (N.Y.: Macmillan, 1896).

———. The Psychology of Revolution. (N.Y.: G. P. Putnam, 1913).

LORENZ, KONRAD. On Aggression. (N.Y.: Harcourt, Brace; 1966).

MACK, RAYMOND W. "The Components of Social Conflict." Social Problems, XII (Spring, 1965).

McNEIL, ELTON B., ed. The Nature of Human Conflict. (Englewood Cliffs, N.J.: Prentice-Hall, 1965).

MOMBOISSE, RAYMOND M. Riots, Revolts and Insurrections. (Springfield, Ill.: Charles C. Thomas, 1967).

MOORE, BARRINGTON. Social Origins of Dictatorship and Democracy: Lord and Peasant in the Making of the Modern World. (Boston: Beacon Press, 1966).

NIEBURG, H. L. "The Threat of Violence and Social Change." Am. Polit. Sci. Rev., LVI (Dec., 1962).

———. "Uses of Violence." J. Conflict Resolution (March, 1963).

OLSON, MANCUR. "Rapid Growth as a Destabilizing Force." J. Econ. History, XXIII (Dec., 1963).

PETTEE, G. S. The Process of Revolution. (N.Y.: Harper, 1938).

RAPOPORT, ANATOL. Fights, Games and Debates. (Ann Arbor: Univ. of Michigan Press, 1961).

———. "Violence in American Fantasy." Liberation (Nov., 1966).

ROUCEK, J. S. "The Sociology of Violence." J. Human Relations, V (Spring, 1957).

RUMMEL, R. J. "A Field Theory of Social Action with Application to Conflict Within Nations." Yearbook of the Society for General Systems (Bedford, Mass., 1965).

SCOTT, J. P. Aggression. (Chicago: Univ. of Chicago Press, 1958).

SETON-WATSON, HUGH. "Twentieth-Century Revolutions." Polit. Q., XII (July-Sept., 1951).

SMELSER, NEIL J. Theory of Collective Behavior. (N.Y.: Free Press, 1963).

SOREL, GEORGES. Reflections on Violence. (N.Y.: Collier Books, 1961).

SOROKIN, PITIRIM. The Sociology of Revolution. (Philadelphia: J. B. Lippincott, 1925).

STONE, LAWRENCE. "Theories of Revolution." World Politics (Jan., 1966).

TANTER, RAYMOND. "Dimensions of Conflict Within and Between Nations, 1958-60." J. Conflict Resolution, X (March, 1966).

——— and MIDLARSKY, MANUS. "A Theory of Revolution." J. Conflict Resolution (Sept., 1967).

TURNER, RALPH H. and SURACE, SAMUEL J. "Zoot-Suiters and Mexicans: Symbols in Crowd Behavior." Am. J. Sociol., LXII (1956).

——— and KILLIAN, LEWIS M. Collective Behavior. (Englewood Cliffs, N.J.: Prentice-Hall, 1957).

WADA, G. and DAVIES, JAMES C. "Riots and Rioters." Western Polit. Q. (Dec., 1957).

WALTER, E. V. "Power and Violence." Am. Polit. Sci. Rev., LVIII (June, 1964).

———. "Violence and the Process of Terror." Am. Sociol. Rev. (April, 1964).

WERTHAM, FREDERIC. The Show of Violence. (N.Y.: Doubleday, 1949).

———. A Sign for Cain: An Exploration of Human Violence. (N.Y.: Macmillan, 1966).

WILLER, DAVID and ZOLLSCHAN, GEORGE K. "Prolegomenon to a Theory of Revolutions." Explorations in

Social Change. Edited by George K.
Zollschan and Walter Hersch. (Boston:
Houghton Mifflin, 1964).

WOLFGANG, MARVIN and FERRA-

CUTTI, F. Subculture of Violence.
(N.Y.: Barnes and Noble, 1967).
WOLIN, SHELDON S. "Violence and the
Western Political Tradition." Am. J.
Orthopsychiatry, XXIII (Jan., 1963).

B. U. S. "Race" Riots

BENNETT, LERONE, JR. Confrontation:
Black and White. (Baltimore: Penguin
Books, 1965).
CERVANTES, ALFONSON J. "To Pre-
vent a Chain of Super-Watts." Harv. Bus.
Rev. (Sept.-Oct., 1967).
CLARK, KENNETH B. "The Wonder Is
There Have Been So Few Riots." N. Y.
Times Magazine, (Sept. 5, 1966).
Congressional Quarterly. Urban Problems
and Civil Disorder. (Washington: Con-
gressional Quarterly, Inc., Sept., 1967).
CONOT, ROBERT. Rivers of Blood, Years
of Darkness: Rebellion in the Streets.
(N.Y.: Bantam Books, 1967).
DYCKMAN, JOHN W. "Some Conditions
of Civic Order in an Urbanized World."
Daedalus (Summer, 1967).
GRIMSHAW, ALLEN. "Changing Patterns
of Racial Violence in the U. S."
(Symposium). Notre Dame Lawyer, XL
(1965).
———. "Lawlessness and Violence in Amer-
ica and Their Special Manifestations in
Changing Negro-White Relationships." J.
Negro History, XLIV (Jan., 1959).
———. "Negro-White Relations in the
Urban North: Two Areas of High Con-
flict Potential." J. Intergroup Relations
(Spring, 1962).
———. "Relationship Among Prejudice,
Discrimination, Social Tension and
Social Violence." J. Intergroup Rela-
tions (Autumn, 1961).
———. "Urban Racial Violence in the
United States: Changing Ecological Con-
siderations." Am. J. Sociol., LXVI
(Sept., 1960).
HANDLIN, OSCAR. Fire-Bell in the
Night: The Crisis in Civil Rights. (Bos-
ton: Beacon Press, 1964).
HENDRICH, WILLIAM C. "Race Riots—
Segregated Slums." Current History, V
(Sept., 1943).
JOHNSON, GUY B. "Patterns of Race
Conflict." Race Relations and the Race
Problem. Edited by Edgar T. Thompson.
(Durham, N.C.: Duke Univ. Press,
1939).

KILLIAN, LEWIS and GRIGG,
CHARLES. Racial Crisis in America:
Leadership in Conflict. (Englewood
Cliffs, N.J.: Prentice-Hall, Spectrum
Books; 1964).
LEO, JOHN. "America's Problem of Vio-
lence." Fellowship (March, 1967).
LEWIS, ANTHONY and New York Times.
Portrait of a Decade: The Second Amer-
ican Revolution. (N.Y.: Random House,
1964).
LIEBERSON, S. "The Meaning of the
Race Riots." Race, VII (April, 1966).
——— and SILVERMAN, ARNOLD R.
"The Precipitants and Underlying Con-
ditions of Race Riots." Am. Sociol.
Rev., XXX (Dec., 1965).
MEIER, RICHARD L. "Some Thoughts
on Conflict and Violence in the Urban
Setting" (summary). Am. Behav. Scien-
tist, X (Sept., 1966).
MYRDAL, GUNNAR. An American
Dilemma. (N.Y.: Harper-Row, 1944).
ROBINSON, BERNARD F. "The Sociol-
ogy of Race Riots." Phylon, II, 2
(1941).
———. "War and Race Conflicts in the U.
S." Phylon, IV, 4 (1943).
RUSTIN, BAYARD. "The Lessons of the
Long Hot Summer." Commentary (Oct.,
1967).
SHAW, JOHN. "Can Cleveland Escape
Burning?" Sat. Eve. Post (July 29,
1967).
SILBERMAN, CHARLES E. Crisis in
Black and White. (N.Y.: Vintage 1964).
SPEAR, ALAN. "The Changing Nature of
Racial Violence." New Politics (March,
1967).
STAHL, DAVID, et al., eds. The Com-
munity and Racial Crisis. (N.Y.: Practicing
Law Institute, 1966).
WASKOW, ARTHUR I. From Race Riot
to Sit-In: 1919 and the 1960's. (N.Y.:
Doubleday, 1966).
WILLS, GARY. "The Second Civil War."
Esquire (March, 1968).

II. EMPIRICAL STUDIES OF POLITICAL VIOLENCE

A. Racial Violence—The American South

AMES, JESSE DANIEL. The Changing Character of Lynching. (Atlanta: Commission on Interracial Cooperation, 1942).

APTHEKER, HERBERT. "American Negro Slave Revolts." Science and Society, I (Summer, 1937).

CARPENTER, JOHN A. "Atrocities in the Reconstruction Period." J. Negro History, XLIV (April, 1960).

CARROLL, JOSEPH C. Slave Insurrections in the United States, 1800-1865. (Boston: Chapman and Grimes, 1938).

CROMWELL, JOHN W. "The Aftermath of Nat Turner's Insurrection." J. Negro History, V (April, 1920).

DOLLARD, JOHN. Caste and Class in a Southern Town. (New Haven: Yale Univ. Press, 1937).

ELKINS, STANLEY M. Slavery: A Problem in American Institutional and Intellectual Life. (Chicago: Univ. of Chicago Press, 1959).

FRANKLIN, JOHN HOPE. The Militant South. (Cambridge, Mass.: Harvard Univ. Press, 1956).

FRAZIER, E. FRANKLIN. The Negro in the U.S. (N.Y.: Macmillan, 1949).

HARRINGTON, FRED H. "The Fort Jackson Mutiny." J. Negro History, XXVII (Oct., 1942).

HOVLAND, CARL I. and SEARS, R. R. "Correlation of Economic Indexes with Lynchings." J. Psy., IX (April, 1940).

LOGAN, RAYFERD W. The Betrayal of the Negro. (N.Y.: Collier Books, 1965).

McKIBBEN, DAVIDSON BURNS. "Negro Slave Insurrection in Mississippi, 1800-1865." J. Negro History, XXXIV (Jan., 1949).

MILES, EDWIN A. "Mississippi Slave Insurrection Scare of 1835." J. Negro History, XLII (1957).

MINTZ, ALEXANDER. "Re-examination of Correlations Between Lynchings and Economic Indices." J. Ab. Soc. Psy., XLI (April, 1946).

RAPER, ARTHUR F. The Tragedy of Lynching. (Chapel Hill: Univ. of North Carolina Press, 1933).

SCHULER, EDGAR A. "The Houston Race Riot, 1917." J. Negro History, XXIX (July, 1944).

WATTERS, PAT and ROUGEAU, WELDON. Events at Orangeburg. (Atlanta: Southern Regional Council, 1968).

B. Racial Violence—The American Urban Community

AKERS, ELMER R. and FOX, VERNON. "The Detroit Rioters and Looters Committed to Prison." J. Criminal Law., Criminology and Police Science, XXXV (July-Aug., 1944).

BARTIMOLE, ROLDO S. and GRUBER, MURRAY. "Cleveland: Recipe for Violence." Nation (June 26, 1967).

BERSON, LEONORA E. Case Study of a Riot: The Philadelphia Story. (N.Y.: Institute of Human Relations Press, 1966).

BOSKIN, JOSEPH and FELDMAN, FRANCES LOMAS. Riots in the City. (Addendum to McCone Commission Report, Los Angeles, 1967).

BROWN, EARL. Why Race Riots? Lessons from Detroit. (Washington: Public Affairs Institute, 1944).

Chicago Commission on Race Relations. The Negro in Chicago. (1922).

Citizens Protective League. Story of the Riot. (N.Y.: Citizens Protective League, 1900).

CLARK, KENNETH B. "Group Violence: A Preliminary Study of the Attitudinal Pattern of Its Acceptance and Rejection: A Study of the 1943 Harlem Race Riot." J. Soc. Psy., XIX (May, 1944).

COHEN, JERRY and MURPHY, WILLIAM S. Burn, Baby, Burn: The Watts Riot. (N.Y.: E. P. Dutton, 1966).

CRUMP, SPENCER. Black Riots in Los Angeles. (Los Angeles: Trans-Anglo Books, 1966).

FISHER, LLOYD H. The Problem of Violence: Observations on Race Conflict in Los Angeles. (Los Angeles: The American Council on Race Relations, 1947).

GRIMSHAW, ALLEN. "Three Major Cases of Colour Violence in the United States." Race, V (July, 1963).

HAYDEN, TOM. Rebellion in Newark: Official Violence and Ghetto Response. (N.Y.: Vintage 1967).

LEE, ALFRED McCLUNG and HUMPHREY, NORMAN D. Race Riot. (N.Y.: Dryden Press, 1943).

National Urban League. Racial Conflict—A Home-Front Danger, Lessons on the Detroit Riot. (New York, 1943).

ORLANSKY, HAROLD. Harlem Riot: A Study in Mass Frustration. Report 1 (N.Y.: Social Analyses, 1943).

RUDWICK, ELLIOTT M. Race Riot at East St. Louis, July 2, 1917. (Carbondale: Southern Illinois Univ. Press, 1964).

SHAPIRO, FRED C. and SULLIVAN, JAMES W. Race Riots in New York, 1964. (N.Y.: Thomas Y. Crowell, 1965).

SKOGAN, ROBERT and CRAIG, TOM. The Detroit Race Riot. (Philadelphia: Chilton Books, 1964).

WHITE, WALTER. "Behind the Harlem Riots." New Republic (Aug. 16, 1943).

——— and MARSHALL, THURGOOD. What Caused the Detroit Riot? (New York, 1943).

C. Other—United States

ADAMIC, LOUIS. Dynamite: The Story of Class Violence in America. (N.Y.: Viking, 1934).

BROOKS, THOMAS. Toil and Trouble: A History of American Labor. (Boulder, Colo.: Delta Publishing, 1965).

BRUCE, ROBERT V. 1877: Year of Violence. (Indianapolis: Bobbs-Merrill, 1959).

HEADLEY, J. T. The Great Riots in New York, 1712-1873. (N.Y.: E. B. Treat, 1873).

HEAPS, WILLARD A. Riots USA, 1765-1965. (N.Y.: Seabury Press, 1966).

HENSEL, PHILIP and GOTTLIEB. "Violence in the Schools." Fellowship (March, 1966).

LINDSEY, ALMONT. The Pullman Strike. (Chicago: Univ. of Chicago Press, 1960).

LOFTON, WILLISTON H. "Northern Labor and the Negro during the Civil War." J. Negro History, XXXIV (July, 1949).

WERSTEIN, IRVING. July 1863: The Incredible Story of the Bloody New York City Draft Riots. (N.Y.: Julian Messner, 1957).

D. Other—Non-United States

COBB, R. C. "The People in the French Revolution." Past and Present (April, 1959).

DARVELL, FRANK. Popular Disturbances and Public Order in Regency England. (Oxford, Eng.: Oxford Univ. Press, 1934).

JENKINS, GEORGE. "Urban Violence in Africa," Am. Behav. Scientist (March-April, 1968).

LeVINE, ROBERT A. "Anti-European Violence in Africa: A Comparative Analysis." J. Conflict Resolution, III (Dec., 1959).

ROSE, R. B. "Eighteenth Century Price Riots and Public Policy in England." Internatl. Rev. Soc. Hist., VI, 2 (1961).

RUDE, GEORGE. The Crowd in the French Revolution. (Oxford, Eng.: Oxford Univ. Press, 1959).

———. The Crowd in History. (N.Y.: Wiley, 1964).

TOCQUEVILLE, ALEXIS De. The Old Regime and the French Revolution. Trans. by S. Gilbert. (N.Y.: Doubleday, 1955).

VITTACHI, TARZIE. Emergency '58: The Story of the Ceylon Race Riots. (London: Andre Deutsch, 1958).

III. THE WHITE RESPONSE TO RACIAL VIOLENCE

A. Attitudes of White Citizens

ABRAMS, CHARLES. Forbidden Neighbors: A Study of Prejudice in Housing. (N.Y.: Harper, 1955).

FAGER, CHARLES E. White Reflections on Black Power. (Grand Rapids, Mich.: Wm. B. Eerdmans, 1967).

GREMLY, WILLIAM. "Social Control in Cicero." British J. of Sociol., III (Dec., 1952).

HARPER, DEAN. "Aftermath of a Long Hot Summer." Trans-action (July-Aug., 1965).

POWLEDGE, FRED. Black Power/White Resistance: Notes on the New Civil War. (Cleveland: World Pub., 1967).

STERBA, RICHARD. "Some Psychological Factors in Negro Race Hatred and in Anti-Negro Riots." Psychoanalysis and the Social Sciences, Vol. I. Edited by Geza Roheim. Internatl. Univ. Press, 1947).

WILLIAMS, ROBIN, JR. Strangers Next Door. (Englewood Cliffs, N.J.: Prentice-Hall, 1964).

B. The Police

BANTON, MICHAEL. The Policeman and the Community. (N.Y.: Basic Books, 1965).

California Department of Justice, Bureau of Criminal Statistics. "Watts Riot Arrests." (Sacramento: Calif. State Printing Office, 1966).

California National Guard. "Military Support of Law Enforcement During Civil Disturbances." (Sacramento: Calif. State Printing Office, 1966).

COATES, J. Non-Lethal Weapons for Use by U.S. Law Enforcement Officers. [Study Memorandum] (Washington: Institute for Defense Analysis, Nov., 1967).

CRAY, E. The Big Blue Line: Police versus Human Rights. (N.Y.: Coward-McCann, 1967).

Federal Bureau of Investigation. Prevention and Control of Mobs and Riots. (Washington, 1965).

———. Report on the 1964 Riots. (Washington, 1964).

GRIMSHAW, ALLEN D. "Action of Police and Military in American Race Riots." Phylon, XXIV (Fall, 1963).

———. "Police Agencies and the Prevention of Racial Violence." J. Criminal Law, Criminology and Police Sciences, LIV (March, 1963).

HOLDEN, MATTHEW, JR. "The Modernization of Urban Law and Order." Urban Affairs Q. (Dec., 1966).

LOHMAN, JOSEPH D. The Police and Minority Groups. (Chicago Park District, 1947).

KEPHART, WILLIAM N. Racial Factors and Urban Law Enforcement. (Philadelphia: Univ. of Pa. Press, 1957).

McMILLAN, GEORGE. Racial Violence and Law Enforcement. (Atlanta: Southern Regional Council, 1960).

NIEDERHOFFER, ARTHUR. Behind the Shield: The Police in Urban Society. (N.Y.: Doubleday, 1967).

President's Commission on Law Enforcement and Administration of Justice. Task Force Report: The Police. (Washington, 1967).

———. National Survey of Police and Community Relations. (Washington, 1967).

SILVER, ALLAN. "The Demand for Order in Civil Society: A Review of Some Themes in the History of Urban Crime, Police and Riots." The Police. Edited by David J. Bordua. (N.Y.: Wiley, 1967).

SKOLNICK, JEROME H. Justice Without Trial: Law Enforcement in Democratic Society. (N.Y.: Wiley, 1966).

U.S. Civil Rights Commission. Hearing [Cleveland]. (Washington, 1965).

———. Law Enforcement: A Report on Equal Protection in the South. (Washington, 1965).

WECKLER, JOSEPH E. and HALL, THEODORE E. The Police and Minority Groups: A Program to Prevent Disorder and to Improve Relations Between Different Racial, Religious and National Groups. (Chicago: Internatl. City Managers Assoc., 1944).

WESTLEY, WILLIAM A. "Violence and the Police." Am. J. Sociol., IL (Aug., 1953).

———. The Formation, Nature and Control of Crowds. (Ottowa: Canadian Defense Research Board, 1955).

C. Governmental Reports and "Findings"

BLAUNER, ROBERT. "Whitewash Over Watts." Trans-action (March-April, 1966).

California Governor's Commission on the Los Angeles Riots. Violence in the City—An End or a Beginning? (Los Angeles: McCone Commission Report, 1965).

Cuyahoga County Grand Jury. Report on the Hough Riots. (Cleveland: August 9, 1966).

FOGELSON, R. "White on Black: A Critique of the McCone Commission Report on the Los Angeles Riots." Polit. Sci. Q., LXXXII (Sept., 1967.

GRIMSHAW, ALLEN D. "Government and Social Violence: The Complexity of Guilt." The Minnesota Review, III (Winter, 1963).

Mayor's Commission on Conditions in Harlem. "The Negro in Harlem: A Report on Social and Economic Conditions Responsible for the Outbreak of March 19, 1935." (New York City Archives).

National Advisory Commission on Civil Disorders (Kerner Commission). Report. (N.Y.: Bantam Books, 1968).

New Jersey Governor's Select Commission on Civil Disorder. Lilley Commission Report. (Trenton, N.J.: 1968).

Ohio House of Representatives Judiciary Committee. Report on Riot Control Bill. (Columbus: Jan., 1968).

Permanent Subcommittee on Investigations, Committee on Governmental Operations, U.S. Senate (McClellan Committee). Riots, Civil and Criminal Disorders. (Washington: U.S. Government Printing Office, 1967).

U.S. Civil Rights Commission. A Time to Listen... A Time to Act: Voices from the Ghettos of the Nation's Cities. (Washington: Nov., 1967).

WEISBERG, B. "Racial Violence and Civil Rights Law Enforcement." Univ. of Chicago Law Rev., XVIII (1951).

LOS ANGELES RIOT STUDY. A series of reports prepared for the Office of Economic Opportunity. NATHAN E. COHEN, Coordinator. Institute of Government and Public Affairs, U.C.L.A., June 1, 1967:

T. M. TOMLINSON and DAVID O. SEARS, "Negro Attitudes Toward the Riots."

RICHARD T. MORRIS and VINCENT JEFFRIES, "The White Reaction Study."

JEROME COHEN, "A Descriptive Study of the Availability and Useability of Social Services in the South Central Area of Los Angeles."

HARRY M. SCOBLE, "Negro Politics in Los Angeles: The Quest for Power."

RAYMOND J. MURPHY and JAMES N. WATSON, "The Structure of Discontent: The Relationship Between Social Structure, Grievance, and Support for the Los Angeles Riot."

NATHAN E. COHEN, "The Context of the Curfew Area."

T. M. TOMLINSON and DIANA L. TEN HOUTEN, "Method: Negro Reaction Survey."

DAVID O. SEARS, "Political Attitudes of Los Angeles Negroes."

WALTER J. RAINE, "The Ghetto Merchant Survey."

DAVID O. SEARS and JOHN B. McCONAHAY, "Riot Participation."

IV. THE VOICE OF THE GHETTO VIOLENT

A. Biography and Narrative

BALDWIN, JAMES. The Fire Next Time. (N.Y.: Dial Press, 1963).

BREITMAN, GEORDE, ed. Malcolm X Speaks. (N.Y.: Grove Press, 1965).

BRODERICK, FRANCIS L. and MEIER, AUGUST, eds. Negro Protest Thought in the Twentieth Century. (Indianapolis: Bobbs-Merrill, 1965).

BROWN, CLAUDE. Manchild in the Promised Land. (N.Y.: Macmillan, 1965).

CARMICHAEL, STOKELY and HAMILTON, CHARLES V. Black Power. (N.Y.: Vintage, 1967).

Civil Rights Congress. We Charge Genocide: The Crime of Government Against the Negro People. (New York, 1951).

DuBOIS, W. E. B. Black Reconstruction in America. (Cleveland: World, Meridian; 1964).

ELLISON, RALPH. The Invisible Man. (N.Y.: Random House, 1952).

ESSIEN-UDOM, E. U. Black Nationalism: A Search for Identify in America. (Chicago: Univ. of Chicago Press, 1962).

FANON, FRANTZ. The Wretched of the Earth. (N.Y.: Grove Press, 1966).

GRANT, JOANNE, ed. Black Protest: Documents and Analysis, 1619 to the Present. (N.Y.: Premier/Fawcett, 1968).

JOHNSON, JAMES WELDON. Black Manhattan. (N.Y.: A. A. Knopf, 1930).

JONES, JAMES A. and BAILEY, LINDA. A Report on the Race Riots. (N.Y.: Haryou-Act, 1966).

LASCH, CHRISTOPHER. "The Trouble with Black Power." N.Y. Review of Books (Feb. 29, 1968).

LUKAS, ANTHONY. "Whitey Hasn't Got the Message." N.Y. Times Magazine (Aug. 27, 1967).

McGRAW, JAMES R. "An Interview with Andrew J. Young." Christianity and Crisis (Jan. 22, 1968).

STYRON, WILLIAM. The Confessions of Nat Turner. (N.Y.: Random House, 1967).

ZOLBERG, A. and V. "The Americanization of Frantz Fannon." Public Interest (Fall, 1967).

B. Research Data

BEARDWOOD, R. "The New Negro Mood." Fortune (Jan., 1968).

BRINK, W. and HARRIS, LOUIS. Black and White: A Study of U.S. Racial Attitudes Today. (N.Y.: Simon and Schuster, 1967).

———. The Negro Revolution in America. (N.Y.: Simon and Schuster, 1964).

CLARK, KENNETH. Dark Ghetto (N.Y.: Harper-Row, 1966).

COLES, ROBERT. Children of Crisis: A Study of Courage and Fear. (Boston: Little, Brown; 1967).

KEPHART, WILLIAM M. "The Negro Offender: An Urban Research Project." Am. J. Sociol., LX (July, 1954).

Lemberg Center for the Study of Violence, Brandeis University. Six-City Study: A Survey of Racial Attitudes in the Six Northern Cities—Preliminary Findings. (July, 1967).

LINCOLN, C. ERIC. The Black Muslims in America. (Boston: Beacon Press, 1961).

LOMAX, LOUIS E. The Negro Revolt. (N.Y.: Signet Books, 1963).

MARX, GARY. Protest and Prejudice. (N.Y.: Harper-Row, 1967).

MEYER, P. A Survey of Attitudes of Detroit Negroes After the Riot of 1967. (Detroit: Urban League, 1967).

PARKER, SEYMOUR and KLEIN, ROBERT J. Mental Illness in the Urban Negro Community. (N.Y.: Free Press, 1966).

WRIGHT, NATHAN. Black Power and Urban Unrest. (N.Y.: Hawthorn Books, 1967).

V. COLLECTIONS IN JOURNAL ISSUES

Am. Behavioral Scientist, XI 4 (March-April, 1968). "Urban Violence and Disorder." Edited by Louis H. Masotti.

Annals of the Am. Acad. of Political and Social Science, CCCLXIV(March, 1966). "Patterns of Violence." Edited by Marvin E. Wolfgang.

City, II, 1 (Jan., 1968). "Riot Reader."

Freedomways, I, 1 (Winter, 1967).

The Nation, CCV, 4 (August 14, 1967). "The Violence." A symposium on violence in the cities with contributions by Paul Good, Phil Kerby, Lewis Moroze, Bennett Cremers, and B. J. Widick.

Negro Year Book, 1947. "Race Riots." (N.Y.: Wm. H. Wise and Co., Inc.).

———, 1952. "Lynching." (N.Y.: Wm. H. Wise and Co., Inc.).

Social Problems, XV, 3 (Winter, 1968). "Race, Collective Behavior, and Politics." A symposium with contributions by Irving L. Horowitz and Martin Lie-

bowitz, Marvin E. Olsen, Carl J. Couch, Anthony Oberschall, and Carl F. Grindstaff.

Trans-action, IV, 9 (Sept., 1967). "Detroit: Violence in the Urban Frontier." A symposium with contributions by Irving L. Horowitz, R. Montgomery, T. Parmenter, and Lee Rainwater.

VI. UNPUBLISHED MATERIALS

BOSKIN, JOSEPH. "A History of Race Riots in Urban Areas, 1917-1964." Report prepared for McCone Commission, 1966.

Citizens Hearings. "Report of the Panel Hearings on the Superior and Hough Disturbances." (Cleveland, Ohio: Sept., 1966).

CLARK, KENNETH B. "The Present Dilemma of the Negro." Address before the Annual Meeting of the Southern Regional Council. (Atlanta, Ga.: Nov. 2, 1967).

Cleveland Council of Churches, Metropolitan Affairs Commission. "Preliminary Analysis of Riots in Hough Area of Cleveland, July 19, 1966."

FISHER, LLOYD H. "The Problem of Violence: Observations on Race Conflict in Los Angeles." (Los Angeles: The American Council on Race Relations, April, 1947).

FOGELSON, ROBERT M. "Violence as Protest: A Definition of the 1960 Riots." Columbia University: History Dept.

GRIMSHAW, ALLEN D. "A Study in Social Violence: Urban Race Riots in the United States." Unpub. Ph.D. Dissertation (Univ. of Pa., 1959).

GURR, TED. "The Genesis of Violence: A Multivariate Theory of the Preconditions for Civil Strife." Unpub. Ph.D. Dissertation (New York Univ., 1965).

HADDEN, JEFFREY K. "Paradox of Clergy Response to Riots." Case Western Reserve Univ.

HOLLOWAY, RALPH. "The Nature of Human Aggression." Paper read at the 66th Annual AAA meetings, Washington, Dec., 1967.

HOMER, P. W. "Report to the Rochester City Council on the Riots of July,

1964." Report of Rochester City Manager, April 7, 1965.

HUNDLEY, JAMES R. "Dynamics of Recent Ghetto Riots." Paper read at meeting of Am. Sociol. Assoc., San Francisco, Aug. 29, 1967.

Industrial Relations Research Association. "Industrial Conflict and Race Conflict: Parallels Between the 1930's and the 1960's." Proceedings of the 1967 Annual Spring Meeting, Detroit, May, 1967.

JANSYN, LEON R. "Violence, Ceremony and Substance in Street Corner Society." Southern Illinois Univ.: Center for the Study of Crime, Delinquency, and Corrections.

JUSTICE, DAVID B. "An Inquiry into Negro Identity and a Methodology for Investigating Potential Racial Violence." Unpub. Ph.D. Dissertation (Rice Univ., 1966).

KATZ, DANIEL. "Strategies of Conflict Resolution." Univ. of Michigan: Conflict Resolution Center (June 28, 1965).

LAKEY, GEORGE. "Civil Disobedience and Nonviolent Action." Conference on Violence in Contemporary American Society, Pa. State Univ., May 25-27, 1967.

LAUE, JAMES H. "Conciliation of Racial Conflict." U.S. Department of Commerce: Community Relations Service.

LEVY, BURTON, "Police-Negro Tensions." Michigan Civil Rights Commission: Community Services Division.

MacNAMARA, DONALD E. J. "Inter-Acting Patterns of Hate and Violence: The Dynamics of Police-Minority Group Tensions in American Cities." Conference on Violence in Contemporary American Society, Pa. State Univ., May 25-27, 1967.

MASOTTI, LOUIS H. "Reaction to the Hough Riots in the Peripheral White Cosmopolitan Areas." Case Western Reserve Univ.

———. "Violent Protest in Urban Society: A Conceptual Framework." Paper read at the 1967 meeting of the Am. Assoc. for the Advancement of Science.

SPEIGEL, JOHN P. "The Media in the Riot City." Brandeis Univ.: Lemberg Center for the Study of Violence (Nov., 1967).

———. "The Social and Psychological Dynamics of Militant Negro Activism: A Preliminary Report." Brandeis Univ.: Lemberg Center for the Study of Violence (1967).

———. "Social-Psychological Impressions of a Riot." Brandeis Univ.: Lemberg Center for the Study of Violence (Sept., 1967).

———. "Violence, the Law, and Race Relalations." Brandeis Univ.: Lemberg Center for the Study of Violence (1966).

SPERGEL, IRVING. "Gang Warfare and Agency Response." Univ. of Chicago: School of Social Service Administration.

About the Authors

LOUIS H. MASOTTI is Associate Professor of Political Science, and Director of the Civil Violence Research Center, Case Western Reserve University, Cleveland, Ohio. He is co-editor (with Jeffrey K. Hadden) of *Metropolis in Crisis* (1967), author of *Education and Politics in Suburbia* (1967) and articles on a variety of urban topics. He is Editor-in-Chief of the new journal, *Education and Urban Society,* and co-editor of Volume IV in the Urban Affairs Annual Reviews. Dr. Masotti's current research includes a study of the political attitudes of the urban poor. A book on the urban riots, past and present, is in process (with Jeffrey K. Hadden, Jerome Corsi, and Kenneth Semenatore). He has served as a consultant to the National Advisory Commission on Civil Disorders (the Kerner Commission), as well as the National Institute of Mental Health's Center for the Study of Social Problems.

DON R. BOWEN is Associate Professor of Political Science and Research Associate in the Civil Violence Research Center, Case Western Reserve University. He is currently co-directing (with L. H. Masotti) a study on the political attitudes and behavior of the poor, and developing a judicial simulation model. He is author of *Political Behavior of the American Public* (1968) and a number of articles on judicial behavior. He has served as a consultant to the National Commission on Civil Disorders.

LEONARD BERKOWITZ is Professor of Psychology, University of Wisconsin (Madison). He is the author of *Aggression: A Social Psychological Analysis* (1962) and numerous articles on aggression and frustration. Dr. Berkowitz is editor of a series entitled "Advances in Experimental Social Psychology." His current research is focused on the consequences and control of aggressive reactions.

ELINOR R. BOWEN is Assistant Professor of Political Science at The Cleveland State University, currently analyzing the results of a study on the political attitudes and behavior of the urban poor. SHELDON GAWISER is an NDEA Fellow and a doctoral candidate in political science, Case Western Reserve University. He is a research associate on the attitudes and behavior of the poor project.

DOUGLAS BWY is Assistant Professor of Political Science at Case Western Reserve University. He is author of the research monograph *Social Conflict: A Keyword-in-Context Bibliography on the Literature of Developing Areas* and of the forthcoming article "Political Instability in Latin America: The Preliminary Test of a Causal Model" (*Latin American Research Rev.,* Spring,

1968). His current interests lie in investigating the explanatory utility of comparative theories of domestic conflict in urban disorders in the United States.

EVERETT F. CATALDO, RICHARD M. JOHNSON and LYMAN A. KELLSTEDT are all Assistant Professors of Political Science at S.U.N.Y., Buffalo. They are part of a team of social scientists engaged in an on-going study of the agencies of social change and the problems of race and poverty in the Buffalo area. Kellstedt has published on community elites in the *Public Administration Review* and Johnson is author of *The Dynamics of Compliance: Supreme Court Decision-Making from a New Perspective* (1967).

E. S. EVANS is a political reporter for the *St. Louis Post-Dispatch.* In 1967 he was awarded a Russell Sage Foundation Fellowship for graduate study and urban affairs at Washington University (St. Louis), where he also served as a part-time staff member of *Trans-Action* magazine. He has published several articles on urban affairs in *FOCUS/Midwest.*

MARILYN GITTELL is Professor of Political Science and Director of the Institute for Community Studies, Queens College of the City University of New York. She has published widely on the politics of education including two recent books (*Participants and Participation: A Study of School Policy in New York City* and *Six Urban School Systems: A Comparative Study of Institutional Response*). She is currently working on a book on the American mayor. SHERMAN KRUPP, Professor of Sociology at Queens College, is currently finishing a book on organizational theory. He is author of *Patterns in Organizational Analysis* and editor of *The Structure of Economic Science.*

ALLEN D. GRIMSHAW is Associate Professor of Sociology and Director of the Institute for Comparative Sociology at Indiana University. He is an acknowledged expert on race relations and urban violence, and has published more than ten articles on the subject in a variety of American and British journals. He is currently reevaluating his earlier theoretical work in terms of the "new" urban riots, based on his research in Detroit.

TED GURR is Assistant Professor of Politics and Associate Director of the Workshop in Comparative Politics at Princeton University. His recent publications include *The Conditions of Civil Violence: First Tests of a Causal Model* (1967). Currently, he is working on a general theory of civil strife and a series of cross-national and longitudinal studies of the causes of civil strife.

JEFFREY K. HADDEN is Associate Professor of Sociology and Research Associate in the Civil Violence Research Center, Case Western Reserve University. His books include *The American City* (with E. Borgotta), *Metropolis in Crisis* (co-editor with L. H. Masotti), and *A House Divided,* a forthcoming study of the role of the clergy in the civil rights movement. In addition, he has published widely on urban issues in a variety of social science journals.

DEAN H. HARPER is Acting Chairman, Department of Sociology, University of Rochester. In addition to his interest in the field of race relations, he is doing research on the sociology of mental illness. An earlier analysis of the Rochester riot data appeared in *Trans-action.*

JOHN R. KRAUSE, JR., is an NDEA Fellow and doctoral candidate in the Department of Political Science and the Civil Violence Research Center, Case Western Reserve University.

KURT LANG is Professor of Sociology at the State University of New York, Stony Brook. GLADYS ENGEL LANG is Assistant Director of the Center for Urban Education, New York City. They co-authored *Collective Dynamics* (1961) and the article on "Collective Behavior" in the new *International Encyclopedia of the Social Sciences* (1968).

JAMES H. LAUE, who holds a Ph.D. in Sociology from Harvard, is currently serving as Chief of Evaluation, Community Relations Service, U. S. Department of Justice. His published work on race relations and social change appears in a number of professional journals and books. A book, co-authored with Martin Oppenheimer, entitled *Black Protest: Toward a Theory of Movements* is forthcoming. His research in process focuses on intergroup conflict, change processes, and conflict resolution.

BURTON LEVY is Director of the Community Relations Division of the Michigan Civil Rights Commission, which operates the Police-Community Relations and Tension Control program. In 1966-67, he served as a consultant to the Community Relations Service of the U. S. Department of Justice in establishing that agency's nationwide police-community relations program.

JOSEPH D. LOHMAN (deceased) served as Professor and Dean of the School of Criminology, University of California (Berkeley) until his untimely death in the Spring of 1968. He had published articles on police-community relations and the problems of law and society in a variety of journals. At the time of his death, he was in the process of completing a study of middle and upper class delinquency in San Francisco.

WILLIAM McCORD is Lena Gohlman Fox Professor of Sociology at Rice University. His most recent books are *The Springtime of Freedom* and *Mississippi: The Long Hot Summer.* JOHN HOWARD is Assistant Professor of Sociology at the University of Oregon, and has done extensive research on the Black Muslims. Jointly, they are preparing a book on the urban Negro.

RICHARD L. MEIER is Professor of Environmental Design at the University of California, Berkeley. His two most recent books are *Developmental Planning* and *Science and Economic Development* (2nd ed.). He has recently published a monograph on Asian urban development, and a number of papers on organization theory, urban simulation, and the impact of science on human affairs. Currect research is focused on simulations of social institutions and the design of resource-conserving urbanism.

H. L. NIEBURG is Professor of Political Science, University of Wisconsin (Milwaukee). The paper in this issue represents his first effort to reevaluate his earlier work ("The Threat of Violence and Social Change," *Am. Polit. Sci. Rev.,* 1962) in light of the urban violence since 1964.

MARTIN OPPENHEIMER is Associate Professor of Sociology, and Chairman of the Department of Sociology and Anthropology, Lincoln University (Pennsylvania). He is co-author, with James Laue, of a forthcoming book on the sociology of black protest.

E. L. QUARANTELLI and RUSSELL R. DYNES are Professors of Sociology and Co-directors of the Disaster Research Center at Ohio State University. Both are involved in continuing research on organizational and community response to disasters and civil disturbances. Professor Quarantelli recently published "Operational Problems of Organizations in Disasters" in the Proceedings of the System Development Corporation Symposium on Emergency Operations; Professor Dynes is author of *Organizational Behavior in Disaster: Analysis and Conceptualization* (forthcoming).

HARRY W. REYNOLDS, JR. is Professor of Political Science at the University of Nebraska at Omaha. He edited "Intergovernmental Relations in the United States" (*Annals,* May, 1965), and is author of a forthcoming volume entitled *The Political Setting of American Public Administration.* In addition to the U. S. Department of Labor sponsored study of unemployment in Omaha, of which the article in this volume is a part, Professor Reynolds is conducting an analysis of the role of the governmental administrator in community policy decisions.

JAY SCHULMAN is Assistant Professor of Sociology at City College of New York. His volume on *Innovation in an Organization* will be published in Fall, 1968. Current research involves a longitudinal study of the Rochester black ghetto and white reactions to black power processes.

HARRY M. SCOBLE is Associate Professor of Political Science, University of California, Los Angeles and formerly a research associate on the Los Angeles Riot Study (LARS). He is the author of *Ideology and Electoral Action* (1968). In addition to his continuing research on Negro politics, Professor Scoble is co-author of forthcoming studies on political culture in middle-size cities and urban poverty.

IRVING A. SPERGEL is Professor and Chairman of the Community Work rogram, School of Social Service Administration, University of Chicago. He is author of *Racketville, Slumtown and Haulburg* (1964), *Street Gang Work: Theory and Practice* (1966), *Community Work with Delinquents* (forthcoming), and editor of *Community Organization Research* (forthcoming). His current research includes an evaluation of The Woodlawn Organization's (Chicago) Youth Manpower Project.

T. M. TOMLINSON, formerly of the Department of Psychology, U.C.L.A., is now a research psychologist with the Office of Economic Opportunity, Research and Plans Division, Washington, D. C. While at U.C.L.A., he was a member of the Interdisciplinary Los Angeles Riot Study research team and co-authored two of the reports ("Negro Attitudes Toward the Riot" and "Method: Negro Reaction Survey").

JOHN G. WHITE is a doctoral candidate in the Department of Political Science, University of Hawaii. His dissertation research is an attempt to examine, systematically, the underlying causes of riots in a number of American cities.